Navigation Signal Processing for GNSS Software Receivers

For a list of recent related titles in the *Artech House GNSS Technology and Applications Series*, please turn to the back of this book.

Navigation Signal Processing for GNSS Software Receivers

Thomas Pany

ARTECH HOUSE
BOSTON | LONDON
artechhouse.com

Library of Congress Cataloging-in-Publication Data
A catalog record for this book is available from the U.S. Library of Congress.

British Library Cataloguing in Publication Data
A catalog record for this book is available from the British Library.

ISBN-13: 978-1-60807-027-5

Cover design by Greg Lamb

Accompanying MATLAB and assembler programs are available at www.artechhouse.com.

685 Canton Street
Norwood, MA 02062

10 9 8 7 6 5 4 3 2 1

Contents

Preface

The continuous developments of software-defined radio technology resulted in the appearance of the first real-time GPS software radios at the beginning of this century. For the first time, it was possible to realize a complete GNSS receiver without going into the depths of cumbersome hardware development that requires development or programming of low-level digital circuitry. The hardware development efforts were indeed so high that only a very limited number of companies or research institutes could afford them. Furthermore, the implementation constraints were so severe, especially for the first generation of GPS receivers, that often crude signal-processing approximations were necessary to allow a real implementation. Currently, software-defined radio technology not only allows receiver implementations by a larger research community, but also drastically increases the signal-processing capabilities. It also has the potential to become, in certain navigation areas, a commercial success.

Software radio technology provides an opportunity to design a new class of GNSS receivers, being more flexible and easier to develop than their FPGA- or ASIC-based counterparts. Therefore, this text reviews navigation signal detection and estimation algorithms and their implementation in a software radio. A focus is put on high-precision applications for GNSS signals and an innovative RTK receiver concept based on difference correlators is proposed.

This text makes extensive use of the least-squares principle. The least-squares principle is the typical basis for the calculation of a navigation solution. An adjustment or a Kalman filter calculates positions from pseudorange observations in virtually any GNSS receiver. Within this text, the least-squares principle is consistently extended to also allow signal samples as observations. In contrast to the pseudorange-observation equation, the signal sample model is highly non-linear, causing a number of difficulties that are discussed. Furthermore, signal sample observations can be complex-valued.

In the author's opinion, the development of a navigation receiver does not necessarily require an in-depth theoretical knowledge of signal-estimation and signal-detection theory. The basic algorithms like correlation and tracking can also be understood on an intuitive basis. Indeed many textbooks skip the highly theoretical signal-estimation and signal-detection framework and focus on engineering aspects. The question arises: What can we learn from the theoretical treatment that is presented here? First, the theory allows a performance assessment without building a receiver. By providing benchmarks like the Cramér-Rao lower bound or the clairvoyant detector, the theory serves also as a reference with which to compare a real implementation. This text attempts to generalize the existing theory for arbitrary navigation signal waveforms, going beyond existing GNSS signals. The theoretical treatment also gives hints for optimal algorithms; useful examples that are discussed are spectral whitening and the least-squares-based multipath-estimating discriminator. Efficient algorithms are found in the frequency domain for signal acquisition, which itself would justify the effort of going into theoretical details. Furthermore, the theoretical analysis points out that new developments could be expected in the field of direct-position estimation (in a single-step procedure, instead of estimating a position via pseudoranges), which should give advantages in terms of interference robustness and sensitivity. Sensitivity might be further increased by using Bayesian techniques (like a particle filter) that do not rely on a linearized signal model.

Unfortunately, the existing navigation signal-processing theory has limits and does not always provide an optimal algorithm for detection or estimation. Examples are the nonexistence of a uniform most powerful detector for acquisition and the nonexistence of a minimum variance unbiased code-phase or Doppler estimator for finitely received signal power. In addition, the practical usability of Bayesian techniques within signal processing (apart from the Kalman filter) is not completely assessed. Overall, it seems that a theoretically optimal navigation receiver is out-of-reach today, even if only signal processing is considered. However, software radio technology closes the gap between existing theory and real implementation.

Overview

Within this text, the navigation signal processing theory is described for generic navigation signals to allow a broad range of applications, beyond that of GNSS. Requirements for navigation signals are introduced in Chapter 1 and are illustrated with one GPS, one Galileo, and two pulsed signals. Software-defined radio technology will be introduced in Chapter 2, together with the architecture and the data flow of a permanent GNSS reference station in Chapter 3. Chapters 4 and 5 focus on theoretical signal-processing aspects and Chapters 6 through 9 shift the focus to implementation. An innovative high-precision software radio concept is presented in Chapter 10 using double-difference correlators, in addition to double-difference pseudorange and carrier-phase observations to increase carrier-phase tracking stability for real-time kinematic applications. Finally, on the Artech House Web site, www.artechhouse.com, this book has some MATLAB and assembler programs that illustrate the core signal-processing concepts of a navigation receiver. Chapter 11 describes this software.

Summary of Presented Signal-Processing Theory

In Chapter 1, requirements are formulated that a generic navigation signal has to fulfill to allow for the simultaneous estimation of the code phase, the Doppler and the carrier phase. Based on those requirements, Chapter 4 reviews the estimation theory for navigation signals using a consistent mathematical notation and derives the theory from first principles. The presented mathematical derivations are very detailed and with the finite sample rate approach, a reader should be able to adapt the theory for his or her purposes easily. The finite sample rate description is also chosen to enable a software receiver developer to establish a one-to-one correspondence of theory and implementation at every stage of signal processing.

Chapter 4 points out that least-squares code, carrier, and Doppler estimation is optimal under white and colored noise for arbitrary sample rates including sub-Nyquist sampling. The least-squares scheme is affected by the squaring loss, but a numerical evaluation of the true Cramér-Rao lower bound demonstrates that the least-squares scheme is an optimal estimator. In fact, a correct treatment of the carrier phase as a uniformly distributed nuisance parameter shows that the squaring loss is part of the code phase and the Doppler Cramér-Rao lower bound. Linearity conditions are discussed that the proposed least-squares multipath-estimating discriminator has to fulfill to be optimal. Techniques like macroscopic long coherent integration times and vector tracking are pointed out to fulfill the linearity requirements, even for weak signals. Nonparametric (i.e., multipath-mitigating) and parametric (i.e., multipath-estimating) least-squares discriminators are detailed in Chapter 8, making use of the different classes of correlators presented in Chapter 7. An efficient software receiver implementation relies on extra- or interpolation of correlator values, a technique that is also investigated from the parameter-estimation point of view in Chapter 4. Signal preprocessing, including quantization and sampling, is detailed in Chapter 6, as well as the handling of colored noise. The theoretical basis for signal detection or acquisition is laid out in Chapter 5. Signal acquisition is a computational-extensive task that is most efficiently done via FFT methods in a software receiver. Those methods are described in Chapter 9 and are related to time-domain correlation.

Overall, this book covers signal processing, starting from the ADC up to the estimation of the code phase, the carrier phase, and Doppler as depicted in Figure P.1. Tracking loops are covered with respect to their influence on the discriminators. For other navigation receiver aspects, like message decoding or positioning, the reader is referred to other literature. Also, some specific high-precision techniques like semi-codeless tracking and pulse blanking are covered.

The software receiver core algorithms consuming most of the processing time are covered in Chapter 9. A run-time performance analysis is presented, making use of x86 assembly language code snippets. The reader can find those snippets in the exemplary source code on the homepage for this book, http://www.artechhouse.-

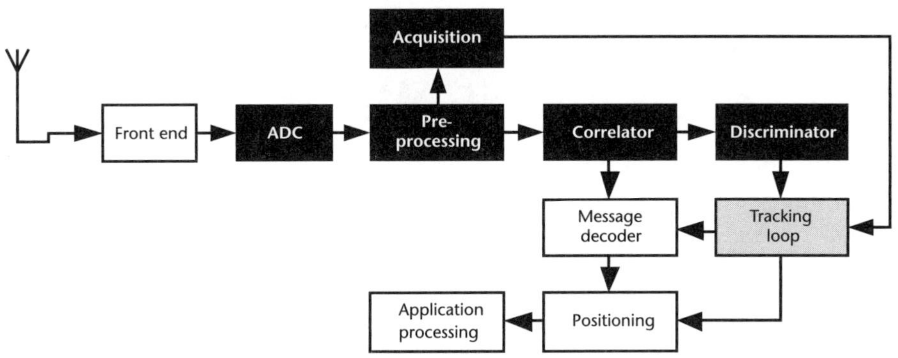

Figure P.1 Covered navigation receiver topics: black—fully covered; gray—partly covered; white—not covered.

com/static/reslib/pany/pany1.html. The source code includes the FFT acquisition methods and multipath-estimating tracking. Furthermore, MATLAB scripts for the true Cramér-Rao lower bound as well as for the thermal-noise analysis of the noncoherent discriminators are included. The scripts run with the four exemplary navigation signals of Chapter 1 and are outlined in Chapter 11.

Book's Usage for Practical Receiver Implementation

This book should help in building advanced navigation software receivers. It is not a beginner's book and the reader should be familiar with the architecture of a GNSS software receiver, which is, for example, excellently described by Borre's book mentioned in Chapter 2. Borre's book also comes with a complete MATLAB receiver and our text may help you to extend this receiver for high-sensitivity applications using efficient FFT techniques or for high-precision applications applying multipath-mitigation schemes or stable double-difference carrier-phase tracking.

To build a navigation software receiver, you need there things: navigation signal samples, a software framework that handles the data flow, and efficient core algorithms.

Signal samples can be obtained by one of the GNSS front ends described in Chapter 2 or you can use the single-channel signal generator of Chapter 11. Sometimes, the front-end manufacturers can provide you with exemplary signal-sample streams, too.

Writing a software receiver framework from scratch can be a quite tedious work. The framework handles the enormous amount of signal samples, synchronizes the different receiver channels, computes the position, and provides some standard output formats. You can short-cut this by adapting the MATLAB source code mentioned above. Another possibility to avoid this cumbersome work is to use a commercial software receiver having an application programming interface, which you can plug into your own source code. Chapter 11 provides you with a single-channel framework that demonstrates how to convert the sample stream into pseudorange measurements.

Finally, the core algorithms can actually be found within this text. They are derived in a way that they can be adapted easily for a specific framework and the assembler code should help to realize them efficiently.

Acknowledgments

This work would not have been possible without the support from numerous coworkers and colleagues. I am especially grateful to the researchers of the University of Federal Armed Forces in Munich, to the researchers at the IFEN GmbH, and to many colleagues from research institutes from all over the world.

I am grateful to Professor Günter W. Hein for continually encouraging me to enter this field, for his uninterrupted belief in technology, and for showing me ways of going beyond limits. Professor Bernd Eissfeller established the basis for GNSS receiver technology research at the Institute of Geodesy and Navigation. His contribution to this work cannot be overvalued. I would also like to thank Professor Jörn Thielecke for fruitful discussions. With his knowledge on communication and navigation signal processing, he showed me several important links between both fields.

CHAPTER 1

Radio Navigation Signals

A radio navigation system analyzes the propagation of electromagnetic signals between a transmitter and a receiver. The signals experience a delay caused by the relative distance between the transmitter and the receiver. The Doppler effect results in a frequency change of the signal and is related to the relative velocity between transmitter and receiver. Navigation signal processing provides the engineer with a number of techniques to estimate the signal delay and Doppler shift; the estimates are ultimately used to calculate the user's position and velocity.

This chapter briefly summarizes the basic framework of radio navigation and derives a generic deterministic signal model plus a stochastic noise model. The models form the basis for the presented navigation signal algorithms in the other chapters. The signal model is illustrated by four radio navigation signals.

1.1 Signal Generation

The typical setup of a navigation system is depicted in Figure 1.1 showing the transmitter with the navigation signal generation unit (NSGU), the high-power amplifier (HPA), the output filter, and the transmit antenna. The signal travels along one or more propagation paths to the receive antenna. The sum of all incoming signals plus noise and interference is received, amplified (by the low-noise amplifier), filtered, and digitized. An optional downconversion is not shown here. The analog-to-digital conversion (ADC) produces streams of digitized signal samples that are processed by the signal processing unit of the navigation receiver.

The term *navigation signal* is defined in a broad sense and may refer to a component of the received electromagnetic field, to the voltage at the receive antenna

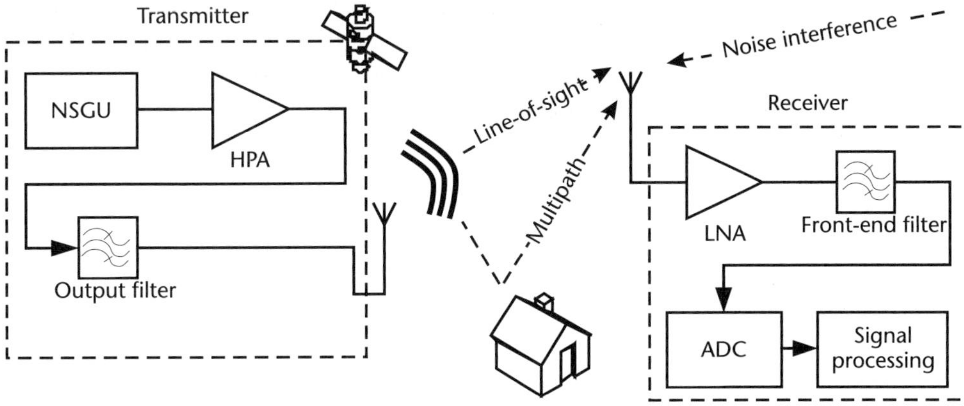

Figure 1.1 Navigation signal generation, propagation, and processing.

output, or to the digitized signal obtained after (optional) downconversion and filtering of the received signal.

The broadcast navigation signal structure is known to the receiver and will be characterized in the following paragraphs. A good way to do this is to merge the frontend filter into the output filter (i.e., into a combined filter) and to describe the signal after the output filter. This simplification is useful for signal processing purposes and is valid, because both antennas, the propagation path, and the frontend filter act as linear systems to the signal. Nonlinear effects of the amplifiers or mixer are not considered.

Without the combined filter (or output filter), the transmitted signal r_S is written as

$$r_S(t) = a_S d(t) c_\infty(t) \cos(2\pi f_{RF} t) \tag{1.1}$$

Here, f_{RF} is the carrier frequency in hertz, $c_\infty(t)$ is the infinite bandwidth signal representation at baseband, and $d(t)$ represents a broadcast navigation message. The transmitted signal is described in the signal-in-space (SIS) interface control document (ICD) [1, 2]. The symbol a_S denotes the signal amplitude in arbitrary units.

Other signals might be transmitted on the same carrier frequency by the same satellite. For example, the Global Positioning System (GPS) C/A signal is broadcast in phase quadrature with the P(Y) signal on the L1 (1,575.42 MHz) carrier frequency. The access to the different signals is controlled via $c_\infty(t)$. Different waveforms of $c_\infty(t)$ can be used to realize code division multiple access (CDMA), time division multiple access (TDMA), or—by including a carrier into $c_\infty(t)$—frequency division multiple access (FDMA) schemes.

With the combined filter, the signal is mathematically written as

$$r_S(t) = -a_S d(t) c_Q(t) \sin(2\pi f_{RF} t) + a_S d(t) c_I(t) \cos(2\pi f_{RF} t) \tag{1.2}$$

with

$$c(t) = c_I(t) + i c_Q(t) \tag{1.3}$$

representing the filtered baseband representation of the signal obtained by applying (1.10). The signal $c_Q(t)$ is an artifact generated by the filter. The filter does not affect $d(t)$, as long as the data rate (e.g., 50 bit/s) is much smaller than the filter bandwidth (e.g., 2 MHz). If the radio frequency (RF) is much higher than the filter bandwidth, $c_Q(t)$ is small and can be ignored.

Using a complex notation, the signal after the combined filter is given as

$$r_S(t) = a_S \operatorname{Re}\{d(t) c(t) \exp(2\pi i f_{RF} t)\} \tag{1.4}$$

1.2 Signal Propagation

According to Fermat's principle, the line-of-sight component of the signal $r_S(t)$ travels along the shortest propagation path from the transmitter to the receiver. Eventually, one or more reflected signals [i.e., copies of $r_S(t)$] are superimposed with the line-of-sight signal. Each copy has a different amplitude, Doppler, carrier phase delay, or code phase delay. Furthermore, the line-of-sight component is delayed by the atmosphere. Accurate modeling of the propagation effects is one of the key elements to obtain a precise position and this topic is covered in many satellite navigation text books [3, 4]. We will not go into the modeling details, but will instead focus on a model for the received signal that can be used as a basis for delay and Doppler estimation.

Delays affect the carrier and the data-modulating signals $c(t)$, $d(t)$ differently. Frequency-independent (nondispersive) delays are caused by the geometric distance and by the electrically neutral part of the atmosphere (troposphere). Frequency-dependent (dispersive) delays are caused by the ionosphere and by the receiver and transmitter hardware. Dispersive and nondispersive delays add up and result in the group delay τ_G, delaying $c(t)$ and $d(t)$. The carrier is delayed by the phase delay τ_P. The delays τ_G and τ_P are not equal because the ionospheric delay contributes to each of them with a different sign. The difference is typically less than 100m and varies with a frequency not more than 0.05 Hz if no scintillations are present. Also, hardware group and phase delay are generally different.

The group delay, τ_G, affects $c(t)$ via

$$c(t) \rightarrow c(t - \tau_G(t)) \approx c(\alpha_D t - \tau_{G,0}) \tag{1.5}$$

and similarly for $d(t)$. The phase delay, τ_P, affects the carrier via

$$\exp(2\pi i f_{RF} t) \rightarrow \exp(2\pi i f_{RF}(t - \tau_P(t))) \approx \exp(2\pi i f_{RF}(\alpha_D t - \tau_{P,0})) \tag{1.6}$$

with

$$\tau_{G/P,0} = \tau_{G/P}(t_0) - (1 - \alpha_D)t_0 \tag{1.7}$$

The coefficient α_D

$$\alpha_D = 1 - \frac{v}{c} \tag{1.8}$$

is the Doppler effect, caused by the change in the group/phase delay, expressed as velocity v in meters per second. The linearization is carried out around the epoch t_0. If the signal duration T_{coh} under consideration is short, the Doppler effect on $c(t)$, $d(t)$ can be ignored, as in

$$\frac{c}{vB} >> T_{coh} \Rightarrow c(\alpha_D t - \tau_{G,0}) \approx c(t - \tau(t_0)) \tag{1.9}$$

and similarly for $d(t)$, where B denotes the signal (or data message) bandwidth in hertz.

1.3 Signal Conditioning

After signal reception by the antenna, the signal is amplified, filtered, and eventually downconverted. Amplification changes the amplitude of the signal (from a_S to a), but leaves the signal structure invariant.

The frontend filter limits the bandwidth of the received navigation signals and of the received noise. It also rejects out-of-band signals. The frontend filter is typically of lower bandwidth than the output filter and neglecting the output filter is a reasonable approximation. The resulting filter is a band pass filter and is described by its baseband equivalent H via

$$c(t) = H(c_\infty(t)) \tag{1.10}$$

For signal estimation and detection, it is largely irrelevant at which center frequency the filter operates; it can be placed at the RF, at the IF, or at baseband. Normally, filters with discrete components or SAW filters operating at the IF are used, but discrete polyphase filters at baseband have also been utilized. Global navigation satellite system (GNSS) receivers normally do not integrate the filter into a chip solution.

Downconversion changes the signal carrier model by

$$\exp(2\pi i f_{RF}(\alpha_D t - \tau_{P,0})) \rightarrow \exp(2\pi i f_{RF}(\alpha_D t - \tau_{P,0}) - 2\pi i f_{LO} t) \tag{1.11}$$

where f_{LO} is the local oscillator frequency in hertz.

For each transmitter and propagation path, a signal of the form

$$r_{rec}(t) = ac(t - \tau(t_0))\exp(2\pi i f_{RF}(\alpha_D t - \tau_{P,0}) - 2\pi i f_{LO} t) \tag{1.12}$$

arrives at the front end's ADC(s). The ADC(s) either quantize(s) the real and the imaginary part of the signal or quantizes only one of them (see Appendix A.2).

1.4 Motivation for a Generic Signal Model

By specifying core signal elements, it is possible to reduce the amount of information that is necessary to analyze a navigation system. The baseband representation of the navigation signal $c(t)$, the time over which $d(t)$ remains constant, and the carrier frequency represent core elements that determine to a large extent how precisely a receiver can estimate code and carrier delay, the Doppler shift, and, consequently, its position. They are summarized in Table 1.1.

The assumptions imposed on the received signals are kept as general as possible to allow the application of the developed theory to a wide range of signals. It focuses mostly on conventional GNSS continuous-time CDMA signals using binary phase shift keying (BPSK), binary offset carrier (BOC), multiplexed BOC (MBOC), alternative BOC (AltBOC), or any other spread-spectrum technique. Furthermore,

Table 1.1 Influence of Navigation Signal Elements on Signal Processing Parameters

Signal Element	*Affected Signal Processing Parameter*
$c(t)$	Code correlation function $R_{c,c}$, Doppler correlation function
$d(t)$	Coherent integration time T_{coh}, Doppler correlation function

pulsed signals are included because they are being used by pseudolites, by LORAN-C, or RADAR-like ranging systems. In principle, the theory can also be adapted for sonar ranging systems and, with some limitations, for optical ranging systems. Only one frequency band is considered (e.g., GPS L1); a generalization from a sampling of one frequency band to multiple frequency bands is obvious, but one should be take care that all bands are sampled synchronously.

No assumption on the relation of the signal bandwidth to the sampling rate is made and, in particular, sub-Nyquist sampling rates can be used. No assumption on the modulation scheme is made as long as the received signal waveform at baseband $c(t)$ is known a priori to the receiver. Eventually, filters influence the theory via the waveform $c(t)$. The more narrow the filter bandwidth, the smoother the waveform will be.

The filter bandwidth and characteristics define how much noise power is being received: the wider the bandwidth, the higher the noise power. For simplicity, a unity noise power is assumed in (1.16) and it is important to keep in mind that only ratios between power levels have meaningful values, as described in Section 1.8.1. In that sense, (1.16) defines the power scale.

The Nyquist criterion does not fully apply because the waveform $c(t)$ is known to the receiver beforehand. Consequently, there is no need to reconstruct the signal waveform from the received samples [5]. As shown in Section 6.5, a good choice for the sample rate is exactly equal to the Nyquist rate (e.g., being equal to the noise bandwidth). Lower sample rates yield less-independent signal samples, thereby generally decreasing the accuracy of the obtained estimates. The accuracy decrease can be modeled as an effective signal power loss. This observation is also true when multiple reflections of the same signal are received.

In the rest of this chapter, we will formulate generic conditions for navigations signals. Later, we illustrate the conditions with two GNSS signals and two pulsed terrestrial navigation signals. In addition to those parameters, the number of received signals, the amplitude, and the geometric placement of the transmitters affect the positioning result (see Sections 4.1.3 and 8.4).

1.5 Sampling

In the following, a number of L signal samples are considered, indexed by μ. The index μ assumes, in general, the values

$$\mu \in \{1,\ldots,L\} \tag{1.13}$$

The sampling epoch t_μ in seconds for the sample μ is given by

$$t_\mu = \frac{\mu}{f_s} \tag{1.14}$$

and f_s defines the ADC sample rate in samples per second. Optionally, the sampling epochs can be offset by a fixed amount of time, which will not explicitly be mentioned here.

For each sampling epoch t_μ, a complex-valued (i.e., I plus Q) signal sample is generated. We chose a complex signal representation to work with more compact expressions compared to working with real-valued signal representations using, for example, only the I-component. In Appendix A.2, it is shown that both signal representations are equivalent. Additionally, it is possible to derive an equation using the complex-valued signal representation and then to adapt the final result for the real-valued signal representation. This can be done by taking the real component of the obtained results or by similar methods. This approach is used in Chapters 7 and 8. The GNSS SDR implementation of Chapter 2 works internally with a real-valued signal representation.

1.6 Deterministic Received Signal Model

A digitized received signal sample is modeled as a complex random variable S_μ being composed of a deterministic part r_μ and a purely stochastic part N_μ,

$$S_\mu = r_\mu + N_\mu = \sum_m r_{m;\mu} + N_\mu \tag{1.15}$$

The deterministic part is the sum of all received signals $r_{m;\mu}$ broadcast from one or more transmitters propagating along one or more paths; for example, the index m distinguishes not only the different emitters, but also the different propagation paths. Furthermore, m may distinguish different signal components (e.g., data and pilot signals) broadcast by the same transmitter.

1.7 Stochastic Noise Model

The baseline assumption for the stochastic component of the received signal is to model it as complex-valued uncorrelated white noise having unity variance in each component

$$\langle N_\mu N_\upsilon \rangle_{\mathrm{N}} = 0, \langle \bar{N}_\mu \bar{N}_\upsilon \rangle_{\mathrm{N}} = 0, \langle \bar{N}_\mu N_\upsilon \rangle_{\mathrm{N}} = 2\delta_{\mu,\upsilon} \tag{1.16}$$

and zero mean

$$\langle N_\mu \rangle_{\mathrm{N}} = 0 \tag{1.17}$$

The white noise originates from the received and internally generated noise.

In general, it is assumed that the amplitudes of the individual noise random variables N_μ are Gaussian distributed. However, important results described in the following chapters also hold for arbitrary amplitude distributions of N_μ, provided that (1.16) and (1.17) hold. A common example of a non-Gaussian noise distribution is the ADC quantization noise, which will be discussed in Section 6.1.

It is important to recognize that the comparably simple white-noise spectral characteristic is sufficient to model many navigation receiver frontends, as long as the sample rate is properly chosen (see Section 6.5). In fact, if (1.16) is not fulfilled, which occurs when oversampling is employed and results in nonwhite noise, then the operation of spectral whitening can be applied as described in Section 6.4. The operation of spectral whitening yields uncorrelated noise samples that allow the operation to work with the simple noise model (1.16) and simultaneously reduces the estimated parameters' variances. If spectral whitening cannot be employed, methods discussed in Section 7.6 show how to adapt the results obtained under the assumption of white noise for the case of colored noise.

1.8 Short-Period Signal Model

The deterministic component of the received signal model is the superposition of a number of signals. Each signal may be different but they are required to share the same structure, which will be defined later. The index m is abandoned in the following.

This structure is valid for short periods of time, typically being related to the coherent integration time and being on the order of a few tens of milliseconds at maximum in a standard GNSS receiver implementation.

We assume that the deterministic part of the received signal samples is modeled during a sufficiently short interval as

$$r_\mu = ac(t_\mu - \tau)\exp\left\{i\left(\omega\left(t_\mu - \frac{L+1}{2f_s}\right) - \varphi\right)\right\} \tag{1.18}$$

Here, a denotes the signal amplitude in arbitrary units, τ is the delay (or code phase) of the signal in seconds, φ is the carrier phase of the signal in radians, and ω is the angular frequency plus Doppler in radians per second. All values are instantaneous values and refer to a particular interval. The relation of the values between the different intervals will not be specified here.

The signal amplitude a is a measure for the received signal power and the delay τ relates to the geometric distance between transmitter and receiver. The carrier phase φ contains information of the geometric distance, but is also used to accommodate a possible broadcast navigation message. The angular frequency ω is given by the nominal frequency (being either zero, an intermediate frequency, or the nominal carrier frequency) plus a Doppler offset caused by the relative velocity between transmitter and receiver.

The signal $c(t)$ is the baseband representation of the broadcast navigation signal waveform (eventually complex-valued). In the example of a GPS C/A-code transmitter, it is a 1-ms pseudorandom noise signal using a BPSK modulation scheme. The signal $c(t)$ is affected by filters located on either the transmitter or the receiver side.

It is important to recognize that the signal $c(t)$ can be quite arbitrary and could assume a CDMA spreading code or a pulsed waveform. The important requirement, however, is that the waveform $c(t)$ must be known to the receiver before receiving it. Additionally, five more requirements are formulated next. Three of these requirements, provided in Sections 1.8.1 through 1.8.3, are mostly of a formal nature (i.e., they relate to the definitions of constants and interval boundaries) and pose only few constraints on the waveform itself. The other two requirements, given in Sections 1.8.4 and 1.8.5, are more strict and need to be fulfilled if the signal will allow separation of Doppler estimates from delay estimates. The two conditions given in Sections 1.8.1 and 1.8.4 are extended to be compatible with the used estimation scheme (complex least-squares adjustment). The extensions are trivially fulfilled if the signal $c(t)$ is real-valued and the extensions ensure that the code phase, the Doppler, and the complex amplitude estimates are uncorrelated in the event that only a line-of-sight signal is present.

It should be noted that the requirements formulated next need to be fulfilled sufficiently, but not necessarily exactly. Overall, they ensure that the Fisher information matrix (4.60) can be sufficiently approximated by a diagonal matrix.

1.8.1 Zeroth-Order Moment of Signal Power

The signal waveform at baseband is required to have an average unity power during the interval of interest, for example,

$$\frac{1}{L}\sum_{\mu=1}^{L}\left|c(t_\mu-\tau)\right|^2 \approx 1 \tag{1.19}$$

This is required so that only the signal amplitude a contains information of the received signal power and the waveform $c(t)$ is assumed to be independent from the actual received power. This equation is approximately valid for all τ values of interest.

The numerical value of a itself is meaningless as it is expressed in arbitrary units. However, it can be related to the ratio between signal power and noise power spectral density C/N_0, which is described in detail in Appendix A.2. Because of (1.19), the following equation holds:

$$C/N_0 = \frac{f_s a^2}{2} \tag{1.20}$$

The value of C/N_0 is expressed in hertz.

Equation (1.19) implies that

$$\operatorname{Re}\left\{\frac{1}{L}\sum_{\mu=1}^{L}\overline{c(t_\mu-\tau)}\,c'(t_\mu-\tau)\right\}\approx 0 \tag{1.21}$$

according to Appendix A.4.2. Note, c' denotes the first derivative of the waveform c. To avoid correlations between the imaginary part of the code phase estimate with the complex signal amplitude, it is required that the imaginary part of the above expression vanishes; overall, we require

$$\frac{1}{L}\sum_{\mu=1}^{L}\overline{c(t_\mu-\tau)}\,c'(t_\mu-\tau)\approx 0 \tag{1.22}$$

Equation (1.22) is trivially fulfilled, if the waveform $c(t)$ is real valued.

1.8.2 First-Order Moment of Signal Power

The first-order moment in time of the signal power, with respect to the midpoint of the interval of interest, is required to vanish; for example,

$$\sum_{\mu=1}^{L}\left(t_\mu-\frac{L+1}{2f_s}\right)\left|c(t_\mu-\tau)\right|^2\approx 0 \tag{1.23}$$

This requirement ensures that estimates for the angular frequency and the carrier phase are uncorrelated with each other, which will be shown in Section 4.3.2. This equation is approximately valid for all τ values of interest. The requirement is fulfilled if the signal power is located symmetrically in time with respect to the midpoint of the interval.

1.8.3 Second-Order Moment of Signal Power

The second-order moment in time of the signal power with respect to the midpoint of the interval of interest evaluates to

$$\sum_{\mu=1}^{L}\left(t_\mu - \frac{L+1}{2f_s}\right)^2 \left|c(t_\mu - \tau)\right|^2 = \sum_{\mu=1}^{L}\left(\frac{2\mu - L - 1}{2f_s}\right)^2 \left|c(t_\mu - \tau)\right|^2 \approx \chi_{freq}\frac{L^3 - L}{12 f_s^2} \tag{1.24}$$

The constant χ_{freq} measures the nonuniformity of the signal power distribution in time. If the signal power is constant during time (i.e., $|c(t)|^2 = 1$), then $\chi_{freq} = 1$. This constant will be much smaller than 1, causing a reduced frequency estimation accuracy, especially for single pulses. This equation is approximately valid for all τ values of interest.

1.8.4 First-Order Moment of Signal Power Variations

Motivated by the requirement of Section 1.8.2, the first-order moment in time of the first derivative of the signal power, with respect to the midpoint of the interval of interest, is required to vanish; for example,

$$\sum_{\mu=1}^{L}\left(t_\mu - \frac{L+1}{2f_s}\right)\frac{\partial}{\partial\tau}\left(\left|c(t_\mu - \tau)\right|^2\right) \approx 0 \tag{1.25}$$

This implies that

$$\sum_{\mu=1}^{L}\left(t_\mu - \frac{L+1}{2f_s}\right)\mathrm{Re}\left\{\overline{c(t_\mu - \tau)}\frac{\partial}{\partial\tau}c(t_\mu - \tau)\right\} \approx 0 \tag{1.26}$$

and, additionally, it is required that the imaginary part vanishes; overall, it is required that

$$\sum_{\mu=1}^{L}\left(t_\mu - \frac{L+1}{2f_s}\right)\overline{c(t_\mu - \tau)}\, c'(t_\mu - \tau) \approx 0 \tag{1.27}$$

This requirement ensures that Doppler estimates and code phase estimates are uncorrelated, which will be shown in Section 4.3.2. This equation is approximately valid for all τ values of interest. This requirement is nontrivial and might not hold for certain waveforms $c(t)$. An example for which it does *not* hold occurs if $c(t)$ assumes a single Gaussian bell-shaped curve. In that case, complex-valued delay and complex-valued Doppler estimates are totally correlated.

1.8.5 Separation of Code and Carrier Correlation

When we consider two waveforms, $c_1(t)$ and $c_2(t)$, and their cross-correlation including Doppler (see also Appendix A.4.4), then we require that the code correlation be sufficiently well separated from the Doppler contribution. More specifically, it is required that

$$\sum_{\mu=1}^{L} c_1(t_\mu - \tau)c_2(t_\mu)\exp\left\{i\omega\left(t_\mu - \frac{L+1}{2f_s}\right)\right\} \approx L\kappa(\omega)R_{c_1,c_2}(\tau) \tag{1.28}$$

is approximately fulfilled for a properly chosen function $\kappa(\omega)$, with $\kappa(0) = 1$. Equation (1.28) is fulfilled for all signals occurring in a certain navigation problem. The function $\kappa(\omega)$ is universal and applies to transmitted, received, or internally generated signal pairs (e.g., baseband signal plus P-, D-, F-, or W-correlator reference signals of Chapter 7) and to any combination of them.

In the case of a timely uniform distributed signal power $|c_1(t)| = |c_2(t)| = \text{const.}$, the function $\kappa(\omega)$ is given as

$$\kappa(\omega) \approx \frac{i\left(e^{-i\omega T_{coh}/2} - e^{i\omega T_{coh}/2}\right)}{\omega T_{coh}} = \frac{2\sin\left(\frac{\omega T_{coh}}{2}\right)}{\omega T_{coh}} = \operatorname{sinc}\left(\frac{\omega T_{coh}}{2}\right) \tag{1.29}$$

Here, $T_{coh} = L/f_s$ denotes the integration time in seconds.

Assuming a signal $c(t)$ that fulfills (1.19), then $\kappa(0) \approx 1$ and the conditions (1.23) and (1.28) require that $\kappa(\omega)$ fulfills

$$\begin{aligned}
&\left(-i\frac{\partial}{\partial\omega}\right)\left(\sum_{\mu=1}^{L}\left|c(t_\mu - \tau)\right|^2 \exp\left\{i\omega\left(t_\mu - \frac{L+1}{2f_s}\right)\right\}\right)\Bigg|_{\omega=0} \\
&= \sum_{\mu=1}^{L}\left|c(t_\mu - \tau)\right|^2\left(t_\mu - \frac{L+1}{2f_s}\right) \approx -iL\kappa'(0) \approx 0 \\
&\Rightarrow \kappa'(0) \approx 0
\end{aligned} \tag{1.30}$$

Furthermore, the conditions (1.24) and (1.28) imply for $\kappa(\omega)$

$$\begin{aligned}
&\left(-i\frac{\partial}{\partial\omega}\right)^2\left(\sum_{\mu=1}^{L}\left|c(t_\mu - \tau)\right|^2 \exp\left\{i\omega\left(t_\mu - \frac{L+1}{2f_s}\right)\right\}\right)\Bigg|_{\omega=0} \\
&= \sum_{\mu=1}^{L}\left|c(t_\mu - \tau)\right|^2\left(t_\mu - \frac{L+1}{2f_s}\right)^2 \approx -L\kappa''(0) \approx \chi_{freq}\frac{L^3 - L}{12f_s^2} \\
&\Rightarrow \kappa''(0) \approx -\chi_{freq}\frac{L^2 - 1}{12f_s^2}
\end{aligned} \tag{1.31}$$

For large L and $T_{coh} \approx L/f_s$, the conditions are summarized as

$$\begin{aligned} \kappa(0) &\approx 1 \\ \kappa'(0) &\approx 0 \\ \kappa''(0) &\approx -\frac{1}{12}\chi_{freq}T_{coh}^2 \end{aligned} \tag{1.32}$$

1.9 Exemplary Signals

This section illustrates the generic signal model with four types of navigation signals. Table 1.2 summarizes signal parameters of the four exemplary navigation signals. For each signal, we chose a typical coherent integration time. The pulsed open-service (OS) pilot signal is introduced in detail in Section 1.9.3 and Chapter 10 and uses T_R = 100 ms and T_P = 4 ms.

1.91 A Model for the GPS C/A-Code Signal

The American GPS is the most popular navigation system worldwide. Its first development satellites were launched in 1978. Currently, 32 GPS satellites orbit the Earth and transmit navigation signals on three different frequencies: L1, L2, and L5. Of all the signals, the C/A-code signal on L1 is best known. All mass-market GNSS receivers make use of it and it is also the workhorse for professional and scientific applications. Originally, the C/A-code signal was forseen for acquisition of the P(Y) code.

The C/A-code signal utilizes the CDMA scheme and the pseudorandom noise (PRN) code sequences are Gold sequences 1 ms in length at a chipping rate of f_c = 1.023 Mchip/s employing the BPSK modulation scheme. Details of the signal can be found in [1].

A baseband model of the signal is given by a convolution of the PRN code sequence c_n with the single-chip waveform $p_\infty(t)$

$$c_\infty(t) = \sum_{n=-\infty}^{\infty} c_n p_\infty(tf_c - n) \tag{1.33}$$

The elements of the spreading code sequence c_n assume values of –1 and +1 and the sequence is periodic with a period of N_{PRN} = 1,023,

Table 1.2 Important Parameters of Exemplary Navigation Signals

	Typical T_{coh} (ms)	$R''_{\tilde{c},c}(0)$ $(\mu s)^{-2}$	$\kappa(\omega)$
GPS C/A	20	–21.37	(1.29) with T_{coh} = 20 ms
Galileo OS Data	4	–76.59	(1.29) with T_{coh} = 4 ms
Galileo OS Pilot	100	–87.58	(1.29) with T_{coh} = 100 ms
Pulsed OS Pilot	100	–87.58	(1.29) with T_{coh} = 4 ms
Gaussian Double Pulse	1	–0.005	(A.124)

$$c_n = c_{n+mN_{PRN}} \qquad m \in \mathbb{Z} \tag{1.34}$$

The navigation message amplitude $d(t)$ assumes values of −1 and +1 and remains constant over the 20-ms bit duration. Spreading code sequences and bits are synchronized.

The signal chip waveform is given by

$$p_\infty(n) = \begin{cases} 1 & 0 \le n < 1 \\ 0 & \text{otherwise} \end{cases} \tag{1.35}$$

As with any PRN code sequence, the C/A-code sequence can be modeled by a sequence of uniform distributed, binary (−1 and +1), and independent random variables with

$$\langle C_n C_m \rangle_C = \delta_{n,m} \tag{1.36}$$

This model is only an approximation to the signal. Especially for short codes like the C/A code, significant deviations (nonvanishing cross-correlation and side autocorrelation peaks) may occur. However, (1.36) is useful because it allows a uniform treatment of the entire GPS C/A-code signal family, independent from the PRN code number.

A filter affects the single-chip waveform and leaves the spreading code sequence untouched,

$$H(c_\infty(t)) = \sum_{n=-\infty}^{\infty} c_n H(p_\infty(tf_c - n)) = \sum_{n=-\infty}^{\infty} c_n p(tf_c - n) \tag{1.37}$$

The filtered single-chip waveform $p(t)$ must be normalized to fulfill the requirement of Section 1.8.1. Using (1.36), the correlation function of the signal can be simplified as

$$\begin{aligned}
R_{\bar{c},c}(\tau) &= \frac{1}{L}\sum_{\mu=1}^{L} \overline{c(t_\mu - \tau)}\, c(t_\mu) = \frac{1}{L}\sum_{\mu=1}^{L}\sum_{m,n=-\infty}^{\infty} c_m \overline{p(t_\mu f_c - m - \tau)}\, c_n p(t_\mu f_c - n) \approx \\
&\approx \left\langle \frac{1}{L}\sum_{\mu=1}^{L}\sum_{m,n=-\infty}^{\infty} C_m \overline{p(t_\mu f_c - m - \tau)}\, C_n p(t_\mu f_c - n) \right\rangle_C \\
&= \frac{1}{L}\sum_{\mu=1}^{L}\sum_{m,n=-\infty}^{\infty} \delta_{n,m} \overline{p(t_\mu f_c - m - \tau)}\, p(t_\mu f_c - n) \\
&= \frac{1}{L}\sum_{n=-\infty}^{\infty}\sum_{\mu=1}^{L} \overline{p(t_\mu f_c - n - \tau)}\, p(t_\mu f_c - n) \\
&\approx \sum_{n=0}^{t_L f_c - 1} \frac{1}{L}\sum_{\mu=1}^{L} \overline{p(t_\mu f_c - n - \tau)}\, p(t_\mu f_c - n) \approx \sum_{n=0}^{t_L f_c - 1} R_{\bar{p},p}(\tau) = t_L f_c R_{\bar{p},p}(\tau)
\end{aligned} \tag{1.38}$$

and $t_L f_c$ is the number of chips contained in the sampling interval.

The left side of Figure 1.2 shows the single-chip C/A waveform after lowpass filtering with a fifth-order Butterworth filter having a one-sided 3-dB bandwidth of 7 MHz.

The right side of Figure 1.2 shows the autocorrelation function $R_{\bar{c},c}(\tau)$ (for PRN1); the second derivative at the origin is listed in Table 1.2.

1.9.2 A Model for the Galileo E1 Open-Service Signal

Galileo is the European GNSS and has been under development since the early 1990s. Two test satellites are currently in orbit; the fully deployed Galileo system will consist of 30 satellites (27 operational + 3 nonactive spares). Galileo will be interoperable and compatible with GPS. The satellites share the same nominal L1 and L5 carrier frequencies as GPS (and also use additional frequencies). On E1 (the Galileo name for L1), the signal design aims to provide a superior ranging performance compared to the GPS C/A code. This is achieved partly by using the CBOC modulation, which is a particular implementation of the MBOC. The modernized GPS block III satellites will transmit TMBOC signals and will also be compatible with the MBOC.

The Galileo E1 OS signal consists of a data-less (pilot) and a data-bearing (data) component [2]. A baseband model of the CBOC(6,1,1/11) signal is given by a convolution of two PRN code sequences $c_{y,n}$ ($y = D$ or P) with two single-chip waveforms $p_y(t)$

$$c_\infty(t) = \sum_{n=-\infty}^{\infty} \left(d(t) c_{D,n} p_{D,\infty}(tf_c - n) + c_{P,n} p_{P,\infty}(tf_c - n)\right) \tag{1.39}$$

The satellites transmit the PRS signal on E1 in phase quadrature plus the intermodulation product. Both components of the Galileo E1 OS signal can be treated separately and the data and pilot signals are modelled similarly to the GPS C/A-code signal.

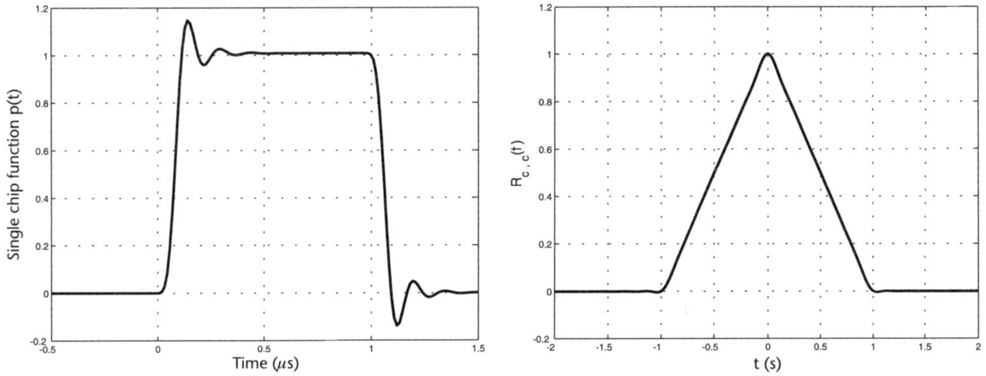

Figure 1.2 GPS C/A single-chip waveform and autocorrelation function.

The infinite bandwidth data single-chip function is

$$p_{D,\infty}(n) = \begin{cases} \sqrt{\frac{10}{11}}\text{sign}(\sin(2\pi n)) + \sqrt{\frac{1}{11}}\text{sign}(\sin(12\pi n)) & 0 \le n < 1 \\ 0 & \text{otherwise} \end{cases} \tag{1.40}$$

and the pilot single-chip function

$$p_{P,\infty}(n) = \begin{cases} \sqrt{\frac{10}{11}}\text{sign}(\sin(2\pi n)) - \sqrt{\frac{1}{11}}\text{sign}(\sin(12\pi n)) & 0 \le n < 1 \\ 0 & \text{otherwise} \end{cases} \tag{1.41}$$

The code rate f_c is 1.023 Mchip/s. For the data component, N_{PRN} = 4,092 and the length of one code sequence is 4 ms, equal to the data symbol duration. For the pilot component, N_{PRN} = 102,300 using a tiered code (see Section 9.5.7). The filtered waveform and the autocorrelation function are shown for the pilot and data components separately in Figure 1.3 using the same filter as for the GPS C/A-code.

1.9.3 Pulsed GNSS Signals

The two GNSS signals discussed so far have a constant envelope over time. A constant envelope signal allows the use of an efficient HPA in the satellite payload with small nonlinearities. If a navigation signal is transmitted from a ground-based station (e.g., from a pseudolite), less power is needed and it is possible to transmit nonconstant envelope signals.

A possible utilization of a nonconstant envelope may consist of periodically switching on and off a GNSS-like signal. This is referred to as *pulsing*. Pulsing is used for pseudolites and is discussed in Chapter 10.

Considering a signal time span from 0 to T_R (the pulse-repetition time), pulsing is mathematically represented as a modification of the baseband waveform via

$$c_\infty(t) \rightarrow h(t)c_\infty(t) \tag{1.42}$$

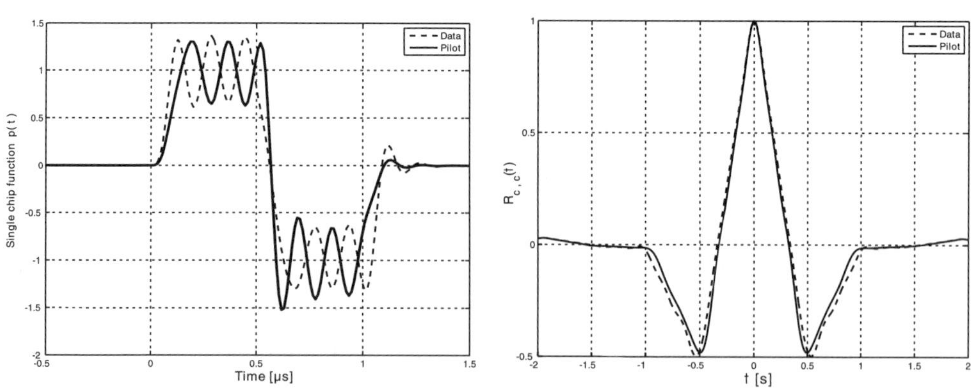

Figure 1.3 Galileo E1 OS single-chip waveform and autocorrelation function.

Table 1.3 Exemplary Gaussian Double-Pulse Signal Parameters

Parameter	*Value*
Repetition period T_R	1 ms
Pulse span b	10 μs
Pulse position a	500 μs
Pulse separation d	250 μs

with

$$h(t) = \begin{cases} \sqrt{\dfrac{T_R}{T_P}} & |2t - T_R| < T_P \\ 0 & \text{otherwise} \end{cases} \tag{1.43}$$

where T_P is the pulse period in seconds. Note, the pulse is located symmetrically within the pulse-repetition interval to fulfill the requirements of Section 1.8.2. Pulsing allows extension of the coherent integration time T_{coh} until T_R, but also maintains a wide Doppler correlation function, making the signal easier to acquire and more stable to track. Also, pulsing represents a TDMA scheme and allows access to the signals in given time slots. The interference caused by a pulsed signal can be easily mitigated by pulse blanking (see Chapter 10).

1.9.4 Gaussian Double Pulse

A simple navigation signal may be realized by broadcasting Gaussian pulses. This signal is approximately obtained by pulsing a pure carrier signal and applying a filter with a bandwidth similar to the inverse of the pulse period.

The Gaussian pulsed navigation signal is described by its waveform at baseband and is given as

$$c(t) = c_0 \exp\left\{-\frac{(t-a-d)^2}{2b^2}\right\} + c_0 \exp\left\{-\frac{(t-a+d)^2}{2b^2}\right\} \tag{1.44}$$

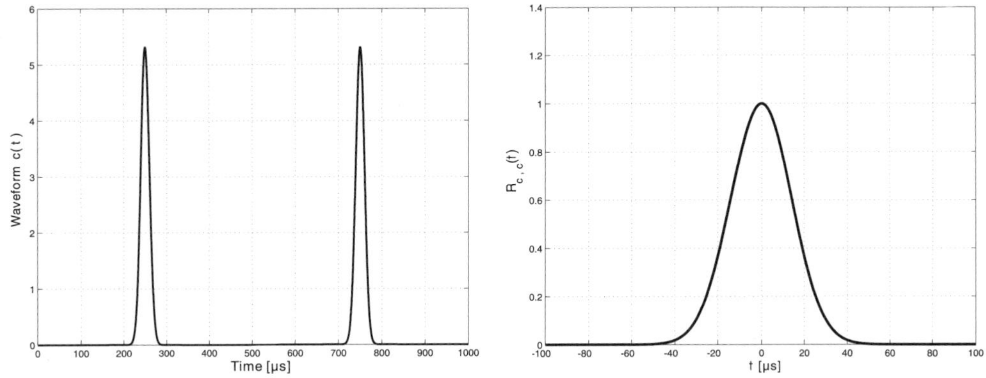

Figure 1.4 Gaussian double-pulse waveform and autocorrelation function.

As shown in Figure 1.4, the signal is composed of two Gaussian pulses, well-separated in time (i.e., $d >> b$). This signal fulfills all the requirements of Section 1.8. The fact that the interval contains two pulses (to fulfill the requirement of Section 1.8.4 and to achieve a reasonable high value of χ_{freq}), and that both pulses are (approximately) located symmetrically in the interval of interest (to fulfill the requirement of Section 1.8.2) is particularly important. The constant c_0 is given by

$$c_0 = \sqrt{\frac{2\sqrt{2\pi b^2}}{T_R}} \quad b \ll d \tag{1.45}$$

where T_R represents the signal duration in seconds. Overlaps between signals emerging from different emitters will not be discussed here. Exemplary signal parameters are summarized in Table 1.3.

The waveform and the autocorrelation function of the signal using the parameters from Table 1.3 are illustrated in Figure 1.4.

References

[1] ARINC Engineering Services, LLC, "NAVSTAR Global Positioning System, Interface Specification," IS-GPS-200, Rev. D, http://www.navcen.uscg.gov/gps/geninfo/IS-GPS-200D.pdf, December 2004.

[2] European Space Agency and European GNSS Supervisory Authority, "Galileo Open Service Signal In Space Interface Control Document," OS SIS ICD, Draft 1, http://www.gsa.europa.eu/go/galileo/os-sis-icd/galileo-open-service-signal-in-space-interface-control-document, 2008.

[3] Hofmann-Wellenhof, B., H. Lichtenegger, and E. Wasle, *GNSS Global Navigation Satellite Systems: GPS, GLONASS, Galileo & More*, New York: Springer, 2008.

[4] Leick, A., *GPS Satellite Surveying*, New York: Wiley, 2004.

[5] Pany, T., and B. Eissfeller, "Code and Phase Tracking of Generic PRN Signals with Sub-Nyquist Sample Rates," *NAVIGATION, Journal of The Institute of Navigation*, Vol. 51, No. 2, 2004, pp. 143–159.

CHAPTER 2

Software-Defined Radio

This chapter revises the software-defined radio concept by providing definitions of terms; it shows the connections between GNSS software radios and developments in the communications area. GNSS software radio applications are presented and it is argued that software radios are suitable for high-end GNSS receivers.

2.1 Definitions

A (true) software radio is a radio consisting of an antenna, an ADC/DAC, and a programmable unit (processor) that runs a piece of software. In contrast, a software-defined radio (SDR), generally, is a design principle to realize one or more components of a radio in a configurable manner. In the extreme case, the term SDR can be applied to radios that support different standards by sharing the same hardware unit (or software modules) but running them with different parameter sets [such as parameter-controlled SDR (PaC-SDR)].

Both radio design methods emphasize the inherent flexibility of software compared to hardware to reduce development efforts and maintain a certain level of flexibility and compatibility in the radio or to realize complex operations efficiently [e.g., a handover between different protocols such as the global system for mobile communication (GSM) and the universal mobile telecommunications system (UMTS)]. As a drawback, SDRs often place higher requirements on the underlying hardware in terms of power consumption and processing performance compared to the radio as an application-specific integrated circuit (ASIC).

J. Mitola presented the software-radio idea to a broad community in 1995. Mitola had previously worked on increasing flexibility and interoperability of radios [1]. A comprehensive summary of SDR, including origins, drivers, and perspectives, was published by Tuttlebee in [2]. Recent aspects of network and user administration that explain flexible mobile communication networks are covered in [3]. These include adaptive protocol layers, organization of different services, security aspects, intelligent (cognitive) radios, spectrum pooling, and automatic modulation detection.

A key characteristic of SDR techniques is the mixture of classical radio technology (e.g., modulation, demodulation, and channel estimation) and modern, mostly object-oriented, software techniques. The hardware of a SDR is required to have increased flexibility for it to be compatible to the SDR software, which follows standards and procedures that mimic the functionally of classical radios.

SDRs can be created using different hardware technologies. The lowest levels of a SDR, which include an antenna and an amplifier, do not differ from conventional hardware radios. At the next level, a SDR might directly digitize the received RF broadband signal, or it may first downconvert and filter it before carrying out digital sampling. It is known that for direct sampling of a broad frequency band (e.g., 0.8–2.2 GHz, which contains all digital civilian mobile phone communication standards) not only is a high sampling rate (e.g., 4.4 GHz) required, a high bit resolution to control signal dynamics of all signals in the frequency range is also required. Assuming a constant sampling rate, the ADC-bit resolution increases by 1.5 bits every 8 years [4]. It is therefore expected that in future SDRs, only limited parts of the RF spectrum will be sampled [3]. Furthermore, power consumption of the ADC and of the subsequent signal-processing units would increase dramatically if they had to operate for a very broad frequency band.

After the signal has been converted to its digital form, it is processed by the radio. Digital signal-processing elements in a SDR exist in configurable ASICs, field-programmable gate arrays (FPGAs), digital signal processors (DSPs), and general purpose processors (GPPs). Note that this work focuses on the receive function of a radio, but that the discussion also applies to the transmission path if the data flow is reversed.

Reconfigurable ASICs are related to the technique called PaC-SDR. For PaC-SDR, several communication standards are investigated with respect to their similarities and differences. Common algorithms are designed, which can then be configured using different parameter sets to work for the investigated communication standards. These algorithms might then be realized as ASICs or software modules. PaC-SDR implies that the standards under consideration do not differ completely. PaC-SDR can be considered as the least flexible form of a SDR as in-field software updates are not possible. In contrast, modular SDR (Mod-SDR) is a technique to run the required algorithms as software modules on exchangeable hardware (i.e., on a processor or on FPGAs). This not only allows reconfiguration of the algorithms with different parameters, but also allows complete reorganization or updating of the algorithm. The main problem with Mod-SDR is that the software and the underlying hardware need to be compatible. Wiesler discusses three methods to run Mod-SDR software [5]:

1. There exists a standardized hardware platform including compilers and libraries. The compiler directly generates binary code for this platform standard. The software is developed specifically for this platform and can be easily downloaded. The disadvantage is that hardware manufacturers have to follow this standard strictly and cannot incorporate their own code or hardware.
2. The compiler generates generic code (similar to the case for JAVA compilers) and the hardware manufacturers provide a translator from the generic code to the binary code of the platform. This allows more freedom for the hardware manufacturers in their hardware development. On the other hand, standardization is still required at the level of the generic language. It is also known that directly generated binary code often runs faster than translated code, which is important for real-time radio applications.
3. Every hardware manufacturer develops its own software specifically for its own hardware radio. There exists no SDR standard, which maintains competitiveness among different manufacturers and protects their intellectual property. Software updates are performed primarily when upgrading to new standards or when errors have to be corrected.

In the context of GNSS SDR, the third solution dominates although some efforts have been undertaken to realize a GNSS radio under the JTRS SCA [6] or as a GNU radio [7]. It is also known that GNSS SDRs often serve as prototype platforms for later ASIC designs or the investigation of different signal-processing algorithms. The use of prototype SDRs is common in research and development but is usually not relevant for deployed radios. Overall, the term SDR covers an enormous number of techniques.

2.2 Communication Radios

A vast number of communication software radio implementations exist because the often small bandwidth of a communication channel allows for an efficient computational implementation on a DSP. The examples in Sections 2.2.1 and 2.2.2 provide insight into the underlying architecture of high-complexity software radios:

2.2.1 GNU Radio

GNU Radio is an open-source project consisting of a library of signal-processing blocks written in C++ and Python scripts used to connect the signal-processing blocks [8]. The GNU radio software runs on the PC platform under various operating systems. The Universal Software Radio Peripheral (USRP) is a dedicated front-end hardware, and allows receiving and transmitting radio signals in the frequency range from direct current (DC) to 2.9 GHz [9]. The USRP can be connected to the GNU radio running on the PC via a USB 2.0 connector. The GNU radio project started in 1998 as a toolkit for learning about, building, and deploying SDRs. It is currently an official GNU project.

The C++ building blocks process an infinite stream of data flowing from their input ports to their output ports. A building block can, for example, implement an FIR filter or perform FM demodulation. Some building blocks connect to hardware like the PC speaker, microphone, or screen on the user side, or to the USRP on the RF side. The Python script defines the sample rate of the radio, selects the C++ building blocks, and defines the connections between the building blocks.

Because the GNU Radio project provides full source-code access, new applications can be developed with relative ease. However, the software does not follow any industry standards and seems to focus on the research sector. GNU Radio applications are numerous and include an HDTV receiver. The USRP includes a small FPGA, which performs minor tasks (e.g., digital upconversions and downconversions) on the FPGA instead of on the PC.

Initial attempts have been undertaken to process GPS signals with the GNU Radio [7]. A GLONASS receiver has been developed that uses the USRP as a front end but uses its own software framework [10].

2.2.2 Joint Tactical Radio System

The largest field of application for SDRs is currently in the military domain. Apart from reduced development costs and increased flexibility, a military SDR approach offers the possibility of reprogramming radios in the field, allowing changes in cryptographic methods to be performed easily. Of importance is the potential to form ad-hoc networks in the field that allow the exchange of various data (e.g., speech, text, maps) between soldiers and the command center.

Several programs exist in different nations to replace legacy radios with SDRs. Due to the size of those programs, great efforts have been undertaken to standardize and coordinate the development efforts. The US Department of Defense (DoD) program Joint Tactical Radio System (JTRS), briefly introduced in Section 2.3, is based on the work of the Joint Program Executive Office for the JTRS [2, 11]. It is important to note that other programs exist as well [12].

The JTRS program was initiated in early 1997 to replace existing legacy radios in the DoD inventory. It evolved from separate radio replacement programs to an integrated effort to network multiple weapon system platforms focusing especially on the last tactical mile. JTRS is intended to link the global information grid (GIG) to the military personnel. This goal would be achieved by developing a family of interoperable SDRs that operate as nodes in a network to ensure secure wireless communication and networking services for mobile and fixed forces.

Within the context of JTRS, waveform is a technical term of importance. A waveform is the entire set of radio and communication functions that occur from the user's input to the radio frequency output and vice versa. A JTRS-waveform implementation consists of a waveform application code, radio set devices, and radio system applications. Originally, there were 32 JTRS waveforms; that number has been reduced to nine: Wideband Networking Waveform (WNW), Soldier Radio Waveform (SRW), Joint Airborne Networking–Tactical Edge (JAN-TE), Mobile User Objective System (MUOS), Single Channel Ground and Airborne Radio System (SINCGARS), Link-16, Enhanced Position Location Reporting System (EPLRS), HF, and UHF SATCOM [13]. The JTRS also considered different form factors that are defined as the linear dimensions and configuration of a device.

The JTRS program tries to exploit key SDR elements to achieve the development of the system within the allocated budget. The key element—government-purpose rights—ensures software reusability among different product lines. This is achieved partly by the JTRS information repository, which is available to industry vendors. Currently, it consists of 3.5 million lines of code, including 15 waveforms and two operating environments.

The next key element—the open-systems architecture approach—focuses on an overarching systems-engineering model. This model directs performance, design specifications, and standards for the operation of the system. It is based on the freely available software communication architecture (SCA) [13]. The SCA is based on a standardized operating environment implementing portable operating system interface (POSIX) and middleware (software that ties together other software blocks) implementing the Common Object Requesting Broker Architecture (CORBA) standard.

The key feature, as shown in Figure 2.1, is that the receiver is a multiprocessor system. The software communications architecture (SCA) requires the underlying operation of the system to realize POSIX application programming interfaces (APIs) between the operating system and the applications. The communication of different parts of the software (objects) running on different processors (GPPs, DSPs, and FPGAs) is of the utmost importance. This is facilitated by the CORBA middleware.

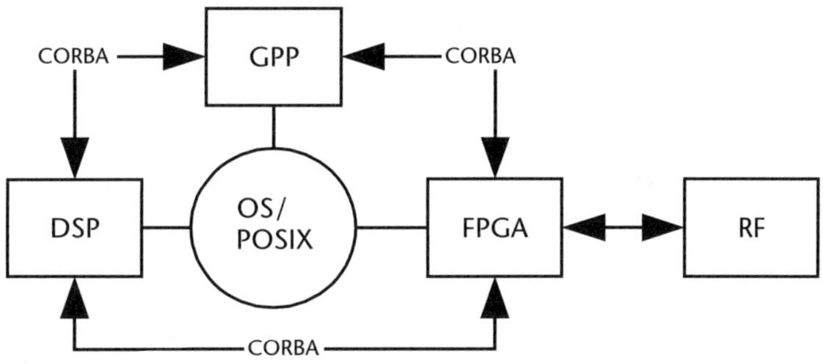

Figure 2.1 Simplified JTRS receiver layout.

CORBA is a specification of middleware whose core is the object request broker (ORB). By using an interface definition language, the CORBA compiler automatically generates the necessary code to launch the objects on a specified processor (or FPGA) and the necessary code for interobject communication. CORBA implementations are hardware specific and come with design tools and specific compilers.

The SCA allows designing and writing code that, to a high degree, can be ported among different hardware platforms. By using an existing SCA implementation, the programmer can concentrate on the radio-specific code instead of developing the whole application framework. It should be noted that, compared to conventional PC software applications, the SDR applications are much more complex because they run simultaneously on different processors (types) and have stringent requirements on the real-time behavior. In Figure 2.2, the different software and hardware layers are shown; on the left side of the figure, interfaces among different software components are illustrated.

It should be noted that the JTRS program experienced significant schedule and cost overruns and was restructured in 2005. The program had to perform in the face of changing requirements and unexpected technical difficulties. In 2007, the first production contracts were issued for next generation SDRs to Thales Communications and Harris.

Regarding GNSS functionality of JTRS radios, it should be mentioned that this can be included as an external hardware block to the radio, not as part of the

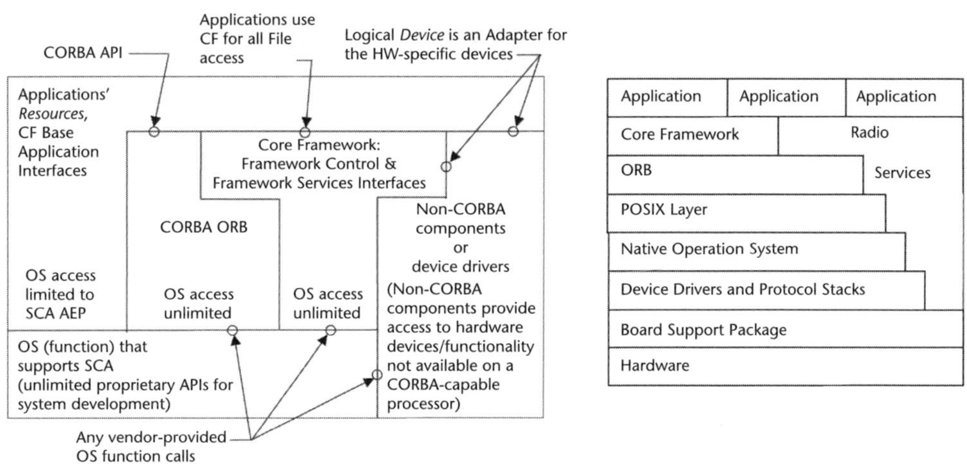

Figure 2.2 SCA horizontal layers [14].

SCA. However, investigations to implement the GPS waveform under the SCA are underway [6].

2.3 GNSS Software Receivers

The idea of a GNSS SDR gained broad attention with the Ph.D. thesis of D. Akos [15]. In that work, the core concepts were implemented and 30 seconds of GPS data was processed until a SPS position fix was achieved. To demonstrate the flexibility, GLONASS signals were successfully acquired and tracked.

The work by Akos performed all the signal processing in the postprocessing mode, but soon thereafter a real-time implementation was achieved on a DSP and on a general-purpose PC [16].

There are many pieces of software from different research groups which do one part (e.g., fast Fourier transform (FFT) acquisition of CDMA signals) or all of the GNSS processing in postprocessing and do not claim to be a SDR [17]. Because of the inherent flexibility of SDR, it is difficult to give a precise definition for a GNSS software radio. For example, a GNSS receiver always has a processor that is responsible for the user interface and the position, velocity, and time (PVT) calculation. Often this processor is used for tracking-loop control. In contrast, simple communication receivers (e.g., an FM radio) can be built without any programmable elements. In a strict sense, even very old GPS receivers were SDRs because at least one programmable element was involved. Additional confusion arose because GNSS receiver prototyping is commonly done using an FPGA for signal correlation. Whereas this approach clearly uses a SDR, within the GNSS community there is a tendency to call only an operational DSP/GPP receiver a *software receiver*.

Within this work, we define a GNSS software receiver as a real-time (capable) GNSS receiver that does all the signal processing after downconversion and sampling on a general-purpose processor. This processor can either be part of an embedded platform or part of a standalone computer. Eventually, some small part of the signal conditioning (e.g., IF filtering or sample rate reduction) can be done by a programmable logic device (PLD) or FPGA. In contrast if the major part of the signal correlation is done in an FPGA, then it shall be called an *FPGA receiver*. If the signal correlation is done on an ASIC, the receiver will be called a *hardware receiver*. Furthermore, we use the terms *receiver* and *radio* synonymously.

2.3.1 Front Ends

From the very beginning of GNSS software receivers, the software development was accompanied by dedicated front-end development. As pointed in an earlier article, three types of software receiver front ends emerged, which are distinguished by their hardware interfaces [18]:

- USB 2.0 front ends, mostly for the PC sector, see Figure 2.3;
- Front ends with proprietary (e.g., serial SPI, SSP) interfaces at the digital-signal level for the embedded sector;

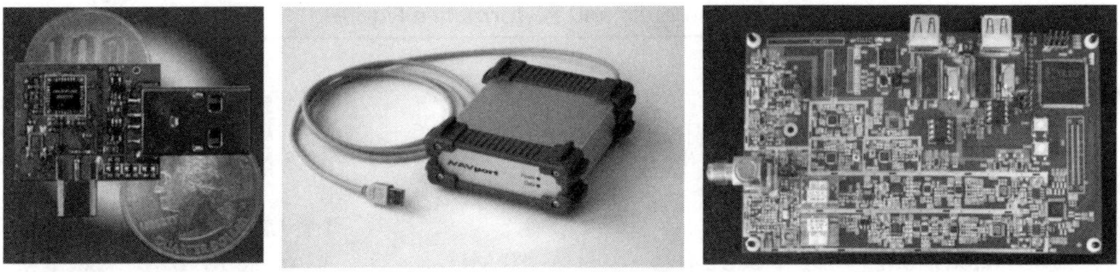

Figure 2.3 Different software receiver front ends, all with USB connectors, from left: embedded [19] (copyright Maxim Integrated Products), R&D L1 [20] (copyright IFEN GmbH), and reference station GNSS receiver [21]. (Copyright Fraunhofer IIS; all images reprinted with permission.)

- Front ends using COTS ADC converters using their own interface or industry standards such as PCI or PCI64.

The number of frequencies, the bandwidth of the sampled frequency bands, the sample rate, and the number of ADC bits influence, to a large extent, the architecture of the software receiver. This is mainly because, with an increasing amount of incoming data, the processing power requirements and IF sample interface specifications change. On the other hand, several front-end architectures such as (super) heterodyne, direct baseband, low-IF, and direct-RF sampling have been used in conjunction with GNSS software receivers, but the influence of the front-end architecture on the software is rather low. It is essentially sufficient to adapt the IF within the software and eventually work with complex (I/Q) or real (I) input samples.

A good front-end design is crucial to achieving high receiver performance because analog design mistakes can often not be compensated by the processing software. Good front ends are, of course, also required for hardware receivers. The only difference between a hardware receiver and a software receiver front end is that the software receiver tries to keep the sample rate as low as possible to reduce the computational load. Typically, the sample rate is chosen slightly above or even below the Nyquist rate, as will be discussed in Section 6.5. In contrast, the hardware receiver might use a higher sample rate to achieve, for instance, a small correlator spacing. For example, a sample rate of 16 MHz for a GPS C/A-code hardware receiver allows a correlator spacing down to 0.064 chip, whereas a sample rate of 4 MHz limits the spacing to 0.25 chip (assuming that different correlators are obtained by shifting the generated PRN code signal by one sample).

To illustrate typical front-end designs for software receivers, consider the four types of front ends whose key parameters are listed in Table 2.1. The first is used for an embedded system focusing on mass-market applications. The second is a research and development USB-based front end for L1 signals. The third has been partly specified at the University FAF Munich and is designed for a GNSS reference station receiver. The fourth is a flexible signal analysis system (created as a laboratory setup at the University FAF Munich) that is able to receive two L-band frequency bands with arbitrary center frequencies.

The information gathered for the embedded system is based on commercial information, and it should be noted that no specification of the IF filter could be found [22]. In fact, the data sheet states that the IF filter order should be tailored

Table 2.1 Front-End Parameters and Performance Figures

Name	*Embedded*	*R&D L1*	*Reference Station*	*Signal Analysis System*
Architecture	Superheterodyne	Heterodyne	Low IF	Heterodyne
Final IF	3.78 MHz	4.348 MHz	53.8 MHz	1–25 MHz
Number of bands	1	1	3	2
Carrier frequency	L1	L1	L1, L2, L5	1–4.2 GHz
Bandwidth	< 3 MHz	10 MHz	13 or 18 MHz	See text
Sample rate	6.5 MHz	23.104 MHz	40.96 MHz	105 MHz
Internal buffer	—	32 MB	None	64 MB
Number of bits	1.5–2.5	1.5	2 or 4	14
Filter characteristic	Discrete L/C or SAW, ceramic	Active discrete	Discrete (Bessel type)	—
Spectral purity	—	< 2.5 dB	< 1 dB	Generally worse
Gain	32–83 dB	36–106 dB	40–90 dB	24 dB
Noise figure	4.7 dB	> 1.6 dB	1.3 dB	2.5 dB
Analog stability	—	High	High	Low
Oscillator	TCXO/crystal	TCXO	TCXO/external	External
IF sample interface	Serial (SPI, SSP)	USB 2.0	2 × USB 2.0	PCI64

to the specific application and that, for example, a standalone GPS receiver will not require stop-band attenuation as strong as that required by a GPS receiver for an integrated wireless handset [19]. Furthermore, the spectral purity [i.e., the maximum height of spikes within the IF power spectral density (PSD)] has not been verified experimentally.

The embedded system uses two downconversion stages to bring the GPS L1 signal down to 3.78 MHz. Eventually, this allows for the use of a wider range of reference oscillator frequencies compared to a single downconversion stage. It includes a SPI serial interface. The SPI is a de facto standard; many processors for embedded systems include SPI controllers.

The R&D L1 front end converts the RF to the IF of 96 MHz in a single step and then uses bandpass sampling to further downconvert it to 4.348 MHz [20]. The sampled IF signal is buffered before transfer over the USB port.

The reference station front end uses a heterodyne architecture. Each RF frequency band is (separately) downconverted to 53.8 MHz. A discrete bandpass filter limits the bandwidth before sampling and a real-valued IF signal is digitized. The front end uses a 2- or 4-bit ADC. It has been demonstrated that the 4-bit ADC provides a significant performance gain if interference is present by being less sensitive to saturation effects.

The highest flexibility can be achieved with a laboratory setup consisting of a commercial ADC card (an ICS-572B is used here) with commercial off-the-shelf (COTS) mixers and local oscillators. The components are connected through SMA connectors. COTS mixers and low-noise amplifiers (LNAs) are available up to frequencies of several GHz. The RF is downconverted to a selectable IF with a single mixer. The ADC Nyquist bandwidth is 52.5 MHz. COTS lowpass filters are used to discard higher-frequency components. The design requires that the antenna LNA already includes an RF filter to suppress the image frequency at RF-2xIF. Furthermore, the experimental setup does not give a completely clean digital IF signal and several spikes appear. The high number of bits results in a high dynamic range of over 80 dB, making the system well suited for interference/jamming applications.

When comparing the four front ends, one clearly sees the increase in the number of frequency bands, bandwidth, bits, and sample rate from the embedded solution to the signal analysis system. The noise figure is best for the reference station front end, but it should be emphasized that the total receiver noise figure is mostly determined by the low-noise amplifier (LNA) that is either integrated into the antenna or located directly behind the antenna. The noise figure of the front end is less important. All four front ends sample a real signal. This avoids the use of a second ADC and simplifies the data transfer.

2.3.2 Illustrative Applications

Sections 2.3.2.1 through 2.3.2.4 will present applications where GNSS SDRs have already been established. Software receivers are excellent teaching tools; two books have already been published on this topic [23, 24].

2.3.2.1 ASIC Replacement

The most obvious application of a GNSS SDR receiver is to replace existing hardware chip solutions. For laptops/mobile phones/PNDs (or in general, for embedded systems), there already exists laboratory software as well as commercial solutions. Software-GPS-enabled laptops have been on the market since 2008 [25].

On an embedded system or, more specifically, on a mobile phone or personal navigation device (PND), a general-purpose processor is available. Therefore, the possibility exists to run a GNSS SDR on this processor. The advantages are that a GNSS baseband chip can be avoided and that the GNSS receiver interacts more flexibly with other software on the same system. Provision of GNSS-aiding data via the mobile phone data link, for example, is more easily achieved. Also, integration with other sensors (e.g., WLAN) or any other user software provides potential advantages in obtaining an integrated position solution. The GNSS software can later be upgraded for upcoming systems and services. An RF front end and an antenna are still required and have to fulfill the same requirements as a hardware receiver. The disadvantages of a software solution compared to a hardware solution include increased processing load and power consumption.

Some design ideas exist for creating a complete GNSS SDR as an embedded system, where a dedicated processor not running the application software is used for the GNSS functions; that is, the GNSS SDR has the same form factor as a hardware receiver. The processor basically replaces the baseband chip. The module itself provides the same interfaces as a hardware receiver, but retains SDR flexibility.

The most critical issues for a SDR embedded receiver are the limited resources in terms of processing power, memory size, and memory bandwidth. Commercial solutions such as SiRFSoft are explicitly optimized for market-leading platforms like the Intel XScale [26]. Those platforms provide dedicated instructions to speed up signal-processing algorithms, like multiply-and-add commands. There are also other vendors of commercial software receivers [22, 27–29]. The processing load necessary to track one satellite signal varies between 3 and 20 MIPS, depending on the power of the received signal.

2.3.2.2 Server-Side GNSS Radio

For a certain area of GNSS applications, completely separating the GNSS front end from the signal processing and navigation solution proves useful. The latter two operations run on a high-performance computer (server) that is installed at a fixed site. The server and front end are connected via an analog or digital data (radio) link, as shown in Figure 2.4; they can be located at separate locations.

The GNSS SDR running on the server retrieves the GNSS signal from one or more front ends and provides the position solution. The advantage is that the server potentially has access to assistance or differential correction data and can thus provide a better position solution. Furthermore, the server is able to store and reprocess the GNSS signal in a postmission mode to further improve the position solution.

Both analog and digital data links are used. For analog transmission, the received analog GNSS signal is converted to another frequency band (e.g., S-band) before transmission. The analog link requires, at a minimum, the same bandwidth as the GNSS signal(s) to be translated. At the server, the translated signal is downconverted and sampled. Digital translation provides the advantage of using existing digital data links that can be implemented more efficiently (e.g., with a lower bandwidth) than an analog link. It should be noted that, in either case (analog or digital), the link itself acts like a long cable and normally has no influence on the positioning result. Furthermore, the data link does not need to be real-time and can be created as a storage device. The server-side GNSS radio has found applications in the areas listed in Table 2.2.

For example, in the article by Brown, a GPS tracking device is presented that periodically records small portions (52 ms) of the GPS IF signal [30]. The device sends this signal's snapshots over a radio network to a server that acquires and tracks the signals and computes and visualizes the position solution. The high-sensitivity signal-processing algorithms utilize assistance data and are run on the server instead of on the mobile device. According to Brown, capturing the GPS signal requires an energy of 15 mJ and sending it over the network requires 70–231 mJ. In contrast, a high-sensitivity chipset solution requires between 100–1,340 mJ per fix. Thus, a server-side GPS SDR reduces power consumption at the client and increases the system's lifetime.

The missile tracking system outlined by Won uses an S-band link to transmit the digital IF signal of the GPS front end attached to the missile [31]. The server-

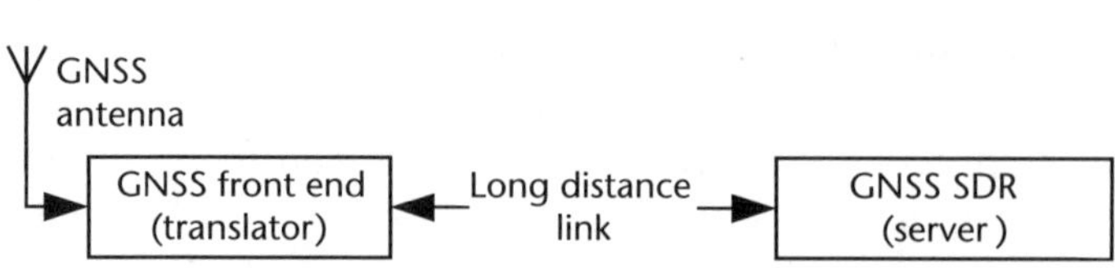

Figure 2.4 Block diagram of a GNSS translator system employing a server-side SDR.

Table 2.2 Server-Side GNSS Radio Applications

Name	*Performance Measure*	*Aiding Data Utilization*	*SDR Advantage*
Missile tracking	TTFF	Ephemeris, coarse PVT	High dynamics
Snapshot for E911	TTFF, accuracy	Ephemeris, coarse PVT	Reduced mobile unit complexity
Asset tracking, image tagging	Sensitivity	Ephemeris, navigation data bits	Increased sensitivity, reliability, low power consumption
GNSS science	—	All	Realization of new algorithms
Signal post processing	Sensitivity, accuracy, availability	All	Highest possible performance

side GPS receiver has a very short TTFF and is capable of tracking the high signal dynamics.

An asset (e.g., container or car) tracking system could periodically record small portions of IF samples that are temporarily stored in a nonvolatile memory. At some point, the data is transferred to a PC that runs the server-side radio. By utilizing ephemeris data and navigation data bits, the server-side radio can do a data wipe-off to increase the acquisition sensitivity. Depending on the duration of the sample snapshot, even high positioning accuracy can be obtained. A related mass-market application could be to localize images shot by a digital camera. The IF samples would then be stored with the image [32].

It should also be noted that all receiver performance parameters, such as sensitivity, accuracy, and availability can be perfectly optimized if a GNSS signal is recorded and processed in a postprocessing manner. For example, in postprocessing mode, an unlimited number of correlators can be used.

2.3.2.3 GNSS Reflectometry

One of the applications where GNSS SDRs have gained much merit is in the area of bistatic remote sensing. In this case, the GNSS signal is reflected by the earth's surface and is received by a receiver mounted on various platforms, ranging from a tower over an airplane to space-borne receivers. The usefulness of these signals was realized over a decade ago when scientists proposed a new method of ocean altimetry [33]. Meanwhile, the method has been adapted to measure sea roughness, wind speed and direction, ice coverage, and soil moisture and has additional potential applications in other remote-sensing areas.

In the case of a space-borne receiver, the bistatically scattered signal is diffusely radiated from a large area of the earth's surface, often covering tens of kilometers and thousands of Hertz in Doppler as the observation geometry changes over the scattering surface or glistening zone [34]. The receiver tracking the reflected signal has to correlate over a wide code phase and Doppler range involving a multitude of correlators. Furthermore, the reflected signals are attenuated and show strong amplitude fluctuations, such that high-end signal processing is generally required.

The United Kingdom's Disaster Monitoring Constellation satellites are examples of a SDR implementation for remote sensing [35]. These satellites normally use two GPS antennas for rapid positioning. For remote sensing, a third antenna (an L1 high-gain, down-looking, left-handed polarized antenna) was added. The IF signal of the down-looking antenna was sampled with a rate of 5.71 MHz and with 2 bits. The data was stored on an onboard, solid-state data recorder. About 20 seconds of data was sampled and later transmitted to the ground station for offline processing. The reflected signals have been detected and Doppler delay maps have been produced. The data is available for other researchers to ensure that the valuable signal is optimally analyzed [34].

2.3.2.4 GNSS Constellation Signal Generators and Spoofers

SDR technology can also be applied to generate GNSS signals in a flexible manner. To the author's knowledge, all available GNSS signal generators that are able to simulate received GNSS signals of a whole satellite constellation use either FPGAs or GPPs to generate the signals. A GNSS software signal generator based on a GPP with RF output capability has been described by Pósfay [36]. The work [37] describes a GNSS signal recorder and playback system used to collect real signals (with traffic information) in the field and to play them back in the laboratory. Humphreys describes a GPS spoofer that simultaneously receives GPS C/A-code signals and transmits an adaptive spoofing signal. Both functions are created with SDR technology on a DSP basis [38].

2.3.3 High-End GNSS Software Receivers

This work endorses the use of GNSS SDRs as high-end receivers. This is an area where software receivers may gain increased importance in the future. New and optimized signal-processing algorithms may be more easily developed on these platforms and current and future processor technology will provide an abundance of processing power. Software receiver developments are also discussed by the International GNSS Service (IGS), an organization that generates precise ephemeris and terrestrial reference frames based on a network of permanent GNSS receivers [39].

Signal processing (acquisition and tracking) in GNSS chips commonly boils down to the evaluation of the cross-correlation function of the received signal with locally generated replica signals. Signal processing in a chip is limited by the following factors (among others):

- A low number of bits are used to represent the received and the locally generated signal.
- The received signal is processed as a stream (e.g., older signal samples are not available for a reiteration).
- Development efforts are generally high.
- Inclusion of high-rate aiding data such as data wipe-off, (semi-)codeless techniques, or deep GPS/INS integration requires explicit synchronization lines.

The GNSS signal and system design took into account these limitations. For example, the GPS signals at baseband can be represented by one bit. The cross-correlation among different signals is minimized so that a tracking channel can generally ignore the influence of PRN codes broadcast from satellites other than the one whose signal is being tracked.

However, these restrictions still limit the signal-processing capabilities of a GNSS receiver and those limitations become relevant if the receiver performance needs to be optimized. Currently, the receiver core technologies listed in Table 2.3 have been identified as areas where improvements can still be expected.

The presented limitations are not intrinsically hardware-receiver specific, but could be solved more easily with a software solution. A high-end GNSS SDR will run on a high-performance computer that generally consumes increased electrical power. If it is assumed that sufficient electrical power is available (e.g., via a power line), then a high-performance computer can provide nearly unlimited processing power for the radio, simply by using multiple processors or cores. Because these computers are equipped with large storage devices (RAM or hard discs) they overcome nearly all deficiencies mentioned in Table 2.3. This work proposes algorithms that exploit this increased processing power and demonstrates how the resulting GNSS SDR outperforms an ASIC-based solution.

Table 2.3 Current High-End Hardware GNSS Receiver Deficiencies

Technology	*Measure*	*Hardware Receiver Deficiency*
Acquisition	Sensitivity	Still a low number of correlators are used, lack of proper interference handling, no data wipe-off, short (≤ 20 ms) coherent integration times
Tracking	Multipath error	Generally suboptimal multipath-mitigation techniques are used and multipath reflections are not tracked
	Thermal noise error	No data wipe-off, independent channel tracking used (no vector tracking), locally generated signal not band-limited
	Transient errors	No reiteration of signal correlation to follow high dynamic signals
	Interference	Only crude interference mitigation (e.g., pulse blanking) used, no adaptive filtering, no cancellation
	GPS C/A-cycle slips	Short (≤ 20 ms) coherent integration times, no data wipe-off, thus Costas discriminators are used
	GPS/INS integration	Ultratight coupling seldom used due to high implementation and synchronization efforts, loose or close coupling schemes predominate; coherent integration is based on a constant line-of-sight velocity, ignoring user accelerations
Positioning	TTFF	TTFF is long if no aiding data is used
	Accuracy	Propagation channel characteristics (e.g., multipath parameters) not used in positioning solution
	Carrier phase	Double differences are formed with carrier phase pseudoranges; less cycle slips are expected if double differences are formed with correlator values
Signal Level	Interference	Interference is normally neither detected nor mitigated
	Flexibility	Inclusion of new navigation signals is generally difficult
	Reliability	Signal-processing problems (e.g., cycle slips, loss of lock) cannot be traced back because the input signal is discarded after processing

Recent developments in ultramobile PCs will allow high-end GNSS signal-processing algorithms to run on a platform with reduced power consumption. Ultramobile PC components form the basis for the innovative real-time kinematic (RTK) receiver concept discussed in Chapter 10.

2.4 Technology Evaluation and Discussion

Previously, it was demonstrated that no clear separation between a hardware receiver and software receiver could be drawn. Important parts of the signal processing (e.g., tracking-loop closure) are often part of the receiver software (firmware) in a typical GPS-chip solution. The chips may even utilize small processors to calculate an FFT to acquire signals. On the other hand, software-GNSS radios utilize dedicated hardware units to accelerate the signal processing. This may range from vector multiply-and-add units of high-performance processors to FPGAs. These units often have a strict structure and the software utilizing them has to be optimized specifically for them. Consequently, the GNSS SDR gives up some of its flexibility.

The distinction between a software and hardware receiver seems not to be very helpful. However, the amount of work a receiver manufacturer puts into developing the receiver hardware platform is important. This may range from a developer choosing a good laptop, to building his own CPU board, to finally designing an ASIC. The hardware platform determines important real-time performance parameters like size, power consumption, and number of correlators. If only COTS components are used (e.g., a low-end PC), these parameters are normally far from optimal. On the other hand, wisely chosen processors integrated into an optimized board may outperform classical ASIC correlator chip solutions for certain applications. However, developing a multiprocessor system may require similar efforts to that of an ASIC design.

At the moment of writing, it is expected that SDR solutions based on COTS components will continue to have some disadvantages in terms of size and power consumption; these criteria are relevant for many positioning applications. COTS software receivers will enter the market at the high-performance sector as reference stations and in niche markets, like specialized RTK receivers. Because the receiver life cycle is typically several years in the field of navigation, the GNSS SDR technology will be mature when Galileo reaches its final operational capability. We also expect that DSP or embedded processor-based solutions will be found in mass-marketed products in the coming years.

References

[1] Mitola, J., "The Software Radio Architecture," *IEEE Commun. Mag.*, Vol. 33, No. 5, 1995, pp. 26–38.

[2] Tuttlebee, W., *Software Defined Radio: Origins, Drivers and International Perspectives*, New York: Wiley, 2002.

[3] Rhiemeier, A.-R., *Modulares Software Defined Radio*, University Karlsruhe (TH), Kaiserstraße 12, D-76131 Karlsruhe, http://digbib.ubka.uni-karlsruhe.de/volltexte/1000001174, 2004.

[4] Walden, R. H., "Analog-to-Digital Converter Survey," *Areas Commun.*, Vol. 17, No. 4, 1999, pp. 539–550.

[5] Wiesler, A., *Parametergesteuertes Software Radio für Mobilfunksysteme*, University Karlsruhe (TH), Kaiserstraße 12, D-76131 Karlsruhe, http://digbib.ubka.uni-karlsruhe.de/volltexte/3222001, 2001.

[6] Brown, A., and D. Babich, "Implementing a GPS Waveform Under the Software Communications Architecture," *Proc. 19th Int. Technical Meeting of the Satellite Division of the Institute of Navigation (ION-GNSS) 2006*, Fort Worth, TX, September 26–29, 2006, pp. 2334–2345.

[7] Danielsen, T., "Creating a GNSS Receiver From Free Software Components," *Proc. 20th Int. Technical Meeting of the Satellite Division of the Institute of Navigation (ION-GNSS) 2007*, Fort Worth, TX, September 25–28, 2007, pp. 2731–2741.

[8] Blossom, E., "Exploring GNU Radio," Free Software Foundation, http://www.gnu.org/software/gnuradio/doc/exploring-gnuradio.html, 2004.

[9] Ettus Research, "Homepage of Ettus Research," Ettus Research LLP, http://www.ettus.com, 2007.

[10] Peng, S., and B. M. Ledvina, "A Real-Time Software Receiver for the GLONASS L1 Signal," *Proc. 21st Int. Technical Meeting of the Satellite Division of the Institute of Navigation (ION-GNSS) 2008*, Savannah, GA, September 16–19, 2008, pp. 2268–2279.

[11] Joint Program Executive Office for the Joint Tactical Radio System, "Joint Tactical Radio System," U.S. Department of the Navy, http://jpeojtrs.mil/, 2007.

[12] Finish Defense Forces, "Finish Software Radio Project," Finnish Defense Forces Telecommunication Laboratory and Centre for Wireless Communications, http://www.mil.fi/laitokset/pvtt/fsrpbook.pdf, 2007.

[13] Anderson, S., and S. A. Davis, "The Joint Tactical Radio System—Reloaded," *CHIPS - the Department of the Navy Information Technology Magazine*, Vol. July–September, No. 2006, pp. 6–9.

[14] Bard, J. D., "Joint Tactical Radio System," Space Coast Communication Systems, Inc., http://www.spacecoastcomm.com/docs/JTRS.pdf, 2003.

[15] Akos, D., *A Software Radio Approach to Global Navigation Satellite System Receiver Design*, Athens, OH: Ohio University, 1997.

[16] Akos, D. M., et al., "Real-Time GPS Software Radio Receiver," *Proc. Institute of Navigation National Technical Meeting (ION-NTM) 2001*, Long Beach, CA, January 22–24, 2001, pp. 809–816.

[17] Cheng, U., W. J. Hurd, and J. I. Statman, "Spread-Spectrum Code Acquisition in the Presence of Doppler Shift and Data Modulation," *IEEE Trans. Commun.*, Vol. 38, No. 2, 1990, pp. 241–250.

[18] Pany, T., J.-H. Won, and G. Hein, "GNSS Software Radio, Real Receivers or Just a Tool for Experts?" *InsideGNSS*, Vol. 1, No. 5, 2006, pp. 48–56.

[19] Maxim Integrated Products, "MAX2741, Maxim Integrated L1-Band GPS Receiver (19-3559; Rev 0)," Maxim Integrated Products, Inc., http://www.maxim-ic.com/products/wireless/gps/max2741.cfm?CMP=490, 2005.

[20] IFEN GmbH, "NavX®-NSR - GPS/Galileo Navigation Software Receiver, Brochure," IFEN GmbH, http://www.ifen.com/content/flyer/NavX-NSR-Flyer.pdf, 2007.

[21] Fraunhofer Institut für Integrierte Schaltungen, "Triband Frontend L1, L2 and L5 with USB," Fraunhofer Institut für Integrierte Schaltungen, http://www.iis.fraunhofer.de, 2008.

[22] NXP Software, "swGPS™ Personal, Software GPS solution for PNDs, PMPs and smartphones," NXP Software, http://www.software.nxp.com/assets/Downloadablefile/swGPS-personal_vs3-13467.pdf, 2007.

[23] Tsui, J. B. Y., *Fundamentals of Global Positioning System Receivers: A Software Approach*, 2nd ed., New York: Wiley, 2005.

[24] Borre, K., et al., *A Software-Defined GPS and Galileo Receiver: A Single Frequency Approach*, Boston, MA: Birkhäuser, 2007.

[25] PR Newswire, "ASUS Selects NXP Software's swGPS™ for World's First Mainstream GPS-Enabled Notebook PC," http://www.prnewswire.com/cgi-bin/stories.pl?ACCT=109&STORY=/www/story/09-13-2007/0004662288&EDATE=, 2008.

[26] SiRF Technology, "Company homepage," SiRF Technology, Inc., http://www.sirf.com, 2008.

[27] Fastrax, "White paper on Fastrax Software GPS receiver: Smart Positioning with Fastrax Software GPS Receiver," Fastrax, Ltd., http://www.fastraxgps.com, 2008.

[28] CSR plc, "LC7830FM: CSR's combined solution for GPS and FM," CSR plc, http://www.csr.com, 2008.

[29] Trimble Navigation, Ltd., "Trimble News Release: Trimble and u-Nav Offer High Performance, Low Power GPS Chipset Solutions, TrimCore NEu," Trimble Navigation, Ltd., http://www.trimble.com/news/release.aspx?id=030806a, 2006.

[30] Brown, A., P. Brown, and J. Griesbach, "GeoZigBee: A Wireless GPS Wristwatch Tracking Solution," *Proc. 19th Int. Technical Meeting of the Satellite Division of the Institute of Navigation (ION-GNSS) 2006*, Fort Worth, TX, September 26–29, 2006, pp. 2883–2888.

[31] Won, J. H., S. J. Ko, and J. S. Lee, "Implementation of External Aiding and Novel Pseudorange Generation Algorithms for Fast TTFF of GPS Translator System," *Proc. 12th GNSS Workshop*, Jeju Island, Korea, December 1–2, 2005.

[32] NXP Software, "NXP SnapSpot GPS Technology and JOBO photoGPS capture a location in an instant," NXP Software, http://www.software.nxp.com/?pageid=139, 2007.

[33] Martin-Neira, M., "A Passive Reflectometry and Interferometry System (PARIS)—Application to Ocean Altimetry," *ESA Journal*, Vol. 17, No. 4, 1993, pp. 331–355.

[34] Gleason, S., "An Open Source Software Receiver for Bistatic Remote Sensing," *Proc. 20th Int. Technical Meeting of the Satellite Division of the Institute of Navigation (ION-GNSS) 2007*, Fort Worth, TX, September 25–28, 2007, pp. 2742–2748.

[35] Gleason, S., et al., "Detection and Processing of Bistatically Reflected GPS Signals from Low Earth Orbit for the Purpose of Ocean Remote Sensing," *IEEE Trans. on Geoscience and Remote Sensing*, Vol. 43, No. 6, 2005, pp. 1229–1241.

[36] Pósfay, A., T. Pany, and B. Eissfeller, "First Results of a GNSS Signal Generator Using a PC and a Digital-to-Analog Converter," *Proc. 18th Int. Technical Meeting of the Satellite Division of the Institute of Navigation (ION-GNSS) 2005*, Long Beach, CA, September 13–16, 2005, pp. 1861–1870.

[37] Averna, The Test Engineering Company, "URT, Universal Receiver Tester," Averna, Inc., http://www.averna.com, 2008.

[38] Humphreys, T. E., et al., "Assessing the Spoofing Threat: Development of a Portable GPS Civilian Spoofer," *Proc. 21st Int. Technical Meeting of the Satellite Division of the Institute of Navigation (ION-GNSS) 2008*, Savannah, GA, September 16–19, 2008, pp. 2314–2325.

[39] Humphreys, T. E., L. E. Young, and T. Pany, "Considerations for Future IGS Receivers," *Proc. American Geophysical Union Fall Meeting*, San Francisco, CA, December 15–19, 2008.

CHAPTER 3

GNSS Receiver Structure and Dataflow

The architecture of a GNSS software receiver determines, to a large extent, the potential receiver applications; it is shaped by the underlying hardware capabilities. The envisaged processing modes (real-time and postprocessing) are important to the architecture as well as to the targeted-use cases, such as a standalone reference-receiver operation, snapshot mode, or receiver development environment.

In this chapter, the receiver architecture of the ipexSR is outlined. It targets a real-time-capable, multifrequency GNSS receiver running under a nonreal-time operating system like Microsoft Windows. It exploits the capabilities of multicore CPUs by separating the tasks into several threads; it is prepared to accommodate external sensor data in addition to the GNSS signals. A screenshot of this software GNSS receiver is shown in Figure 3.1. Its real-time behavior will be described in Section 3.3.1 and performance results are included in Section 3.4.

The architecture of the receiver is governed by the receiver block diagram that describes the constituting receiver modules and the exchanged data. The execution diagram shows the timeline and the synchronization of the different operations performed by the receiver. The real-time requirement dictates, to a large extent, the data flow within the receiver. An optimal implementation of the receiver core operations, correlation, and FFT (described in Chapter 9) is achieved by grouping the GNSS samples in fixed-length units instead of processing them on a sample-by-sample basis (as would be done using a hardware receiver). Because this GNSS sample arrangement influences the block diagram as well as the execution flow, it shall be described first.

3.1 GNSS Sample Handling

In the following section, the GNSS sample handling shall be described. The focus will be on the real-time mode. In addition, several possibilities for postprocessing shall be described.

It should be mentioned that by counting the GNSS signal samples, the receiver's *internal time base* is realized, which shall also be described below.

For GNSS receivers, there exist several methods of how IF signal samples can be accessed by the signal processing unit. An overview is shown in Figure 3.2. In the simplest case, an incoming IF sample is processed immediately after its output by the ADC.

This method is used in most hardware-based receivers that do not buffer the IF samples and have the possibility of processing the samples instantaneously within multiple channels (this implies, for example, that all code generators and correlators run at the ADC rate or faster). If the hardware receiver uses a massive parallel

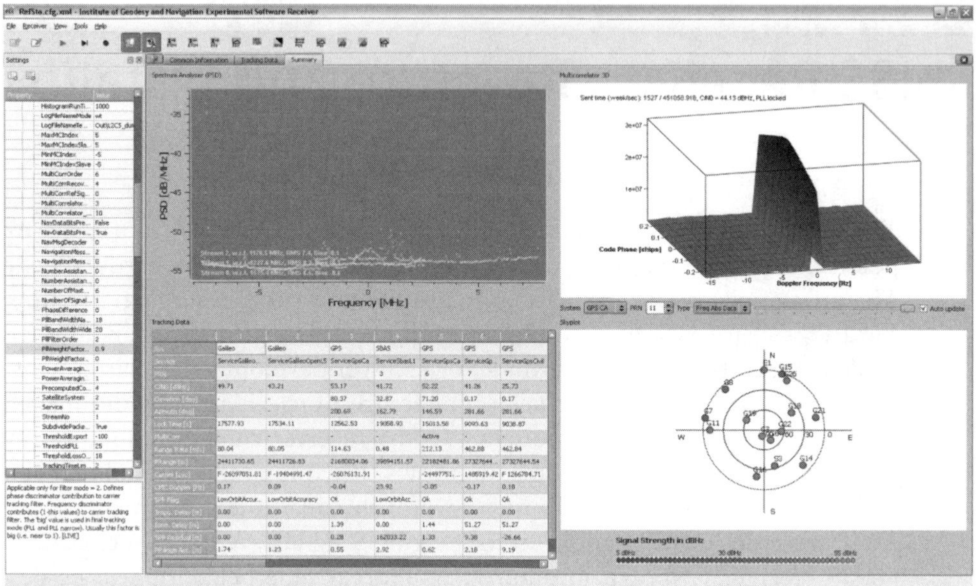

Figure 3.1 Screenshot of the ipex software receiver.

correlator structure for signal acquisition, a small sample buffer may already be required to organize the data flow. If an FFT unit is used in a hardware receiver, a small sample buffer that has to be filled up with samples before starting the FFT is definitely required. In the ipexSR, samples are grouped into batches and the signal processing unit can randomly access the samples within a batch. Once processing of one batch is finished, the samples are discarded. The most powerful method is, however, to allow random access to all samples. This targets a postprocessing application, but even in this case a signal processing algorithm consequently exploiting the random access possibilities is very difficult to realize, simply due to the large amount of data involved (see also Section 4.2.3 and the remark on snapshot acquisition).

3.1.1 Real-Time Mode

In a real-time GNSS SDR, the received GNSS signals are amplified, filtered, and eventually down converted by the front end. Then the signal is digitized and the

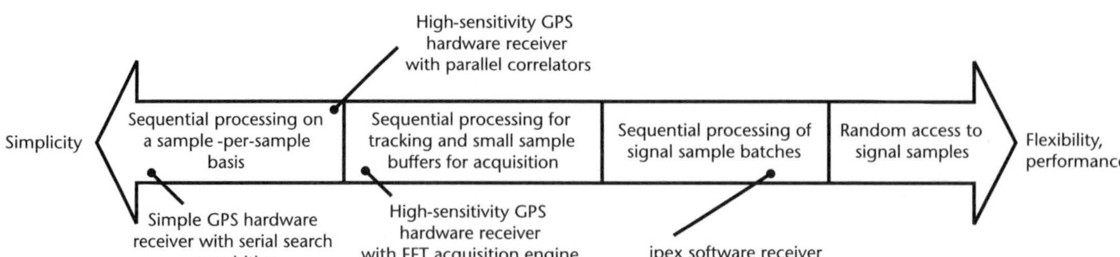

Figure 3.2 IF sample access methods.

digitized samples are transferred to the processing platform running the software receiver. The platform sees the received signal as a continuous stream of samples. There may exist multiple synchronous streams of samples if multiple frequency bands are received (e.g., GPS L1 and L2).

3.1.1.1 Timing

The streams are required to be synchronously derived from the ADC output; no sample must be lost when the data is transferred from the ADC to the processing platform. The ADC itself is required to be synchronized to the front-end oscillator, which also controls signal down conversion. As a consequence, the internal time base can be realized by counting the number of received GNSS signal samples. The internal receiver time is determined by a startup epoch value plus the number of received samples divided by the sample rate. The value of the startup epoch can, for example, be read from the PC clock, but can also be set to zero. If multiple streams are present, then every stream can be used for time-based calculations because all streams are required to be synchronous. After reading the startup epoch, external clocks (like the PC clock) are no longer used and the internal time base is completely based on counting the received samples. After a PVT solution is available, the receiver clock error is obtained; it estimates the difference between the GNSS time scale and the internal receiver time scale. It is composed of a constant part related to the startup epoch reading and a time variable part related to the front-end oscillator stability.

It should be noted that reading and adjusting the PC clock and establishing a relationship to the IF sample stream can be quite imprecise (e.g., on the order of several tenths of milliseconds under the Windows operating system due to the 10-ms granularity of the Windows task scheduler). There exist special items of hardware, such as an IEEE1588-enabled PCI network interface card, that allow sample synchronization down to 100-ns accuracy, provided that a hardware connection between the front end and the interface card is established [1]. If the IEEE1588 time server is properly configured, the samples can be directly synchronized to a GNSS time scale without the necessity to compute a PVT solution. This could be important for indoor positioning, where precise synchronization with approximate coordinates can shrink the acquisition search space drastically.

3.1.1.2 Batch-Processing

The GNSS sample streams are based on a relatively high sample rate being, at minimum, several megahertz. Because of overhead operations (loop control, data movement) direct processing of each single sample on a sample per sample basis is computationally too demanding for real-time operations in a GNSS SDR (in contrast to hardware receivers based either on ASICs of FPGAs). Instead, the samples are grouped into units called batches. Certain processing parameters (e.g., NCO rates) are kept constant for all the samples in one batch. This approach is commonly called *batch-processing*. There exist several possibilities of what to call those batches; in the following, the terms packets and frames will be used. These terms do not have any specific meaning by themselves, but they distinguish several ways of how to group GNSS signal samples. They will be explained later.

One key requirement of batch-processing is to make the batches as long as possible to speed up the algorithms. On the other hand, short batches are required to have, for example, high tracking loop update rates, high positioning output rates, or low latency. There must be a trade-off regarding both requirements.

Within the ipexSR, the received GNSS signal samples are first grouped into *packets*, as shown in Figure 3.3. A packet represents a relatively large number of GNSS signal samples covering a time span of approximately 0.01–0.4 second. The received GNSS signal samples are collected by the USB front-end driver (see Figure 3.4). After a packet is filled up with samples it is passed to the IF sample buffer. The master receiver (see Figure 3.4) is informed when a packet is ready and then retrieves the packet from the buffer. The packet size ultimately determines the *latency* of the GNSS receiver because the processing only starts if a packet is completely filled up with samples. Furthermore, the packet size also determines the position output rate of the GNSS receiver; at maximum, one position solution can be calculated per packet (note that this is particular to the ipexSR and might also be solved differently).

GNSS signal samples are passed to the signal acquisition unit in the form of packets because signal acquisition typically operates on a rather large number of samples to achieve high acquisition sensitivity. On the contrary, signal tracking is synchronized to the GNSS signal PRN code structure and has to follow the (eventually rapid) antenna movement. Therefore, signal tracking operates on shorter batches of IF samples, called *frames*. The size of a frame is specific for a receiver (see Figure 3.4) or, in other words, the size is specific for a GNSS service. It is required that the packet size is an integer multiple of all frame sizes, as shown in Figure 3.3. Additionally, it is required that the frame size is shorter than the primary coherent integration period. The primary coherent integration period is defined by the PRN code length divided by a user-definable integer (usually 1 for most civil GNSS signals, with the exception of L2CL, where 75, for example, might be used). At maximum one correlation value is produced per frame. More specifically, in most cases one frame gives one correlation value. There may, however, be cases where a frame does not contain an integrate and dump epoch for a specific GNSS signal. In that case, no correlation value is produced by this frame. The samples contained within one frame are typically correlated with internally generated reference signals whose parameters (code rate, Doppler) are constant during the correlation interval. Therefore, the frame size also influences the maximum GNSS dynamics, which can be tracked by the receiver (i.e., higher signal dynamics may require shorter frames).

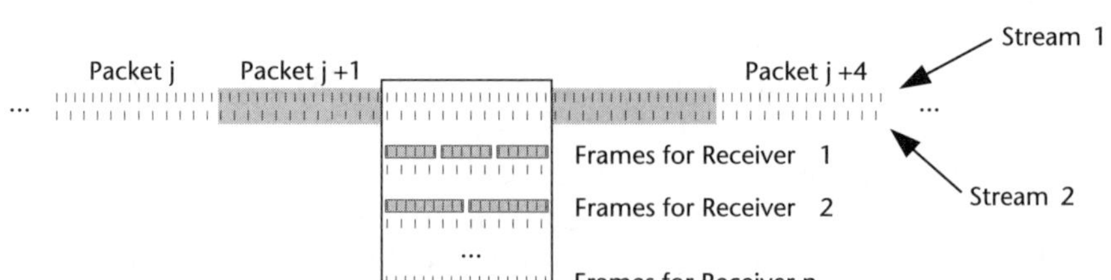

Figure 3.3 IF sample subdivision.

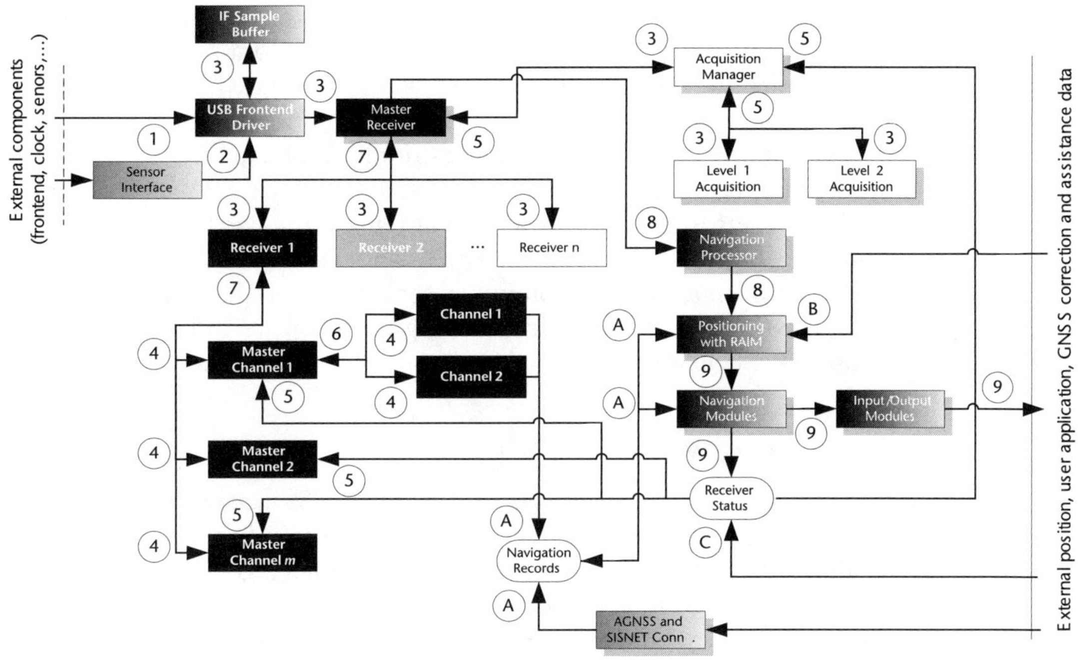

Figure 3.4 GNSS software receiver block diagram (rectangular blocks of the same style run in same thread; rounded blocks represent global data elements)

3.1.1.3 Real-Time Modes

For a GNSS SDR, it is useful to distinguish two different definitions of real-time operation modes: *soft* and *hard* real-time capability.

- Soft real-time capable: The GNSS SDR processes continuously GNSS signals fast enough that the latency of the processing output remains on average, constant.
- Hard real-time capable: The GNSS SDR processes continuously GNSS signals and outputs its processing results within a guaranteed latency with respect to the received signals.

For both cases, the core software requirement is that the software runs quick enough to process one packet (e.g., 0.4 second) of received samples faster than the time span covered by one packet (e.g., 0.4 second). For the first case, this requirement refers to the *average* packet processing time, whereas the second case refers to the *maximum* processing time.

For the second case of hard real-time capability, it is useful to establish a controllable hardware connection between the GNSS signal input (e.g., the ADC) and the GNSS SDR output. This is required because a hard real-time capable system normally interacts with other systems (e.g., a navigation system) and must output their data in "real-time." The hardware connection has to be controlled by the GNSS SDR and the SDR may precisely determine the output values (e.g., a pulse-per-second signal) for a given (future) output epoch. To calculate these values, the SDR may use extrapolation techniques to compensate for the latency caused by the processing and the latency introduced when changing the output state (e.g., changing the pulse-per-second signal state from high to low). The error introduced by the extrapolation can be exactly evaluated (and consequently be minimized) because the receiver itself relies on assumptions on the signal dynamics (e.g., tracking loop filter bandwidths or positioning Kalman filter settings). Also, hardware receivers have latencies of at least 1–20 ms caused by the tracking loop update only being performed after the coherent integration has finished [2].

The ipexSR is soft real-time capable. There is no hardware synchronization between the receiver output (e.g., the computers serial ports) and the GNSS front end, and it would not make much sense in changing the software to be hard real-time capable, a task that would require porting the source code onto a real-time OS.

3.1.2 Postprocessing Mode

In a postprocessing mode (or postmission mode) the GNSS signals are first recorded and stored on a hard disk and are later processed in a second step. Generally, postprocessing allows the implementation of more sophisticated algorithms because there is no time limit for the algorithm execution time. However, recorded GNSS signals can also be processed as if they were captured in real-time (like a replay mode), which is the mode implemented in the ipexSR. The postprocessing modes shall be described later.

3.1.2.1 Real-Time Like Processing

When the ipexSR processes record GNSS signals in a postprocessing mode, it simply reads the signals from the hard disk and subdivides them in the same way into packets and frames, as described in Section 3.1.1. A difference in the real-time operation is given only by the possibility to configure processing modules in a way that other modules wait until a certain operation is completed. For example, the master receiver can be configured to pause its signal tracking until a signal acquisition finishes. No IF sample buffer (see Figure 3.4) is consequently required and obviously much more external data (like precise ephemeris data available several days after the actual measurement) can be used to aid the receiver.

3.1.2.2 Random Access

Postprocessing of GNSS signals has the potential to allow the implementation of completely different signal processing algorithms because, in principle for a given measurement period, all signal samples are available at once. This idea is nicely introduced by van Grass and is compared to sequential processing methods [3]. However, more sophisticated algorithms, as described by van Grass, are possible. In an extreme case, a direct solution of the MLE equations as described in Section 4.2.3 or the method shown by Closas is imaginable [4]. Another example is snapshot positioning as described by Brown, where the position is estimated from a very short duration (< 1 second) piece of IF samples [5].

From the data-handling point of view, the distinguishing feature is that those algorithms have *random access* to the recorded signal samples and there is no need to group them into packets or frames. The time to access those samples is, however, orders of magnitude slower because that data has to be read from the hard disk if the signal duration under consideration is longer than a few seconds.

3.2 Module Diagram

The ipexSR architecture is described as a module diagram shown in Figure 3.4. The single modules have self-explaining names and are described in detail in the following sections. A module is actually realized as a number of C++ classes. The ipexSR is a *multithread* program. Modules that are executed within the same thread have the same style in Figure 3.4.

The data exchanged by the modules is described in Table 3.1. As the software receiver is configurable, the number of modules and the number of threads is configuration-dependent. Specifically, the number of receiver modules in Figure 3.4 depends on the configuration, and only for Receiver 1 in Figure 3.4 is the master channel/channel structure shown. All other receivers share the same structure, with the exception of the last receiver, which acts as a spectrum analyzer.

The main data flow in the software receiver is as follows. The USB front-end driver collects the IF samples from the GNSS front end, groups them into packets and stores them in the IF sample buffer. The packets also contain data from external sensors captured within the same time span covered by the respective packet. The master receiver retrieves a packet from the IF sample buffer and passes it to the acquisition manager and the receivers. The acquisition manager, with the Level 1 and 2 acquisition units, acquires the GNSS signals and the receivers track the GNSS signals. A dedicated receiver acts as an IF spectrum analyzer. Each receiver utilizes several master channels to track different satellites. Each master channel needs one

Table 3.1 Data Types Occurring in Figure 3.4

Code	*Description*
1	USB microframes containing GNSS IF samples
2	Data from other sensors (PC clock reading, IMU measurements, barometric measurements)
3	Time-stamped packets of IF samples plus external sensor data; one packet contains data for time span of about 10–400 ms
4	Frame of preprocessed IF samples; one frame contains data for a time span less than the primary coherent integration interval (e.g., less than 1 ms for the GPS C/A-code)
5	Code phase or pseudorange, Doppler and visibility indicator of GNSS signals
6	Correlator values
7	Raw data (code and phase pseudorange, Doppler, C/N_0, cycle slip flag, signal quality indicators, measurement accuracy)
8	Same as (7) plus time synchronized sensor data
9	Same as (8) plus position estimate
A	Decoded GNSS navigation messages, ephemeris data, clock corrections, atmospheric models, AGNSS data
B	GNSS correction data (differential corrections, data from a GNSS reference station or reference station network)
C	Position estimate from external source

or two channels that do the correlation for the data and for the pilot signal component. The master receiver retrieves from each receiver the pseudorange measurements and passes them to the navigation processor. A PVT solution is calculated, verified, visualized, and output to other user applications. The most recent PVT is kept in memory (called receiver status) and is used to aid signal acquisition (e.g., Doppler prediction) and tracking (e.g., vector tracking).

In the following subsections, a brief functional description of each module shall be given.

3.2.1 USB Front-End Driver

The USB front-end driver establishes a connection to the GNSS front end and retrieves the GNSS samples. The ipexSR allows the use of the isochronous USB 2.0 high-speed mode, where IF sample bits are transferred over the USB link grouped in USB microframes of a size of 1,024 bytes. The ipexSR also allows for the use of other USB transfer modes that will not be described here. After a USB microframe has been received, its samples are bit-converted, as described in Section 9.2. The samples are converted from a 2- or 4-bit format to a 16-bit format. The converted samples are used to fill up a packet of IF samples. A special protocol using time stamps (patent pending) is used to ensure time synchronicity within and in-between IF sample streams. After a packet is filled up with samples, it is put on the IF sample buffer.

The USB front-end driver represents the most time-critical task in the software receiver because it runs before the IF sample buffer. After a USB microframe has been received, the front-end driver has to react immediately. Otherwise, the following USB microframe will be lost, which potentially can cause serious problems in the signal processing. The ipexSR uses two different threads to realize the USB front-end driver. One thread retrieves the USB microframes. The other one does the bit conversion, puts the packet on the IF sample buffer, and optionally writes the sample data into a file on the computer's hard disk.

3.2.2 IF Sample Buffer

The IF sample buffer is part of the USB front-end driver and represents a relatively large amount of main memory used to buffer IF samples. The buffer itself acts as a FIFO buffer. If a buffer overflow occurs (if the software receiver is too slow), the buffer is completely flushed. This is communicated to the master receiver, which initiates a series of actions (similar to reacquisition) to continue tracking after the buffer flush. Although a *buffer flush* should not occur during standard operation, buffer flushes cannot be completely avoided. In the case of a GNSS reference station, a buffer flush may occur once per day and the buffer flush strategy ensures that pseudoranges are continuously tracked with a reduced accuracy, but the buffer flush causes carrier phase cycle slips.

3.2.3 Sensor Interface

If external sensor data is provided to the software receiver, like IMU data for GNSS/INS integration, the sensor interface retrieves this data and passes it to the USB front-end driver, which includes the sensor data in the IF sample packets. Currently, different IMUs, a magnetometer, a barometer, Wi-Fi power readings, and external NMEA strings are supported.

A particularly simple external sensor is the PC clock, which can be (optionally) read on a continuous basis. For each IF sample packet, the corresponding PC clock reading is stored. Reading the PC clock allows the software receiver to also determine the PC clock error with respect to a GNSS time scale. Assuming that the PC clock error varies smoothly, it allows other applications running on the same PC to have access to the GNSS time scale. If the PC is equipped with a special clock, as described by researchers of the Institute of Embedded Systems, then the time transfer could potentially be with submicrosecond accuracy [1]. The internal PC clock allows only an accuracy of several 10 ms.

3.2.4 Postprocessing Mode

If the software receiver runs in postprocessing mode, the USB front-end interface, the IF sample buffer, and the sensor interface are deactivated and are replaced by a data input module that reads the same data from the hard disk.

3.2.5 Master Receiver

The master receiver coordinates the signal processing but does no processing itself. It retrieves one packet of IF samples and passes it to the receivers. It checks whether GNSS signals shall be acquired and, if yes, it collects the necessary number of IF sample packets to be passed to the acquisition manager. The master receiver checks if a pseudorange measurement shall be performed and determines the exact measurement epoch. The measurement epoch is aligned to, for example, full seconds of GPS time. If a pseudorange measurement is to be performed within a given packet, it informs the receiver modules (which make use of the master channels for this task) to read the code and carrier pseudorange (plus auxiliary data) for a given reference sample. The master receiver retrieves the measurements from the receivers, merges them to a single record, and passes them to the navigation processor.

After the master receiver has completed its tasks, it waits until all Receiver modules finish processing the current packet of IF samples. In this way, processing of different GNSS services is synchronized. In the postprocessing mode, the master receiver waits for the acquisition manager to finish, whereas in real-time mode, the master receiver does not wait for the acquisition manager. Results from the acquisition manager are passed to the receiver modules. Finally, the master receiver retrieves the next packet of IF samples.

3.2.6 Receiver

For each tracked GNSS service (e.g., the GPS C/A-code, or the Galileo E5a Open Service) there exists one receiver module that controls a configurable number of master channels to track satellite signals for this service.

The receiver module does service-specific preprocessing (e.g., pulse blanking for E5/L5/E6 services) and then subdivides the sample packet into smaller frames, as described in Section 3.1.1.2. Within a loop, all frames are processed by one master channel by calling the respective master channel functions. After the loop ends, the receiver advances to the next master channel. This has been found to be more effective than passing one frame of data to all master channels and then advancing to the next frame. All the channel configuration data including PRN codes are kept in the CPU's memory caches. The receiver module also controls the multicorrelator behavior of the master channels, which are later used for signal quality monitoring. If the receiver is configured to work in the multiplexing multicorrelator mode, it sequentially activates and stops the multicorrelator mode of the respective master channels.

Normally there exists a special receiver module—called the spectrum analyzer—that computes and monitors the power spectral density of the received IF sample streams. This receiver module neither uses master channels nor subdivides packets into frames.

3.2.7 Master Channel

The master channels run within the same thread as their controlling receiver module. One master channel maintains the NCO values and can determine the code and phase pseudorange data based on the NCO values. It monitors several status variables like the loss-of-lock or the cycle slip flag. A master channel controls one or two channels, depending upon whether the service contains only a data component or a data plus pilot component.

The Master channel receives one frame of IF samples from the receiver that is passed directly to the channel(s) with the current NCO values. The master channel retrieves the correlation values from the channel(s) and computes the tracking errors. It then updates the NCO variables based on the selected tracking loop scheme.

3.2.8 Channel

The channel run within the same thread as their controlling receiver module. One channel retrieves IF sample frames from the controlling master channel with NCO values. The channel does the reference signal generation, the correlation, the bit synchronization, navigation data bit extraction and navigation data message decoding. The channel returns, at maximum, one set of correlation values to the master channel per frame. Decoded navigation data messages are stored in the navigation record data structure.

3.2.9 Acquisition Manger

The acquisition manager coordinates GNSS signal acquisition in the software receiver. Signal acquisition is by far the computationally most demanding operation in the software receiver; careful control of this algorithm is required to leave enough processing resources for, for example, tracking. The signal acquisition algorithms run within their own thread. Signal acquisition is triggered by the master receiver, in regular, user-configurable intervals.

After an acquisition has been initiated, the acquisition manager first determines which signals can be expected by the receiver to be above a certain elevation cutoff angle and which are already tracked. If a PVT solution is available and if ephemeris or almanac data is available, the code phase and Doppler search range are determined. If both values can be determined very precisely (e.g., less than one chip and a few hertz accuracy) then this signal is considered to be acquired. This is called *vector acquisition* because, similar to a VDLL, the code phase and the Doppler are derived from the navigation solution. If vector acquisition is not possible, the parameters are put into a list. This is repeated for all signals and finally the list is passed to the level-1 (cold-start) acquisition module. This module tries to acquire the signals contained in that list by FFT methods. Signals that cannot be acquired by the level-1 module are then passed to the level-2 (warm-start) acquisition module. Finally, the results from all three acquisition methods (vector, level 1, and level 2) are combined and returned to the master receiver.

3.2.10 Level-1 and Level-2 Acquisitions

Within the ipexSR, the actual signal acquisition is fully flexible and different acquisition algorithms can be used. Normally, the correlation is done in the frequency domain and squaring or differential detectors can be used for further noncoherent integration. The algorithms can be configured and two different configuration sets are possible. They are named level-1 (cold-start) and level-2 (warm-start) acquisitions. The level-1 parameters are applied to all satellite signals, whereas the level-2 parameters are applied only to acquire those signals where it is possible to reduce the Doppler search space based on an (approximate) PVT solution and satellite ephemeris/almanac data. Obviously the level-2 parameters can be chosen to use longer coherent integration times and/or more noncoherent steps.

3.2.11 Navigation Processor

The navigation processor is like the master receiver or the acquisition manager. It is a controlling module that does no processing itself. It receives the code and phase pseudorange measurements and passes them to encapsulated modules described in the following three sections. The navigation processor, with its modules, runs in its own thread.

3.2.12 Positioning with RAIM

Within a GNSS receiver, positioning has a central role. In fact, there may be different positioning algorithms running and operating on the same data, with each positioning algorithm being optimized for certain aspects (e.g., integrity, accuracy, RTK). Within the ipexSR there is one positioning algorithm that has an outstanding role because it is used to determine the receiver internal PVT estimate. This estimate is used to determine satellite visibility or to do vector acquisition. Therefore, the positioning algorithm is optimized for utmost robustness. It operates on an epoch-per-epoch basis instead of relying on a previous position estimate. This avoids linked errors caused by occasional gross positioning errors. Furthermore, the positioning includes RAIM; it also informs the master receiver to stop tracking satellite signals that are identified to have gross errors. If no position at all can be determined for a certain number of epochs, the master receiver is totally reset to acquire all signals from scratch. All signals that can be successfully verified that their code pseudorange and Doppler match the PVT estimates are flagged to be acceptable.

3.2.13 Navigation Modules

The ipexSR includes a large number of navigation modules that all operate on the same pseudorange data on an epoch-per-epoch basis. This includes positioning with EGNOS, positioning with RTCM corrections, ionospheric monitoring, and signal quality monitoring using multicorrelator values [6, 7].

3.2.14 Input and Output Modules

Input and output modules are special navigation modules because they interact with other applications and/or user equipment. The following output modules can be used: RTCM output, RINEX observation file output, commercial GNSS receiver emulation, NMEA output, and tracking-status output. Currently, there is one input module used to read RTCM correction data.

3.2.15 Receiver Status

The receiver status is a special data structure containing the most recently validated PVT estimate with the corresponding measurements. All software receiver modules have access to this data structure. Additionally, it can be set by the user based on an external PVT solution.

3.2.16 Navigation Records

The software receiver contains a database, which contains all navigation messages. The database is called "Navigation Records." Records can be provided via the GNSS signals, they can be read from the hard disk, and they can be obtained via an EGNOS-SISNET connection or from an AGNSS server. The Navigation Record type and content are usually directly derived from the different GNSS message structures, but special records also exist for, for example, tropospheric modeling, or to define fixed pseudolite positions. The management of this database can be quite complex, especially if a permanent operation of the software receiver is considered.

3.2.17 AGNSS and SISNET Connection

The navigation records database can be configured to connect to ESA's SISNET or to establish a connection to an AGNSS server. Both connections use a TCP/IP connection. The SISNET connection provides EGNOS records. The assisted connection uses proprietary formats to exchange navigation records between an AGNSS server and AGNSS receiver. Both are software receivers, but one is configured to act as the server and the other to act as the client. Multiple clients can connect to one server.

3.3 Execution Flow

It is common practice to define an SDR as a multiprocessor system as described in Section 2.2.2. To exploit the capabilities of a multiprocessor system, the software itself needs to be parallel or, in other terms, the software execution path should be split up into several *threads*. Threads can be executed on different processors simultaneously and need to be synchronized to each other to maintain data coherency. If the software uses more threads than available processors (or cores), different threads are executed partly sequentially.

To get more insight into an execution flow of a GNSS SDR, the ipexSR shall be analyzed with a thread profiler [8]. This tool visualizes the activity of all threads and shows the synchronization between the threads. The thread activity is plotted as a function of time and the diagrams are called (thread) time lines (see Figures 3.9 or 3.10). As shown in Figure 3.5, a thread can be *active* (actually doing a computation) or it can *wait* for a certain event (e.g., a receiver thread may wait for the next packets of IF samples to arrive). The thread profiler also indicates if a thread is in a *spin* loop and permanently polling an event that should be set by another thread (and is thereby wasting CPU resources). Thin (yellow) vertical or diagonal lines in the time line indicate thread synchronization (e.g., one thread tells another thread to resume its activity). The time line also shows the most time-consuming path or call sequence originating from the program start as a (red) bold, highlighted horizontal line. Please see this book's homepage (mentioned in Chapter 11) for color

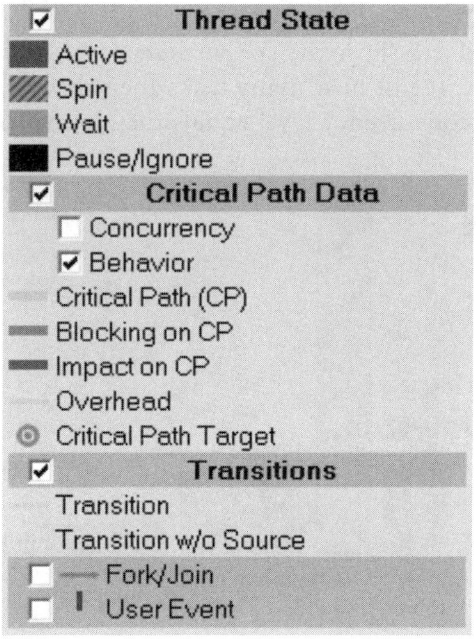

Figure 3.5 Thread time line diagram legend.

versions of Figures 3.9, 3.10, and 3.11. A critical path can flow across different threads at different times during a program's execution. A serial application has a single flow, which corresponds to the default critical path. A multithreaded application has multiple flows. The critical path determines the total execution time of the software; code along the critical path should be optimized. Optimizing code outside the critical path does *not* reduce the software execution time.

It should be mentioned that thread profiling causes a computational overhead of several percentages. The software execution flow thus slightly differs from a program run without profiling.

Two real-time configurations shall be investigated in the following section: the first runs a multifrequency configuration on a multicore computer; the second investigates a single-frequency configuration on a single-core computer.

3.3.1 Computer with Four CPU Cores

The software receiver was configured to track five GNSS services (GPS C/A, GPS L2C, GIOVE-A E1-OS, SBAS on L1, GIOVE-A E5a-OS). A sixth receiver was configured to act as an IF spectrum analyzer. A triple-frequency front end (the reference station front end of Table 2.1) has been used and the time span covered by one packet of IF samples has been 0.41 second. The computer platform was equipped with two Xeon 5140 (2.33 GHz) processors, each having two CPU cores. The overall processing load of the GNSS SDR in this configuration was quite low and much less than 50%. The profiling results of this configuration are shown in Figures 3.6 and 3.9 during signal tracking. Figures 3.7 and 3.10 contain profiling results for a period during which the SDR acquires signals.

3.3.1.1 Concurrency Level

Figures 3.6 and 3.7 show the *concurrency* level of the software. The concurrency level is an indicator of how many tasks the software programs could run simultaneously. If the concurrency level equals the number of cores (in this case, four) the

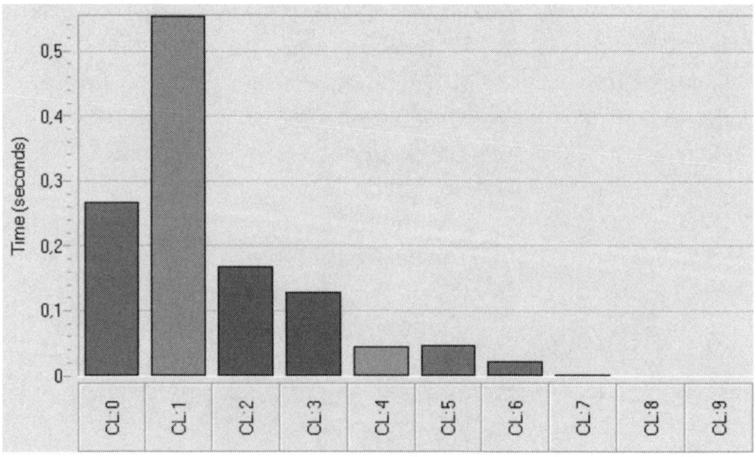

Figure 3.6 Concurrency level distribution corresponding to Figure 3.9.

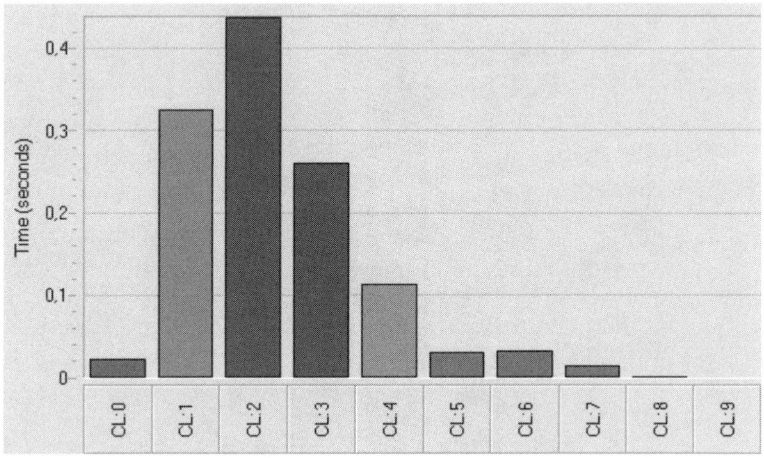

Figure 3.7 Concurrency level distribution corresponding to Figure 3.10.

computer is perfectly loaded. If the concurrency level exceeds this number, the system is overutilized, and if the number is less than the number of cores the system is underutilized. During pure tracking (Figure 3.6), most of the time only one thread is executed. Alternatively, the system may wait for IF samples, which is indicated as a concurrency level of zero.

If signal acquisition is performed, the average concurrency level increases by one, as shown in Figure 3.7. Both figures also demonstrate that the computer is required to run four or more tasks simultaneously only in rare cases.

3.3.1.2 Sample Reception

The SDR runs two threads to retrieve IF samples from the USB front end. They are called "FhGUSBInterface" and "FhGUSBInterfaceDispatcher" in Figures 3.9 and 3.10. The first thread retrieves the USB micropackets, using low-level operating system calls, more or less directly from the USB host located in the computer. It receives this data with a rate faster than 19.2 ms. This thread is extremely time-critical because it must be almost instantaneously ready to receive the next batch of data. Therefore, it puts the raw USB data into memory and informs the FhGUSBInterfaceDispatcher. The FhGUSBInterfaceDispatcher thread brings the raw USB samples into timely correct order and performs a bit conversion from the front end 2/4-bit format to the internal 16-bit format (see Section 9.2). Figures 3.9 and 3.10 show that the dispatcher is active with a rate of slower than 100–200 ms. Every 410 ms (on average) a packet of IF samples is completed and the Master Receiver is informed.

3.3.1.3 Signal Processing

After being triggered, the master receiver starts with preprocessing of the next IF sample packet (actually copying the data within the computer's main memory; see, e.g., the time span of 54.7–54.84 seconds in Figure 3.9). Then the different receivers are triggered to start processing this packet. The master receiver and the receivers can work in overlapping fashion; for example, the master receiver can fetch the next

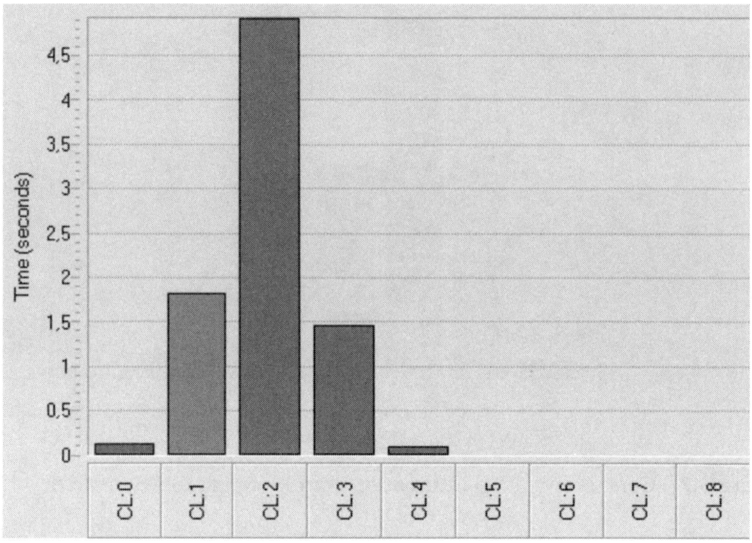

Figure 3.8 Concurrency distribution corresponding to Figure 3.11.

packet (if available) and start its preprocessing while the Receivers work on the old packet. This is, for example, visible during time spans of 55.00 to 55.01 seconds in Figure 3.9. Normally, however, there is no need for overlapping because the receivers finish before the next packet arrives. In Figures 3.9 and 3.10, one sees that different receivers need different execution times. The C/A-code receiver (receiver 0) needs the most time because it tracks the most signals. The spectrum analyzer (receiver 5) is activated only for every second packet.

When an acquisition is running the computer spends fewer resources for tracking and signal processing; in some receivers tracking is slightly delayed (see time span 41.61–41.74 seconds of Figure 3.10).

3.3.1.4 Navigation Processor

Both threading time lines in Figures 3.9 and 3.10 show nicely that indeed most of the time is spent for signal processing. The navigation processor is activated (in this particular configuration) only every second and needs a few milliseconds to complete. This also includes quite extensive data logging and RINEX output. Also the controller (the interface to the user or to the GUI) consumes only marginal computational resources.

3.3.2 Computer with a Single CPU Core

A second real-time threading analysis was done on the test system described in Table 9.1. This is a single-core computer and obviously only one thread can run at a time. The system is less powerful and a single-frequency (GPS C/A-code on Receiver 0) configuration with a spectrum analyzer (Receiver 1) was chosen. An L1 front end (R&D L1 front end of Table 2.1) was used, which has a different software

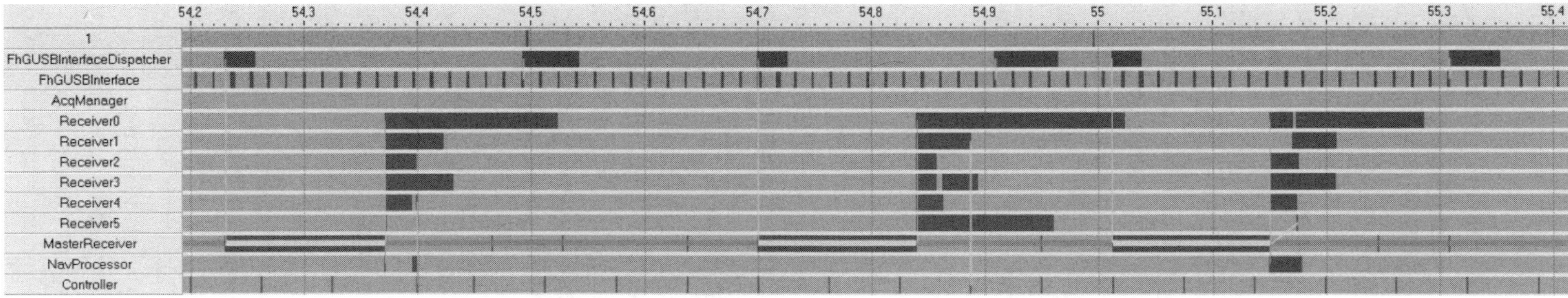

Figure 3.9 Threading time line for the multifrequency GNSS receiver configuration running on four CPU cores (no acquisition).

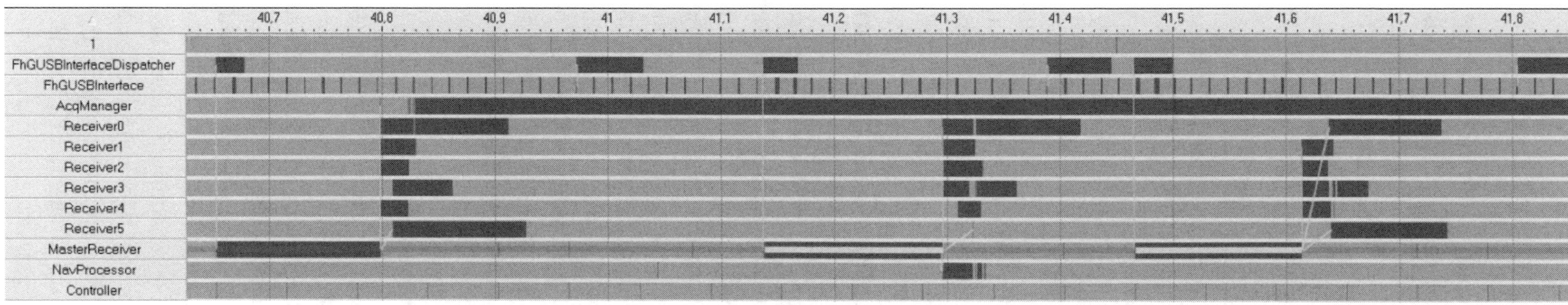

Figure 3.10 Threading time line for the multifrequency GNSS receiver configuration running on four CPU cores (acquisition running).

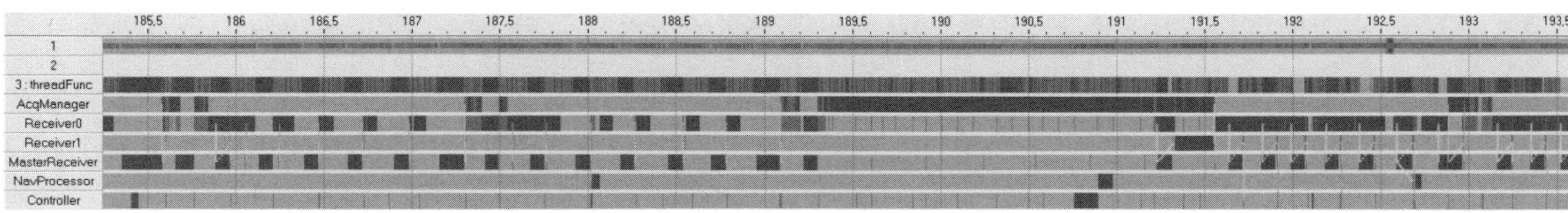

Figure 3.11 Threading time line for the single-frequency GNSS receiver configuration running on one CPU core.

interface compared to Section 3.3.1.2. The time span covered by one packet of IF samples is 91 ms.

The concurrency level is shown in Figure 3.8 for the threading time line of Figure 3.11. Most of the time the software has a need to run two threads simultaneously, but only one can be executed. The system is overutilized. The threading time line in Figure 3.11 is also quite different from the ones of Section 3.3.1. It appears to be more chaotic because the computer switches between the threads. The computer itself has a processing load of nearly 100% when running the GNSS SDR with the thread profiler.

The thread time line in Figure 3.11 shows no visible structure in the IF sample reception thread. The respective thread "threadFunc" is more or less continuously active while receiving data from the USB port. Interestingly, IF sample packets are usually processed in groups of two, as can be seen from the active time line of the Master Receiver or the Receiver 0. One packet contains data for 91 ms and the Master Receiver is active around five times within a period of 1 second. Signal tracking almost completely pauses while an acquisition is running (time span 189.3–191.5 seconds of Figure 3.11). Immediately after acquisition, the signal processing runs faster than in real time to recover the lost time.

3.4 GNSS Reference Station Configuration

The ipex software receiver can be configured to work as a GNSS reference station and the algorithms and parameters required for this operation mode shall be outlined in the following sections.

A GNSS reference station is typically located at a fixed site and the antenna usually has a good field of view to track all GNSS satellites above the horizon. GNSS measurements (code and carrier pseudorange) are produced with a rate of 1 Hz and are used as correction data for roving GNSS receivers, to precisely determine the position of the reference station itself or for other professional and scientific applications. A reference station is required to track all signals in view and to provide measurements of highest accuracy and reliability.

3.4.1 Acquisition Parameters

The reference station configuration uses a single coherent integration (see Section 5.7.1) to acquire signals in a *cold-start* mode. The cold-start mode is used when the receiver is unable to predict the Doppler frequency of a signal. This prediction step would require an available satellite almanac and a coarse receiver position and receiver time information. When this prediction can be done, a *warm start* is performed and the receiver uses multiple coherent integrations (with noncoherent summation) as described in Section 5.7.2.3. The used FFT algorithm will be outline in Section 9.5.4. The presence of data bits or a secondary code is simply ignored. Before applying the FFT acquisition, the signals are resampled to an acquisition sample rate defined by dividing the FFT length by the coherent integration time.

Acquisition parameters are summarized in Table 3.2. For each satellite system (GPS or Galileo) there is one *acquisition signal* that is acquired first. For example,

Table 3.2 GNSS Reference Station Acquisition Parameters

Signal	T_{coh} *(ms)*	*Noncoherent Integrations*	*FFT Length*
Cold start			
L1 C/A	4	1	8,192
E1 C	8	1	65,536
Warm start			
L1 C/A	4	5	8,192
E5a Q	4	25	131,072

the C/A-code on L1 is the acquisition signal for GPS. After it is acquired, code phase and Doppler for L2 (or L5) can be calculated and a *handover* from L1 to the other frequencies is performed.

The single coherent integration method is less memory-consuming than using multiple coherent integrations and is usually sufficient to acquire rising GPS signals as soon as the satellite has an elevation higher than 2–3°. Usually, the signal is tracked continuously until the satellite disappears behind the horizon.

We found it useful to have a backup solution for GIOVE-A E5a that is used when the handover from E1 is not working. The backup solution is realized as warm-start parameters for the Q-channel of E5a (see Table 3.2). During the early GIOVE-A tests, the code phase relationship between E1 and E5a was not totally clear from open literature. It is not possible to do a handover if an unknown code phase offset between E1 and E5a larger than 30m is present. The lack of ephemeris and almanac data for GIOVE-A (remember, this was a test satellite) does not allow for a warm start on E1. Only when E1 is already tracked can the Doppler search range for E5a be limited.

3.4.2 Tracking Parameters

For signal tracking, the ipex software receiver generates references signals, as described in Section 9.3, and correlates them with the method of Section 9.4. Code-phase, carrier-phase, and Doppler discriminators are described in Section 8.1 and are based on one P-, D-, and F-correlator (see Chapter 7). To speed up the numerical performance, already-generated reference signals are reused for multiple correlations, as pointed out in Section 4.3.2.10 and Section 9.6.

The P-correlator reference signal equals the infinite-bandwidth PRN code signal, including the modulation scheme. No carrier-phase multipath mitigation scheme is employed. This choice is the simplest one and the most easy to realize. The F-correlator is approximated by splitting the coherent integration into two parts of equal length, as described in Section 7.3.2. The D-correlator reference signals are partly optimized to mitigate code multipath. For the C/A-code, a double-delta-like reference signal is used that is given by (8.32) with $d = 0.2$. For L2 CM, an early-minus-late reference signal with $d = 0.1$ is used, as defined via (7.116). A difficulty arose when precomputing a reference signal for the L2 CL code (which has a period of 1.5 seconds). It turns out, it needs too much memory if a narrow correlator spacing is required. Consequently, there were three options: disregarding L2 CL, using L2 CL with $d = 1$, or porting the software receiver to a 64-bit operating system. We

Table 3.3 GNSS Reference Station Tracking Parameters

Signal	*Phase Discriminator*	T_{coh} *(ms)*	B_{DLL} *(Hz)*	B_{PLL} *(Hz)*
L1 C/A	atan	20	0.01	18
L2 CM	atan2	20	0.01	18
E1 B+C	atan2	4	0.05	18
E5a I+Q	atan2	20	0.05	18

chose the first option for simplicity. The BOC(1,1) signal on GIOVE-A E1 is tracked using an S-curve shaping correlator (see Section 8.2.4) and the E5a signal is tracked with an early-minus-late $d = 0.66$ correlator.

Tracking loop filters are summarized in Table 3.3, together with the coherent integration time. The GPS C/A carrier phase is tracked with an atan (one-quadrant) phase discriminator, and the full carrier phase is derived by decoding the sign of the GPS C/A NAV message preamble. For all other signals, an atan2 (four-quadrant) phase discriminator is used that works only on the pilot component. At this writing, the GPS L2 CM signal is broadcast as a pilot signal without any navigation message.

Sub-Nyquist sample rates are used to reduce the computational load (see Section 6.5). The front-end samples the IF signals with a rate of 40.96 MHz. All frequency bands have a 3-dB bandwidth of around 13 MHz. For tracking, the sample rate is reduced to 20.48 MHz. For GPS C/A tracking, the sample rate is further reduced to 10.24 MHz, if the C/N_0 lies above 40 dBHz.

3.4.3 Performance Results

Two exemplary spectrums of the received GNSS frequency bands are shown in Figure 3.12. For Figure 3.12(b), an active antenna with a 50-dB integrated LNA was used; Figure 3.12(a) is based on a passive antenna with a 17-dB LNA directly attached to it. The spectra were obtained from analyzing 0.2 second of the received signal. The spectra are offset with respect to the nominal carrier frequencies and are shifted with respect to each other because the nominal IF is different for each frequency band. For L1 and L2, one sees the main lobes of the GPS L1 C/A and the GPS L2 CM+CL signals centered directly at the nominal carrier frequency. Doppler shifts are too small to be visible in these figures. Furthermore, it should be pointed out that the signal power spectral density adds to the noise power spectral density. Therefore, we can actually see the GPS signal power spectral density in Figure 3.12 although the GPS signal amplitude is much smaller than the noise amplitude.

Remarkably, interfering signals are present in Figure 3.12. On L2, several narrowband interferers show up. On L1 and L5, narrowband interferers are present with one being of wider bandwidth (at around +5.2 MHz on L1 and –6.2 MHz on L5). Whereas the interferers on L1 and L5 are well suppressed, when the GNSS signals are despreaded, the strong interfering signal on L2 in Figure 3.12(a) significantly distorts the signal reception and brings the 2-bit ADC into saturation. Generally, we have found interfering signals of low power in many places. However, the strong L2 interferer is a local phenomenon at the campus of the University FAF Munich. It can be well-suppressed by a digital notch filter within the software receiver, but due to ADC saturation effects, an effective signal power loss of around 5 dB is the consequence.

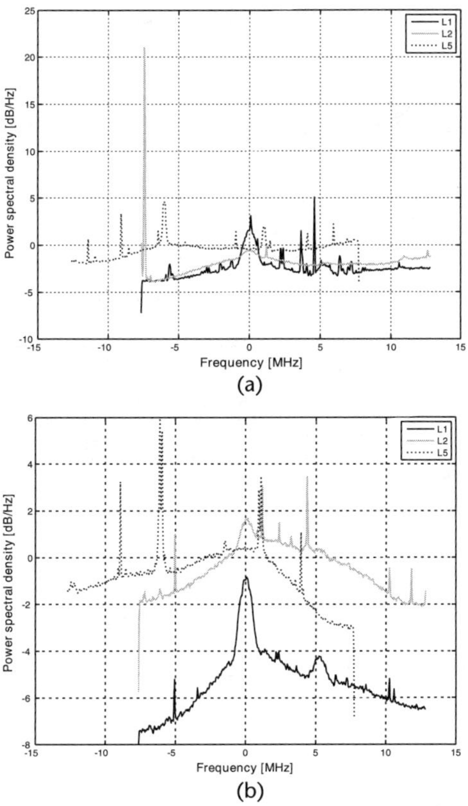

Figure 3.12 (a) L1, L2, and L5 GNSS spectra, at 48.08144° N/11.6349° E on April 11, 2009. (b) At 48.17147° N/11.80860° E on April 28, 2009.

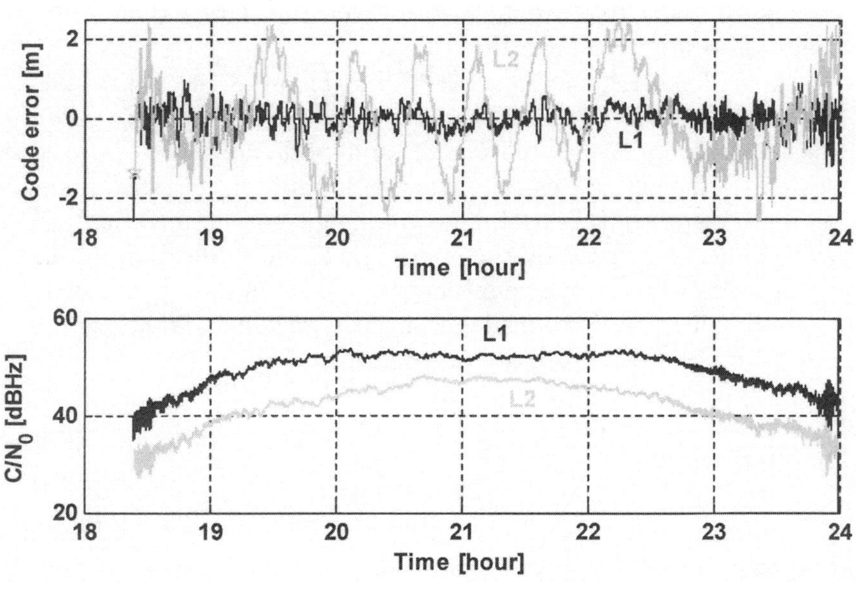

Figure 3.13 Code-minus-carrier and estimated C/N_0 for GPS PRN7, January 1, 2009.

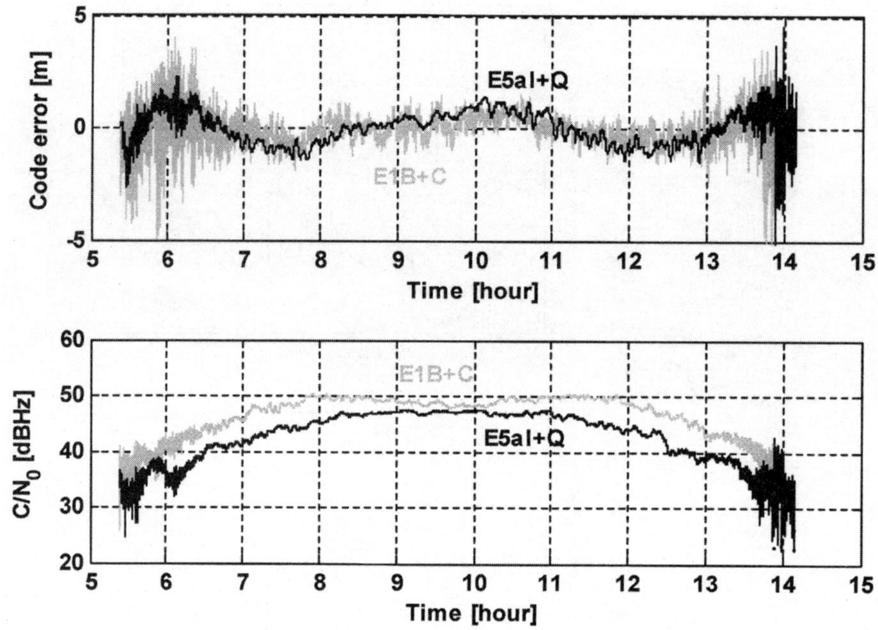

Figure 3.14 Code-minus-carrier and estimated C/N_0 for GIOVE-A, January 1, 2009.

The code and carrier tracking performance is shown in Figure 3.13 and Figure 3.14 for one pass of GPS PRN7 and GIOVE-A. The code error is computed by first subtracting the code pseudorange from the carrier pseudorange (both expressed in meters). Then a fifth-order polynomial is fitted to the difference. The polynomial models the ionospheric delays. The ionospheric delay is removed from the difference and the result is plotted as "code error" in both figures. The carrier phase is totally cycle-slip free if data with an elevation larger than 7° is considered. The figures also show the estimated C/N_0.

Apart from L2, the Figure 3.13 and Figure 3.14 confirm the expected signal performance. GPS C/A code pseudoranges are of decimeter accuracy and E5a has the highest accuracy due to its large bandwidth. For low elevations, all signals show strong C/N_0 fluctuations caused by multipath effects. The antenna was mounted on top a metal roof. For L2, code tracking errors of maximally 2m occur and the C/N_0 value is reduced by the interference. The rather large code tracking errors result from the reference signals not being recomputed frequently enough. The correlation point wanders outside the linearity region and this degrades the code phase estimation accuracy (see Section 4.3.2.10).

3.5 Discussion

Several possibilities for the data flow in a GNSS SDR have been discussed in this chapter. For real-time processing (and in practice for most postprocessing configurations), the immense data volume processed by a GNSS SDR constrains the signal processing to work on only a limited portion of the IF sample data at a time (batch processing). The most striking difference between a hardware and a software receiver is the capability and necessity for a software receiver to buffer IF samples.

A specific receiver architecture has been presented that was used to develop the ipexSR as a multifrequency GNSS SDR. The architecture was carefully chosen and not only influences the resulting software itself, but also relates to the software development process. It is important that the architecture fits the development environment (here C++), the involved developing team members and their specific capabilities. To some extent, the team requirements may have an even larger impact on the architecture than the technical requirements.

Different options exists to run a GNSS SDR in real time. The ipexSR is implemented in a manner to allow its operation in real time (in a loose sense) under Windows (a non real-time operating system). The software itself utilizes multiple processors or cores and is separated into several threads. A threading analysis shows that the receiver behaves very much like a real-time GNSS SDR in a strict sense. It immediately reacts after receiving the IF samples if (and most likely only if) there is a sufficient number of cores available and if the processing load is clearly beyond 100% (e.g., around 50–60% at a maximum). In that case, acquisition and tracking can run in parallel. If the software gets heavily loaded to more than 70–80%, there will be, for example, tracking pauses of several seconds or other phenomena that may occur that should not be present in a real-time GNSS SDR.

References

[1] Institute of Embedded Systems InES, "IEEE1588 Enabled PCI Network Interface Card," http://ines.zhaw.ch/institut/products/ieee-1588-hardware/page.html, 2008.

[2] Trimble Navigation Ltd., "MS750, Dual Frequency RTK Receiver for Precise Dynamic Positioning," http://trl.trimble.com/docushare/dsweb/Get/Document-12640/ms750pp.pdf, 2007.

[3] van Grass, F., et al., "Comparison of Two Approaches for GNSS Receiver Algorithms: Batch Processing and Sequential Processing Considerations," *Proc. 18th Int. Technical Meeting of the Satellite Division of the Institute of Navigation (ION-GNNS) 2005*, Long Beach, CA, September 13–16, 2005, pp. 200–211.

[4] Closas, P., et al., "Bayesian Direct Position Estimation," *Proc. 21st Int. Technical Meeting of the Satellite Division of the Institute of Navigation (ION-GNNS) 2008*, Savannah, GA, September 16–19, 2008, pp. 183–190.

[5] Brown, A., P. Brown, and J. Griesbach, "GeoZigBee: A Wireless GPS Wristwatch Tracking Solution," *Proc. 19th Int. Technical Meeting of the Satellite Division of the Institute of Navigation (ION-GNNS) 2006,* Fort Worth, TX, September 26–29, 2006, pp. 2883–2888.

[6] Anghileri, M., et al., "Performance Evaluation of a Multi-Frequency GPS/Galileo/SBAS Software Receiver," *Proc. 20th Int. Technical Meeting of the Satellite Division of the Institute of Navigation (ION-GNNS) 2007,* Fort Worth, TX, September 25–28, 2007, pp. 2749–2761.

[7] Stöber, C., et al., "Implementing Real-time Signal Monitoring within a GNSS Software Receiver," *Proc. European Navigation Conference (ION-GNNS) 2008*, Toulouse, April 22–25, 2008.

[8] Intel Corp., "Thread Profiler 3.1 Build:0.25466," http://www.intel.com, 2007.

CHAPTER 4

Signal Estimation

This chapter will introduce signal-estimation techniques that are used by a navigation receiver. The basic goal of the receiver is to obtain an estimate of its position; this will be optimized with respect to application-specific criteria. One option is to look for unbiased estimators having minimum variances. Another option could be to minimize the mean-squared error if a priori stochastic information on the distribution of the true position (e.g., a user-motion model) or of any related quantity is available. The first approach is called *nonrandom parameter estimation* because the position is regarded as an unknown but otherwise deterministic quantity. In a GNSS receiver, for example, pseudorange estimation (an intermediate step before position estimation) is usually treated as a nonrandom parameter-estimation problem. The second approach—*Bayesian parameter estimation*—treats the position as a random variable. Kalman filtering, being a Bayesian technique, is commonly used to estimate user positions in a navigation receiver.

The chapter starts with nonrandom parameter-estimation techniques, which are regarded in this context to be more fundamental, simply because no stochastic models are needed. Furthermore, the nonrandom parameter-estimation approach allows the use of powerful analytical tools, like the Cramér–Rao lower bound (CRLB). The described estimators try to achieve this lower bound as much as possible. Nonrandom parameter estimation is especially useful if GNSS signal parameters such as code phase, Doppler, frequency, or amplitude are considered, because no (or only very crude) stochastic models are available for them. Bayesian estimation (or more specifically, a Kalman filter) is a suitable approach if the estimated parameters are the user position or trajectory. In that case, better stochastic models are available. In other words, it is possible to provide useful stochastic information for the movement of a vehicle, but range-measurement errors from the vehicle to a satellite, including multipath, atmospheric delays or satellite orbit/clock errors, may require sophisticated modeling that is too complex for practical use within the signal-processing stage.

Reviewing the algorithmic design of a navigation receiver is of special importance for a SDR receiver because the SDR concept potentially provides the possibility to redesign well-known navigation signal-processing techniques to implement more complex algorithms as compared to hardware receivers.

4.1 Parameters of Interest

Before starting with the description of the different estimation techniques, it is useful to look at the parameters to be estimated in a navigation receiver. Three classes of parameters will be identified: position, low-rate pseudorange data, and high-rate

Table 4.1 Generalized Useful Parameters

Name	*Symbol*	*Description*
Position	x	End product of the estimation scheme. Normally includes the position estimate in a given coordinate system plus velocity and clock error/drift, but may also include atmospheric delays, transmitter positions, or any other quantity correlated with position estimate.
High-rate pseudorange	q	Short period signal parameters (typically delay, Doppler, carrier phase, amplitude) as introduced in Section 1.8 for line-of-sight and multipath signals.
Low-rate pseudorange	p	Reduced and filtered subset of high-rate data (e.g., only delay and Doppler) for a given measurement rate of for example, every 1 seconds.

pseudorange data. They are summarized in Table 4.1 and will be explained below. *Nuisance parameters* are introduced for formal reasons and extend the useful parameters to have a complete model for the deterministic part of the received signal.

All parameters can either be constant during the signal-estimation process or be time-dependent. In the latter case, we assume that the time dependence can be described through a finite number of parameters. For example, if a moving user is considered, we assume that his or her trajectory can be modeled sufficiently well by a spline or similar curve, which is determined by a finite number of parameters.

4.1.1 Useful Parameters

The position is understood as the vector of parameters, that is, the final output of the navigation receiver. In the simplest case, this is the estimated position within a predefined-coordinate system. It may, however, include other parameters of interest, such as the velocity, clock error, or atmospheric delays. Furthermore, the position vector may include parameters that are related to the aforementioned variables (e.g., multipath-signal parameters) and, in the most general case, it includes all parameters necessary to define the signal model. Therefore, the term position is used here in a general way, but it defines the end product of the estimation scheme.

The term *high-rate pseudorange data* summarizes all parameters that are directly related to the received signal model. These parameters are those introduced in Section 1.8: delay, Doppler, carrier phase, and amplitude.

The term *low-rate pseudorange data* is used to describe all data being passed from the signal-processing unit to the navigation filter within a GNSS receiver. Low-rate pseudorange data is derived directly from the high-rate pseudorange data, usually via a filtering procedure that reduces the amount of data to be processed by the navigation filter.

4.1.2 Nuisance Parameters

The deterministic part of the received signal may depend, in addition to the useful parameters, upon further unknown parameters—the nuisance parameters ξ, which do not stand in any relation to the position; we are therefore not interested in their estimated values. Typically, the nuisance parameters are a subset of the high-rate pseudorange data. Formally, the nuisance parameters can be estimated along with the other parameters.

For certain applications, the carrier phase and the signal amplitude can be treated as nuisance parameters. Another example of nuisance parameters includes signal parameters of possibly present multipath signals. However, for both examples it may be reasonable to consider them as useful parameters for a different set of applications, especially if correlations between them and the position estimates are important.

The nuisance parameters ξ are treated as random variables, described by a probability density function $p(\xi)$. This implies that any parameter for which we are not able to give a probability density function must be treated as a useful parameter, even if we are not interested in its estimated value.

4.1.3 Relationship Between the Parameters

The deterministic part of the received signal, r_μ, contains all available information for position estimation. It is reasonable to assume that the deterministic model depends upon the position, the low- or high-rate pseudorange data, and on the nuisance parameters as

$$r_\mu = r_\mu(\mathbf{x},\xi) = r_\mu(\mathbf{q},\xi) = r_\mu(\mathbf{p},\xi) \tag{4.1}$$

In each of these three cases, the same signal model is meant; it is just expressed as a function of different parameters. The nuisance parameters ξ are identical for all three cases.

For the useful parameters, we assume that they represent different ways of expressing the user's position. Therefore, high- or low-rate pseudorange data is uniquely defined by the position:

$$\mathbf{q} = \mathbf{q}(\mathbf{p}(\mathbf{x})) \tag{4.2}$$

Equation (4.1), together with (4.2), states that the signal does not directly depend on the position, but only through the pseudorange $\mathbf{q}$, respectively through $\mathbf{p}$ and $\mathbf{q}$. The vector $\mathbf{q}$ is typically of a higher dimension than $\mathbf{p}$, which is itself of a higher dimension than $\mathbf{x}$. Fulfilling (4.2) in a real-world situation can be complicated; many modeling efforts are necessary to account for user dynamics, multipath, atmospheric effects, transmitter position/clock errors, and other effects. Nevertheless, for the theoretical discussion we assume that the condition (4.2)—called *sufficient modeling*—is fulfilled.

4.2 Nonrandom Parameter Estimation

This section introduces several concepts of nonrandom parameter estimation that are illustrated by the positioning problem. The CRLB and some modified bounds for position estimators will be evaluated. It will be shown that, under certain assumptions, the estimation problem can be subdivided into three-steps: high-rate pseudorange estimation, filtering, and positioning. The CRLB for the position estimate will be maintained.

Because the discussion in this chapter is more theoretical, a simplified picture of the receiver will be assumed (see Figure 4.1). It is seen as a device that accepts as input discrete samples of the received navigation signal and outputs the estimated position. Without loss of generality, we assume that only a single frequency is processed. The signal samples are considered to be a vector of complex-valued random variables ($\mathbf{S}$, or equivalently, S_μ) whose distribution is given by $p_\mathbf{x}(\mathbf{s})$. The distribution $p_\mathbf{x}(\mathbf{s})$ depends on the position $\mathbf{x}$. For simplicity, we presently consider only a finite (although large) number of samples and assume all samples are available to the estimator simultaneously [in contrast to an (in principle) unlimited number of samples being used by an estimation filter, like a Kalman filter].

The position $\mathbf{x}$ acts as a parameter to the distribution of the received samples and is assumed to be from an open interval Λ of the n-dimensional space $\mathbb{R}^n$ of real numbers (called *admissible range* in Figure 4.1),

$$\mathbf{x} \in \Lambda \subseteq \mathbb{R}^n \tag{4.3}$$

The exact choice of Λ represents a priori information on the position, but otherwise no information (i.e., the statistical distribution for $\mathbf{x}$) is available.

The navigation receiver combines the received signal samples properly to obtain an estimate of the position, which is written as $\hat{\mathbf{x}} = \hat{\mathbf{x}}(\mathbf{S})$. Assuming that the true position assumes the value $\mathbf{x}$, the expected value $\langle \hat{\mathbf{x}} \rangle_\mathbf{S}$ and the variance $\operatorname{var}\langle \hat{\mathbf{x}} \rangle_\mathbf{S}$ of the estimated position shall be optimal. In the context of nonrandom parameter estimation, it is understood that no bias shall be present:

$$\langle \hat{\mathbf{x}} \rangle_\mathbf{S} = \mathbf{x} \tag{4.4}$$

It is also understood that the estimated position $\hat{\mathbf{x}}$ shall have minimum variance

$$\operatorname{var}\langle \hat{\mathbf{x}} \rangle_\mathbf{S} = \left\langle (\hat{\mathbf{x}} - \langle \hat{\mathbf{x}} \rangle_\mathbf{S})(\hat{\mathbf{x}} - \langle \hat{\mathbf{x}} \rangle_\mathbf{S})^* \right\rangle_\mathbf{S} \to \min \tag{4.5}$$

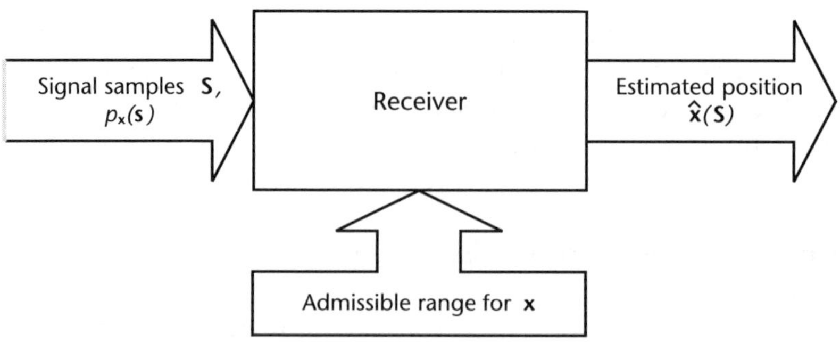

Figure 4.1 Simplified GNSS receiver used to describe signal-estimation techniques.

in the sense that there should be no estimator having a smaller variance matrix. For a definition on how to compare two matrices, see (4.15). Consequently, the position estimator shall be a minimum variance unbiased estimator (MVUE). It is known that an MVUE is, in general, not optimal if the MSE (or RMS) value

$$\mathrm{MSE}\langle\hat{\mathbf{x}}\rangle_{\mathbf{S},\mathbf{x}} = \left\langle(\hat{\mathbf{x}}(\mathbf{S})-\mathbf{x})^*(\hat{\mathbf{x}}(\mathbf{S})-\mathbf{x})\right\rangle_{\mathbf{S},\mathbf{x}} \tag{4.6}$$

is considered; that is, biased estimators may exist whose MSE is below the MSE of an MVUE (e.g., noise-variance estimation as described in IV.D.2 of [1] illustrates this case). Minimum MSE estimators are Bayesian estimators; they require a priori knowledge on the distribution of **x** and are discussed in Section 4.5.

Additionally, it should be noted that unbiased estimators are important in high-precision navigation problems (see Table 4.2). There, the variance can, in general, be reduced by collecting more data by, or by using a longer time span to determine the position. Another option is to increase the bandwidth of the received navigation signal (e.g., change from a low-bandwidth front end to a high-bandwidth front end) and thus obtain more independent samples within the same observation period.

The signal samples comprise a deterministic part, r_μ, and an additive stochastic component N_μ,

$$S_\mu(\mathbf{x},\xi) = r_\mu(\mathbf{x},\xi) + N_\mu \tag{4.7}$$

This type of estimation problem is called a *signal-in-noise* problem.

For a fixed value of **x** and ξ, the expected value with respect to the signal samples equals the expected value with respect to the noise:

$$\mathbf{x},\xi = \text{const.} \Rightarrow \langle\bullet\rangle_{\mathbf{S}} = \langle\bullet\rangle_{\mathbf{N}} \tag{4.8}$$

If ξ is treated as a random variable, then the expected value has to be computed with respect to the probability density functions of the noise *and* of the nuisance parameters:

$$\mathbf{x} = \text{const.} \Rightarrow \langle\bullet\rangle_{\mathbf{S}} = \langle\bullet\rangle_{\mathbf{N},\xi} \tag{4.9}$$

The stochastic component N_μ represents the received noise and is modeled as a vector of complex-valued i.i.d. random variables. For purposes pursued in this section, it is sufficient to consider a comparable simple Gaussian model; that is,

$$\mathrm{Re}\{N_\mu\} \sim N(0,1), \quad \mathrm{Im}\{N_\mu\} \sim N(0,1) \tag{4.10}$$

Table 4.2 Desired Receiver Properties for High-Precision Applications

Property	*Importance*	*Motivation*
Position estimate is unbiased	High	To avoid systematic errors in positioning that cannot be reduced by further averaging and are not modeled by the position-variance matrix.
Small variance, small MSE	Medium	Should generally be as low as possible.

where $N(0,1)$ denotes a Gaussian distribution with zero mean and unit variance, such that (1.16) holds.

A Gaussian noise model is considered because only in that case can comparable simple analytical expressions for the CRLB be derived. The evaluation of the CRLB in non-Gaussian noise is cumbersome and, in most cases, analytically impossible [2]. Furthermore, thermal noise is well-described by a Gaussian model. A detailed discussion of non-Gaussian noise occurring during signal quantization is given in Section 6.1.

4.2.1 Position CRLB Without Nuisance Parameters

Making all those assumptions, the received signal is modeled as a sequence of complex-valued random variables whose distribution depends on the true position **x**. Assuming for the following that either x completely describes the received signal or that the nuisance parameters have fixed known values [both assumption imply that (4.8) is valid], then the probability density function of the signal is given as (see Appendix A.1.2)

$$p_{\mathbf{x}}(\mathbf{s}) = \frac{1}{(2\pi)^L} \exp\left\{ -\frac{1}{2} \sum_{\mu=1}^{L} \left| s_\mu - r_\mu(\mathbf{x}) \right|^2 \right\} \tag{4.11}$$

where L is the number of available complex-valued signal samples. It is known that the variance of any MVUE for the position is bounded from below by the inverse of the Fisher information matrix (see Section IV.E.1 from the book by Poor [1]):

$$I_{\mathbf{x};i,j} = \left\langle \left(\frac{\partial}{\partial x_i} \log p_{\mathbf{x}}(\mathbf{S}) \right) \left(\frac{\partial}{\partial x_j} \log p_{\mathbf{x}}(\mathbf{S}) \right) \right\rangle_{\mathbf{N}} \tag{4.12}$$

The derivative is evaluated as

$$\begin{aligned} \frac{\partial}{\partial x_i} \log p_{\mathbf{x}}(\mathbf{S}) &= \frac{1}{2} \sum_{\mu} \left(\overline{(S_\mu - r_\mu(\mathbf{x}))} \frac{\partial r_\mu(\mathbf{x})}{\partial x_i} + (S_\mu - r_\mu(\mathbf{x})) \frac{\partial \overline{r_\mu}(\mathbf{x})}{\partial x_i} \right) \\ &= \mathrm{Re}\left\{ \sum_{\mu} \left((S_\mu - r_\mu(\mathbf{x})) \frac{\partial \overline{r_\mu}(\mathbf{x})}{\partial x_i} \right) \right\} = \mathrm{Re}\left\{ \sum_{\mu} \left(N_\mu \frac{\partial \overline{r_\mu}(\mathbf{x})}{\partial x_i} \right) \right\} \\ &= \mathrm{Re}\left\{ \sum_{\mu} \left(\overline{N}_\mu \frac{\partial r_\mu(\mathbf{x})}{\partial x_i} \right) \right\} \end{aligned} \tag{4.13}$$

Thus, the Fisher information matrix evaluates to

$$
\begin{aligned}
I_{\mathbf{x};i,j} &= \left\langle \mathrm{Re}\left\{\sum_{\mu}\left(\bar{N}_{\mu}\frac{\partial r_{\mu}(\mathbf{x})}{\partial x_i}\right)\right\}\mathrm{Re}\left\{\sum_{\upsilon}\left(N_{\upsilon}\frac{\partial \bar{r}_{\upsilon}(\mathbf{x})}{\partial x_j}\right)\right\}\right\rangle_{\mathbf{N}} \\
&= \frac{1}{4}\left\langle \sum_{\mu}\left(\bar{N}_{\mu}\frac{\partial r_{\mu}(\mathbf{x})}{\partial x_i} + N_{\mu}\frac{\partial \bar{r}_{\mu}(\mathbf{x})}{\partial x_i}\right)\sum_{\upsilon}\left(N_{\upsilon}\frac{\partial \bar{r}_{\upsilon}(\mathbf{x})}{\partial x_j} + \bar{N}_{\upsilon}\frac{\partial r_{\upsilon}(\mathbf{x})}{\partial x_j}\right)\right\rangle_{\mathbf{N}} \\
&= \frac{1}{4}\left\langle \sum_{\mu,\upsilon}\left(\bar{N}_{\mu}N_{\upsilon}\frac{\partial r_{\mu}(\mathbf{x})}{\partial x_i}\frac{\partial \bar{r}_{\upsilon}(\mathbf{x})}{\partial x_j} + N_{\mu}\bar{N}_{\upsilon}\frac{\partial \bar{r}_{\mu}(\mathbf{x})}{\partial x_i}\frac{\partial r_{\upsilon}(\mathbf{x})}{\partial x_j}\right)\right\rangle_{\mathbf{N}} \\
&= \mathrm{Re}\left\{\sum_{\mu}\frac{\partial r_{\mu}(\mathbf{x})}{\partial x_i}\frac{\partial \bar{r}_{\mu}(\mathbf{x})}{\partial x_j}\right\}
\end{aligned}
\tag{4.14}
$$

because of (1.16). A similar expression can be found in Section 15.7, Equation (15.52) in the book by Lehmann [3].

Equation (4.14) represents the best possible accuracy a navigation receiver can achieve in the sense that the variance of any unbiased estimator $\hat{\mathbf{x}}(\mathbf{S})$ is bounded from below by

$$
\mathrm{var}\left\langle \hat{\mathbf{x}}(\mathbf{S})\right\rangle_{\mathbf{N}} = \left\langle (\hat{\mathbf{x}}(\mathbf{S}) - \mathbf{x})(\hat{\mathbf{x}}(\mathbf{S}) - \mathbf{x})^{*}\right\rangle_{\mathbf{N}} \geq I_{\mathbf{x}}^{-1} = \mathrm{CRLB}(\mathbf{x}) \tag{4.15}
$$

The difference matrix D,

$$
D = \mathrm{var}\left\langle \hat{\mathbf{x}}(\mathbf{S})\right\rangle_{\mathbf{N}} - I_{\mathbf{x}}^{-1} \tag{4.16}
$$

is nonnegative definite (i.e., all the eigenvalues of D are greater than or equal to zero).

Any unbiased estimator that achieves the CRLB is called an *efficient* estimator. An MVUE is not necessarily efficient, because "minimum" does not mean equality holds on the Cramér–Rao inequality (4.15).

The equality to the CRLB of an unbiased estimate $\hat{\mathbf{x}}(\mathbf{S})$ is achieved if and only if

$$
\hat{\mathbf{x}}(\mathbf{s}) - \mathbf{x} = I_{\mathbf{x}}^{-1}\frac{\partial \log p_{\mathbf{x}}(\mathbf{s})}{\partial \mathbf{x}} \tag{4.17}
$$

at least somewhere around $\mathbf{x}$; see Theorem 3.4 from the book by Porat [4].

4.2.2 Position Estimation with Nuisance Parameters

If nuisance parameters are present, the probability density function of the received signal samples takes the form

$$
p_{\mathbf{x},\xi}(\mathbf{s}) = \frac{1}{(2\pi)^{L}}\exp\left\{-\frac{1}{2}\sum_{\mu=1}^{L}\left|s_{\mu} - r_{\mu}(\mathbf{x},\xi)\right|^{2}\right\} \tag{4.18}
$$

Formally, the dependency on the nuisance parameters can be eliminated if we calculate the expected value over the nuisance parameters

$$\begin{aligned} p_{\mathbf{x}}(\mathbf{s}) &= \left\langle \frac{1}{(2\pi)^L} \exp\left\{ -\frac{1}{2} \sum_{\mu=1}^{L} \left| s_\mu - r_\mu(\mathbf{x},\xi) \right|^2 \right\} \right\rangle_\xi \\ &= \frac{1}{(2\pi)^L} \int \exp\left\{ -\frac{1}{2} \sum_{\mu=1}^{L} \left| s_\mu - r_\mu(\mathbf{x},\xi) \right|^2 \right\} p(\xi) d\xi \end{aligned} \tag{4.19}$$

This is the *marginal distribution* for disregarding the nuisance parameters. We could motivate this step as an emphasis that the nuisance parameters cannot be well estimated and modeling them as a random variable is a reasonable approximation. However, it should be made clear that the marginal distribution is *not* an approximation (as long as the a priori nuisance parameter distribution is exact) and working with (4.19) gives exact results. We will show that (4.19) models the coupling between the different high rate pseudorange parameters better than (4.18) if the CRLB is considered.

Equation (4.19) allows calculation of the *true* CRLB via (4.12); in practice, however, this calculation is impracticable because a strict evaluation of (4.19) is, in most cases, impossible. Therefore, several less-strict bounds have been proposed in the literature that are much easier to calculate.

4.2.2.1 Modified Cramér–Rao Lower Bound

A scalar lower bound of the error variance of an unbiased parameter estimator with the presence of nuisance parameters has been proposed, with the name *modified Cramér–Rao lower bound* (MCRLB) [5]. This bound can be easily calculated and it is generally looser than the CRLB. In some cases of practical interest, it approaches the CRLB as computed for known nuisance parameters.

Gini et al. extended the MCRLB to the estimation of a vector of nonrandom parameters [6]. Like the conventional CRLB, the MCRLB relies on the definition of a properly modified Fisher information matrix, which is the expectation (with respect to the nuisance parameters) of the conventional Fisher information matrix as computed for fixed nuisance parameters. In the following, the results obtained by Gini et al. shall be summarized.

The modified Fisher information matrix $\tilde{I}_{\mathbf{x};i,j}$ is defined as

$$\tilde{I}_{\mathbf{x};i,j} = \left\langle I_{\mathbf{x};i,j}(\xi) \right\rangle_\xi = \int I_{\mathbf{x};i,j}(\xi) p(\xi) d\xi \tag{4.20}$$

where the conventional Fisher information matrix $I_{\mathbf{x};I,,j}(\xi)$ is given by (4.12) using fixed values for the nuisance parameters ξ. A particularly simple case occurs if the conventional Fisher information matrix does not depend on ξ, in which case the MCRLB equals the CRLB.

The variance of any unbiased estimator $\hat{\mathbf{x}}(\mathbf{S})$ is bounded from below

$$\mathrm{var}\left\langle \hat{\mathbf{x}}(\mathbf{S})\right\rangle_{\mathbf{N},\xi} = \left\langle (\hat{\mathbf{x}}(\mathbf{S})-\mathbf{x})(\hat{\mathbf{x}}(\mathbf{S})-\mathbf{x})^{*}\right\rangle_{\mathbf{N},\xi} \geq \tilde{I}_{\mathbf{x}}^{-1} = \mathrm{MCRLB}(\mathbf{x}) \tag{4.21}$$

This is the mathematical formulation of the MCRLB. The MCRLB is generally looser than the true CRLB, calculated via (4.19). This is expressed as

$$I_{\mathbf{x}}^{-1} \geq \tilde{I}_{\mathbf{x}}^{-1} \quad \Leftrightarrow \quad \mathrm{CRLB}(\mathbf{x}) \geq \mathrm{MCRLB}(\mathbf{x}) \tag{4.22}$$

It should be noted that the CRLB of Sections 4.2.1 and 4.2.2 are not identical; only in the second case are nuisance parameters considered at all.

The modified Fisher information matrix satisfies Fisher's five properties and thus qualifies as an information quantity (see page 60 in the book by Porat [4]). Fisher's idea was to express the information carried by the density $p_{\mathbf{x}}(\mathbf{S})$ about the parameters $\mathbf{x}$ in quantitative terms. Most important, the larger the sensitivity of $p_{\mathbf{x}}(\mathbf{S})$ to changes in $\mathbf{x}$, the larger should be the information.

A necessary and sufficient condition for the equality in (4.21) of an unbiased estimate $\hat{\mathbf{x}}(\mathbf{S})$ to hold is

$$\hat{\mathbf{x}}(\mathbf{s}) - \mathbf{x} = \tilde{I}_{\mathbf{x}}^{-1} \frac{\partial \log p_{\mathbf{x},\xi}(\mathbf{s})}{\partial \mathbf{x}} \tag{4.23}$$

which corresponds to (4.17).

4.2.2.2 Asymptotic Cramér-Rao Lower Bound

In the work of Moeneclaey, the CRLB has been investigated for signal-in-noise problems in the limit of infinite high signal-to-noise ratios [7]. A single useful scalar parameter in the presence of a vector of nuisance parameters has been considered. The obtained bound is called *asymptotic* CRLB (ACRLB). The ACRLB is more fundamental than the MCRLB because the ACRLB is directly derived from the CRLB via a series expansion in the inverse signal-to-noise ratio, whereas the derivation of the MCRLB is more motivated by computational simplicity. However, Moeneclaey has shown that the ACRLB equals the MCRLB in the case where the parameter of interest and the nuisance parameters are uncoupled. In the following, the results of Moeneclaey shall be summarized.

The derivation of the ARCLB is based on a combined Fisher information matrix containing the information for the useful parameter x (here only a scalar random variable is considered) and the nuisance parameters ξ. This Fisher information matrix shall be represented as $I_{x,\xi}$ and is obtained by (4.12) using (4.18). We assume that the parameters x together with ξ are ordered such that x is the first element in the vector used for the definition of $I_{x,\xi}$.

The derivation first expands the second derivative with respect to x of the logarithm of the averaged probability density function (4.19) into a second-order polynomial and then evaluates the expected value with respect to $\mathbf{N}$ for high signal-to-noise ratios.

The ACRLB reads as

$$\left\langle (\hat{x}(\mathbf{S}) - \mathbf{x})(\hat{x}(\mathbf{S}) - x)^* \right\rangle_{\mathrm{N},\xi} \underset{\mathrm{SNR} \geq \infty}{\geq} \frac{1}{\left\langle ((I_{x,\xi}{}^{-1})_{1,1})^{-1} \right\rangle_{\xi}} = \mathrm{ACRLB}(x) \tag{4.24}$$

Simplified expressions for the calculation of the first element of the inverse combined Fisher information matrix are given by Moeneclaey [7].

The ACRLB is generally stricter than the MCRLB:

$$\mathrm{ACRLB}(x) \geq \mathrm{MCRLB}(x) \tag{4.25}$$

Equality is achieved if the parameter x is *uncoupled* to the nuisance parameters ξ (i.e., if the coupling elements of the combined Fisher information matrix vanish). This is expressed as

$$I_{x,\xi;1,n} = I_{x,\xi;n,1} = 0 \quad n > 1 \tag{4.26}$$

Equality is also achieved if the nuisance parameters are considered to assume discrete values.

4.2.2.3 Joint Cramér–Rao Lower Bound

The work by Moeneclaey also introduces the *joint* CRLB (JCRLB) [7]. The JCRLB is obtained by: (1) treating the nuisance parameters as nonrandom parameters, (2) jointly estimating x and ξ, (3) computing the CRLB for the useful parameter of interest x using the full combined Fisher information matrix $I_{x,\xi}$ of Section 4.2.2.2, and (4) averaging the obtained bound over the nuisance parameters ξ.

It should be noted that joint estimation—defined by Steps (1) and (2) above—implies a certain way to handle the nuisance parameters. Joint estimators are thus only a subset of all possible useful parameter estimators.

The JCRLB is expressed as

$$\left\langle (\hat{x}_J(\mathbf{S}) - \mathbf{x})(\hat{x}_J(\mathbf{S}) - x)^* \right\rangle_{\mathrm{N},\xi} \geq \left\langle (I_{x,\xi}{}^{-1})_{1,1} \right\rangle_{\xi} = \mathrm{JCRLB}(x) \tag{4.27}$$

The JCRLB is larger than the ACRLB (ACRLB ≤ JCRLB). When the combined Fisher information matrix is not a function of ξ, the ACRLB equals the JCRLB (and to the CRLB). When the combined Fisher information matrix does depend on ξ, the ACRLB is below the average (over ξ) of the JCRLB.

The JCRLB assumes a specific-estimation strategy (the joint estimation of x and ξ), which is indicated by the subscript J in (4.27). The JCRLB is less general than the true, asymptotic, or modified CRLB. The joint estimation does not make use of the a priori probability density function $p(\xi)$. In the examples of Moeneclaey [7], not using this information yields suboptimal estimators, especially for low signal-to-noise ratios.

4.2.2.4 Discussion

With the presence of nuisance parameters, the computation of bounds for the useful parameters becomes analytically difficult because of the complex evaluation of the *true* CRLB based on (4.19). However, it is important to recognize that the provided probability density function for the nuisance parameters represents useful information that should enter the computation of any bound or the estimator design. Ignoring this information and treating the nuisance parameters as nonrandom parameters may incorrectly value the influence of the nuisance parameters and produce bounds or estimators diverging from the true CRLB.

The presented bounds are related to each other in the sense

$$\mathrm{JCRLB}(x) \geq \mathrm{ACRLB}(x) \geq \mathrm{MCRLB}(x) \tag{4.28}$$

and

$$\mathrm{CRLB}(x) \geq \mathrm{MCRLB}(x) \tag{4.29}$$

If the combined Fisher information matrix does not depend on ξ, then

$$\mathrm{JCRLB}(x) = \mathrm{ACRLB}(x) = \mathrm{MCRLB}(x) = \mathrm{CRLB}(x) \tag{4.30}$$

In the limit of infinitely high signal-to-noise ratio, the ACRLB equals the CRLB. The JCRLB equals the CRLB if only a subset of all possible estimators are considered (the jointly estimating ones).

4.2.3 Single-Step Maximum Likelihood Estimation

The ML principle provides a commonly used basis to estimate parameters. Given a set of measured signal samples $\mathbf{s}$, the position parameters $\hat{\mathbf{x}}(\mathbf{s})$ are chosen to maximize the probability density function, which is described as

$$\hat{\mathbf{x}}(\mathbf{s}) = \arg\max_{\mathbf{x}} p_{\mathbf{x}}(\mathbf{s}) \tag{4.31}$$

In this section, we again assume that either $\mathbf{x}$ completely describes the received signal sample distribution or that the nuisance parameters have fixed known values. Both assumptions imply that (4.8) is valid.

For sufficiently smooth probability density functions, a necessary condition for the ML estimate is the *likelihood equation*:

$$\frac{\partial}{\partial \mathbf{x}} \log p_{\mathbf{x}}(\mathbf{s})\Big|_{\mathbf{x}=\hat{\mathbf{x}}(\mathbf{s})} = 0 \tag{4.32}$$

In general, the MLE (4.31) does *not* have the desired optimal properties (4.4) and (4.5). Thus, it is a priori not clear if the MLE is an MVUE. However, the following two facts make it a good candidate: First, finding the positions that make the observations most likely is a legitimate criterion on its own [1]. Second, within the limit of an infinite number of samples, it can be shown, under very general assumptions, that the MLE is *consistent* (i.e., that the estimated position converges in probability to the true value and thus becomes unbiased, see proposition IV.E.1 in the book by Poor [1]). Furthermore, under some regularity conditions for the derivates of $r_\mu(\mathbf{x})$ with respect to $\mathbf{x}$, the variance of the estimated parameters approaches the inverse of Fisher's information matrix (see proposition IV.E.2 of [1]) and the estimated parameter's probability density function approaches a Gaussian density; that is, $\hat{\mathbf{x}} \sim N(\mathbf{x}, {I_{\mathbf{x}}}^{-1})$. This property is called *asymptotic normality* of the MLE. A high number of samples has a similar effect to a high signal-to-noise ratio, thus the statement can be rephrased in the sense that, for signal-in-noise problems like (4.7), the ML estimates are Gaussian and achieve the CRLB for high signal-to-noise ratios (see Example 7.6 in the book by Lehmann [3]).

For a finite number of samples and an arbitrary (not high) signal-to-noise ratio, no strict mathematical theorem is available that would state that the MLE is unbiased or that its variance is optimal.

Consistency and asymptotic normality also hold if the admissible values for $\mathbf{x}$ are limited; that is, $\Lambda \neq \mathbb{R}^n$. It is, however, required that the solutions of the likelihood equation (4.32) converge to an isolated root and cases where the likelihood equation has multiple roots may require a more careful analysis.

In the case of Gaussian noise (4.11), the likelihood equation (4.32) is written as

$$\mathrm{Re}\left\{\sum_{\mu}\left((s_\mu - r_\mu(\hat{\mathbf{x}}))\frac{\partial \overline{r_\mu}(\mathbf{x})}{\partial x_j}\Big|_{\mathbf{x}=\hat{\mathbf{x}}}\right)\right\} = 0 \tag{4.33}$$

which is equal to the *least-squares* (LSQ) estimate of the position $\mathbf{x}$ based on the measured samples $\mathbf{s}$. It minimizes the squared differences of the signal model minus the received samples in the observation space. It should be noted that the solution of the least-squares equations (4.33) retains, in the limit of an infinite number of samples or high signal-to-noise ratio, the properties of consistency and asymptotic normality (in the same sense as for the MLE), even if the noise is *non-Gaussian* distributed (see proposition IV.E.3 in the book by Poor [1]). However, in the non-Gaussian case, the CRLB is not given by an expression like (4.14) and better estimators than the LSQ may exist. In other words, for non-Gaussian noise, the LSQ estimated parameter distribution approaches an unbiased Gaussian distribution whose covariance matrix is given by (4.14); for the non-Gaussian case, (4.14) does, in general, not define the Fisher information matrix. It should also be pointed out that, for the general case of a finite number of samples and a finite signal-to-noise ratio, (4.33) may have no solution at all.

Equation (4.31) can, in principle, be used to realize a special type of GNSS receiver that directly solves this equation by, for example, a brute-force maximization routine. This could be done in a special operation mode, called *snapshot* mode. A number of samples are collected (e.g., several tens of milliseconds) and, starting from an approximate position value, (4.31) is iteratively solved. However, to the author's knowledge, no such receiver has ever been realized, but results with simulated data are presented in works by Closas [8]. Snapshot receivers exist [9], but they determine the pseudorange for single signals first and do not directly relate the received signal samples to the position [10].

Directly estimating the position from the samples provides a number of theoretical advantages because the number of estimated parameters is minimal since no intermediate pseudorange parameters are introduced. Optimum handling of signal interference caused by multiple transmitted signals is ensured because the ML principle tries to match the sum of *all* replica signals to the received signal samples. In contrast a typical GNSS receiver ignores the influence of other signals on parameter estimates of a specific signal. It may even happen that the likelihood function has a clear and unique peak only if the likelihood function is expressed as a function of $\mathbf{x}$, but not if each signal is considered separately. In this sense, the single-step estimation optimizes the signal-to-noise ratio.

4.2.4 Cascaded Estimation

As a direct estimation of the position via (4.31) is virtually impossible, most navigation receivers subdivide the problem into several estimation steps approximating (or even being equivalent to) the full LSQ problem. One speaks of a *cascaded* estimation.

The position estimate of (4.31) is then determined in a three-step procedure:

1. Estimation of high-rate pseudorange data based on the signal samples;
2. Estimation (or filtering) to determine synchronized low-rate pseudorange data;
3. Positioning.

If the relationship between pseudorange and position is sufficiently linear, cascaded-position estimation based on a high-rate pseudorange MVUE yields a MVUE of the position.

4.2.4.1 Least-Squares Estimation of High-Rate Pseudorange Data

To refine the definition of the generalized high-rate pseudorange data given in Section 4.1, we recall that the received signal samples are the sum of the transmitted signals, each propagating along one or more propagation paths. We assume that each signal for each transmitter can be described by the same number of parameters (see Table 4.1 and Section 1.8). Those parameters are assumed to be constant during short intervals. This interval may, for example, be equal to the period of one PRN-code period if a GNSS signal is considered. For each interval and signal, the high-rate pseudorange data usually consists of one value per propagation path each for: pseudorange (or code phase), signal power, carrier phase, and Doppler frequency. Collecting these values for all intervals and signals, we obtain the vector of generalized high-rate pseudorange data $\mathbf{q}$. The vector $\mathbf{q}$ depends on the position $\mathbf{x}$ [i.e., $\mathbf{q} = \mathbf{q}(\mathbf{x})$]. For the following, we assume that this dependence can be sufficiently well-linearized (by contrast, the dependence of r_μ on $\mathbf{q}$ is, in general, highly nonlinear). Furthermore, we make the necessary assumptions to ensure that the high-rate pseudorange MLE is optimal.

The presence of possibly present nuisance parameters (actually a subset of parameters $\mathbf{q}$) is not explicitly discussed here, but the methods of Section 4.2.2 will be adapted properly in the next section.

The index range of signal samples during which the high-rate pseudorange data for a specific transmitter is constant may be different for each signal and is denoted as $L_{a,i}$. The index a denotes the different signal sources and i denotes the successive intervals. Similarly, we denote the corresponding set of high-rate pseudorange data as $\mathbf{q}_{a,i}$.

The LSQ estimate of $\mathbf{q}_{a,i}$ is given by solving for all a,i simultaneously

$$\operatorname{Re}\left\{\sum_{\mu\in L_{a,i}}(s_\mu - r_\mu(\hat{\mathbf{q}}))\frac{\partial \overline{r}_{a,\mu}(\mathbf{q}_{a,i})}{\partial \mathbf{q}_{a,i}}\bigg|_{\mathbf{q}_{a,i}=\hat{\mathbf{q}}_{a,i}}\right\} = 0 \tag{4.35}$$

With a similar argument as given for the LSQ position estimator (4.33), it can be argued that, in the limit of an infinite number of samples or high signal-to-noise ratio, the LSQ high-rate pseudorange data estimate is optimal [i.e., it is a MVUE (Gaussian noise assumed)]. In this case, the probability density function of the high-range pseudorange data estimate approaches the Gaussian multivariate density $N(\mathbf{q}_{a,i}, I_\mathbf{q}^{-1})$ [1]. The symbol $I_\mathbf{q}^{-1}$ denotes the Fisher information matrix for the high-rate pseudorange data $\mathbf{q}$.

To obtain the position estimate from the high-rate pseudorange data via an LSQ adjustment, we start from the following observation equation

$$\hat{\mathbf{q}} = \mathbf{q}(\mathbf{x}_0) + \frac{\partial \mathbf{q}}{\partial \mathbf{x}}\bigg|_{\mathbf{x}=\mathbf{x}_0} \cdot \Delta\mathbf{x}' + \mathbf{v}_\mathbf{q} \tag{4.36}$$

with the following stochastic model for the high-rate pseudorange measurement errors $\mathbf{v}_\mathbf{q}$:

$$\langle \mathbf{v}_\mathbf{q} \rangle_\mathrm{N} = 0, \quad \operatorname{var}\langle \mathbf{v}_\mathbf{q} \rangle_\mathrm{N} = \Sigma_\mathbf{q} \tag{4.37}$$

It should be noted that the LSQ estimate does not require that the distribution for $\mathbf{v}_\mathbf{q}$ is Gaussian. The LSQ is also the best *linear* unbiased estimator for non-Gaussian distributions.

If the high-rate pseudorange estimates are optimal, then $\Sigma_\mathbf{q} = I_\mathbf{q}^{-1}$.

The symbol

$$\left(\frac{\partial \mathbf{q}}{\partial \mathbf{x}}\right)_{i,j} = \frac{\partial q_i}{\partial x_j}\bigg|_{\mathbf{x}=\mathbf{x}_0} \tag{4.38}$$

denotes the design matrix. Note, the linear approximation in (4.36) needs to be a sufficient approximation and the linearization point $\mathbf{x}_0$ needs to be sufficiently near the true value $\mathbf{x}$. The least-squares adjustment of the observations $\mathbf{q}$ with respect to the improvements in the parameters $\Delta\mathbf{x}$ yields the best linear unbiased estimator (see [11]),

$$\Delta\hat{\mathbf{x}}' = J^{-1}\frac{\partial \mathbf{q}}{\partial \mathbf{x}}\Sigma_\mathbf{q}^{-1}(\hat{\mathbf{q}} - \mathbf{q}(\mathbf{x}_0)) \tag{4.39}$$

where J denotes the normal matrix

$$J = \left(\frac{\partial \mathbf{q}}{\partial \mathbf{x}}\right)^T \Sigma_\mathbf{q}^{-1}\frac{\partial \mathbf{q}}{\partial \mathbf{x}} \tag{4.40}$$

By employing this two-step procedure (first estimating $\hat{\mathbf{q}}$ and then $\Delta\hat{\mathbf{x}}'$), we finally obtain an estimate for the position as

$$\hat{\mathbf{x}}' = \mathbf{x}_0 + \Delta\hat{\mathbf{x}}' \tag{4.41}$$

Appendix A.1 shows that

$$\langle \hat{\mathbf{x}}' \rangle_{\mathrm{N}} = \mathbf{x}, \quad \mathrm{cov}\langle \hat{\mathbf{x}}' \rangle_{\mathrm{N}} = J^{-1} \tag{4.42}$$

and the normal matrix can be rewritten as

$$J_{i,j} = \sum_{k,l} \frac{\partial q_k}{\partial x_i} (\Sigma_{\mathrm{q}})^{-1}{}_{k,l} \frac{\partial q_l}{\partial x_j} \tag{4.43}$$

4.2.4.2 Bounds for Cascaded Estimation

We define an accuracy bound for an unbiased cascaded estimator (as defined in Section 4.2.4.1) as the obtained accuracy of the final estimate (i.e., the position estimate), provided that the first estimate (i.e., the high-rate pseudorange estimate) achieves a given accuracy bound. For the moment, we simply assume the existence of a high-rate pseudorange estimator achieving a given accuracy bound (e.g., the CRLB, MCRLB, ACRLB, or the JCRLB). In other words, the bound for the cascaded estimation analyzes the effect of the LSQ adjustment of the previous Section 4.2.4.1 compared to single-step position estimation.

If the high-rate pseudorange estimates achieve the CRLB and the relation between $\mathbf{q}$ and $\mathbf{x}$ is sufficiently linear, the cascaded position estimate achieves the CRLB. This is due to a general formulation of the CRLB that considers bounds of functions of parameters described by Theorem 3.4 in the book by Porat [4]. The reader is invited to prove this statement by relating (4.14) to (4.43) using (4.2).

Furthermore, it can be shown that the LSQ adjustment does not decrease the accuracy when the high-rate pseudorange estimates achieve the MCRLB. For the case of the ACRLB and the JCRLB (when they differ from the CRLB or the MCRLB), accuracy degradation caused by the LSQ adjustment cannot be excluded.

4.2.4.3 Discussion

This section has discussed different strategies to obtain position estimates and has introduced several bounds on them. It has been shown that, under certain assumptions, the single-step estimation of the position and cascaded procedure potentially achieve the same "'performance," (i.e., the CRLB is achieved for all approaches). If one of the assumptions is not valid, the single-step estimation might result in a better positioning solution compared to the cascaded procedure. The question arises: which practical cases may cause a violation of those assumptions?

To answer the question, we recall that for both cases (i.e., single-step and cascaded procedure) the signal model (4.1) must be correct. It relates the signal samples to the position parameters $\mathbf{r} = \mathbf{r}(\mathbf{x})$. For example, if multiple propagation paths are present, they must be included in the model to obtain an unbiased solution. In that sense, the direct ML approach is not superior to the cascaded approach.

More interesting is the case when the assumption of sufficient modeling (4.2) cannot be achieved. This is the case that occurs if the signal samples are not a function of delay, Doppler frequency, and phase, but show a more complex dependence on the position. This is, for example, the case when the propagation of electromagnetic waves emitted by the GNSS transmitters (satellites) cannot be described by geometric optics and when complex wave-phenomena occur. A practical example could be a signal creeping of GNSS signals along an aircraft body [12]. But generally, in that case, the direct solution of (4.33) seems to be extremely demanding.

To achieve the optimum positioning result with the cascaded procedure, it is necessary that the full covariance matrixes are passed from the high-rate pseudorange estimation to the position estimation and that the matrices are used in the LSQ adjustments. This condition, however, can often not be fulfilled in practice.

It has been assumed that the high-rate pseudorange estimators are MVUEs, which is achieved by ML estimation under the assumption of a high signal-to-noise ratio or a large number of samples. Additionally, in Section 4.7.1 we will show that, for uncorrelated and uniformly distributed carrier phases, a modified MLE scheme (a single LSQ step under the assumption that the linearity conditions of Section 4.3.2.10 are fulfilled) achieves the CRLB for *arbitrary* signal-to-noise ratio values.

If the signal model considered here can be linearized, the MLE becomes a MVUE (even for low signal-to-noise ratio). Furthermore, Bayesian techniques (e.g., a Kalman filter) can be used easily after linearization. The prerequisite for linearization is that good approximate values for the parameters are available. If a Kalman filter is used, the approximate values are the predicted values. Then the prerequisite for linearization is that the predicted parameter values—especially the carrier phase if a coherent Kalman filter is used—are precise (see Section 4.5.4).

The discussion of what happens if the MLE of (4.31) or of (4.35) is not a MVUE is difficult. Looking for an alternative to an MLE is not easy; in fact, it is not always guaranteed that a MVUE (or any better estimator than the MLE) exists. For example, the possible strategy to construct a MVUE via sufficient statistics and the Rao–Blackwell–Lehmann–Scheffe theorem is of little practical value [3]. In fact, the vast majority of practical estimators are MLEs [3].

If no optimum estimation scheme can be provided, we may speculate that, in this case, the direct solution of the position via (4.33) is superior to the cascaded procedure; this is due to the fewer degrees of freedom involved, which thereby increases the effective signal-to-noise ratio.

4.3 LSQ Correlators/Discriminators

This section contains a further analysis of the LSQ estimators for high-rate pseudorange data that have been defined by (4.35) in Section 4.2.4.1. Estimators for code phase, Doppler, carrier phase, and signal power will be derived. The effect of the unknown carrier phase—causing the squaring loss—will be discussed. The importance of limiting the admissible range of the estimated values will be emphasized, and various solutions (single-channel tracking, vector tracking, and reiteration) will be compared. After a general introduction, the section concentrates on the case where only one line-of-sight signal is received; in Section 4.3.5, the results will be discussed in further detail for one line-of-sight signal and multiple propagation paths.

It has already been demonstrated that the ML solution can be treated under the assumption of Gaussian noise as a complex LSQ adjustment problem whose solution is described in Appendix A.1. As discussed in Section 4.2.3, we expect the LSQ estimate to be optimal if the signal-to-noise ratio is high or if a sufficiently large number of samples is considered. If neither condition is fulfilled or if the noise is non-Gaussian, the LSQ estimate is still defined via (4.35) but has to be treated as an engineering solution, which cannot be expected to be optimal without further analysis. This is especially true because the signal model defined in Chapter 1 is highly nonlinear. Only if the linearization is carried out sufficiently near or at the true parameter values will the LSQ solution achieve the CRLB due to (4.17). The linearization will be discussed in Section 4.3.2.10.

4.3.1 Model for One or More Propagation Paths

We assume that the deterministic part r_μ of the received signal samples S is modeled during one integration interval, given by the index range μ= 1, ..., L, as (see Sections 1.6 and 1.8)

$$r_\mu = \sum_{m=1}^{M} r_{m;\mu} = \sum_{m=1}^{M} a'_m c_m(t_\mu - \tau_m)\exp\{i(\omega_m t_\mu - \varphi_m)\} \tag{4.44}$$

The signal is a superposition of transmitted signals originating from various transmitters with propagation along one or more propagation paths. The index m uniquely identifies transmitter, propagation path, and, eventually, signal component (e.g., a GNSS data or pilot component emitted from the same transmitter). For each index m, the four fundamental signal parameters a'_m (amplitude), τ_m (code pseudorange in seconds), ω_m (angular frequency in radians per second), and φ_m (carrier-phase delay in radians) are assumed to be constant. The symbol t_μ denotes the sampling epochs given by μ divided by the sample rate plus a fixed-time offset. These signals $c_m(t)$ fulfill the condition of unity power mentioned in Section 1.8.1. The other conditions of Section 1.8 are, in general, (e.g., for signals with a noncontinuous power distribution in time) not fulfilled, as it might be impossible to find a proper symmetric sampling interval μ = 1, ..., L for *all* signals if they are considered over a common interval. For example, the pulses might be located non-symmetrically within this interval.

For simplicity, we consider that I- and Q-components of the signal have been sampled. In Appendix A.2, it is shown that this is mathematically equivalent to sampling only one component (e.g., the I-component) with twice the sample rate. Both representations of the sampled signal can be transformed into each other.

It is convenient to define the complex-valued signal amplitude as

$$a_m = \exp\{-i\varphi_m + i\omega_m t_{mid}\}a'_m \tag{4.45}$$

therefore

$$r_\mu = \sum_{m=1}^{M} a_m c_m(t_\mu - \tau_m)\exp\{i\omega_m(t_\mu - t_{mid})\} = \sum_{m=1}^{M} a_m c_m(t_\mu - \tau_m)\exp\{i\omega_m t'_\mu\} \tag{4.46}$$

with

$$t'_\mu = t_\mu - t_{mid} \quad \text{and} \quad t_{mid} = \frac{t_1 + t_L}{2} \tag{4.47}$$

The vector of high-rate pseudorange data is composed of

$$\mathbf{q} = (\ a_1 \quad \tau_1 \quad \omega_1 \quad \ldots \quad a_M \quad \tau_M \quad \omega_M)^T \tag{4.48}$$

It should be noted that, for the moment, we assume that the number of received signals m is known from a priori; that is, however, not true because the number of multipath signals, in particular, is difficult to determine. Estimation of m will be touched upon later in Section 8.3.7. Furthermore, it should be mentioned that, for the moment, we treat all parameters as useful parameters (see Section 4.1).

Analyzing (4.35), we obtain for the derivatives the following set of equations:

$$\begin{aligned}
\frac{\partial r_\mu}{\partial a_m} &= \frac{\partial r_{m;\mu}}{\partial a_m} = c_m(t_\mu - \tau_m)\exp\{i\omega_m t'_\mu\} \\
\frac{\partial r_\mu}{\partial \tau_m} &= \frac{\partial r_{m;\mu}}{\partial \tau_m} = -a_m c_m{}'(t_\mu - \tau_m)\exp\{i\omega_m t'_\mu\} \\
\frac{\partial r_\mu}{\partial \omega_m} &= \frac{\partial r_{m;\mu}}{\partial \omega_m} = it'_\mu a_m c_m(t_\mu - \tau_m)\exp\{i\omega_m t'_\mu\}
\end{aligned} \tag{4.49}$$

These derivatives are used to set up the design matrix. Please note that the apostrophe denotes a derivate of a function $c'(\ldots)$ and it may also denote a separate symbol t', as in (4.58). The first index (row) of the design matrix is denoted by μ and is used to index the samples; the second index m (the column index) is used to enumerate the vector of unknowns.

To perform the LSQ adjustment, it should be noted that the vector of unknowns is composed of the high-rate pseudorange parameters **q**. In the following, only the symbol **q** is used for the unknowns (and not **x**, as in Appendix A.1).

The design matrix takes the form

$$A = (A_1 \quad \ldots \quad A_M) \tag{4.50}$$

with A_m as submatrices given by

$$A_m = \begin{pmatrix} \frac{\partial r_1}{\partial a_m} & \frac{\partial r_1}{\partial \tau_m} & \frac{\partial r_1}{\partial \omega_m} \\ \frac{\partial r_2}{\partial a_m} & \frac{\partial r_2}{\partial \tau_m} & \frac{\partial r_2}{\partial \omega_m} \\ \vdots & \vdots & \vdots \\ \frac{\partial r_L}{\partial a_m} & \frac{\partial r_L}{\partial \tau_m} & \frac{\partial r_L}{\partial \omega_m} \end{pmatrix} = \begin{pmatrix} \frac{\partial r_{m;1}}{\partial a_m} & \frac{\partial r_{m;1}}{\partial \tau_m} & \frac{\partial r_{m;1}}{\partial \omega_m} \\ \frac{\partial r_{m;2}}{\partial a_m} & \frac{\partial r_{m;2}}{\partial \tau_m} & \frac{\partial r_{m;2}}{\partial \omega_m} \\ \vdots & \vdots & \vdots \\ \frac{\partial r_{m;L}}{\partial a_m} & \frac{\partial r_{m;L}}{\partial \tau_m} & \frac{\partial r_{m;L}}{\partial \omega_m} \end{pmatrix} \tag{4.51}$$

The observations are given by the received signal samples **S** of (1.15), which are modeled as complex-valued random variables, each (the real and imaginary component) having unit variance as defined in (1.16). The samples are composed of the deterministic part r_μ and the thermal noise **N**, which is expressed in vector notation as

$$\mathbf{S} = \mathbf{r} + \mathbf{N} \quad \Leftrightarrow \quad S_\mu = r_\mu + N_\mu \tag{4.52}$$

The covariance matrix for the composite complex-valued observations is (see Section 1.7)

$$Q_{V;\mu,\upsilon} = \left\langle \overline{(S_\mu - \langle S_\mu \rangle_{\mathbf{N}})}(S_\upsilon - \langle S_\upsilon \rangle_{\mathbf{N}}) \right\rangle_{\mathbf{N}} = \left\langle \overline{N_\mu} N_\upsilon \right\rangle_{\mathbf{N}} = 2\delta_{\mu,\upsilon} \tag{4.53}$$

and the normal matrix I takes the form

$$I = A^* Q_V^{-1} A = \frac{1}{2} \begin{pmatrix} A_1^* A_1 & A_2^* A_1 & \cdots \\ A_1^* A_2 & A_2^* A_2 & \\ \vdots & & \ddots \end{pmatrix} \tag{4.54}$$

The normal matrix has the same form as the Fisher information matrix because Gaussian noise is assumed. Therefore, we choose the symbol I for the normal matrix (and not N); this also distinguishes it from the noise symbol.

Because of the nonlinear signal model (4.46), the LSQ problem has to be solved iteratively. The iterative solution of the LSQ problem starts with calculating the derivates of (4.51) at a first linearization point $\mathbf{q}_0$ that shall be denoted as

$$\mathbf{q}_0 = (\, a_{1,0} \quad \tau_{1,0} \quad \omega_{1,0} \;\ldots\; a_{M,0} \quad \tau_{M,0} \quad \omega_{M,0})^T \tag{4.55}$$

The observation equations are linear in the amplitude, such that no iteration is required for amplitude determination alone. However, iterations may be required in Doppler and code-phase dimensions.

For a single iteration, the estimated corrections of the high-rate pseudorange data read as

$$\Delta\hat{\mathbf{q}} = I^{-1} A^* Q_V^{-1} (\mathbf{S} - \mathbf{r}(\mathbf{q}_0)) = \frac{1}{2} I^{-1} A^* (\mathbf{S} - \mathbf{r}(\mathbf{q}_0)) \tag{4.56}$$

As mentioned in Appendix A.1, the estimates are complex-valued random variables and, after computing, code phase, and Doppler are constrained to be real-valued. The complex-valued estimated amplitudes shall be separated into the real-valued amplitude and the estimated carrier phase after the adjustment. A real valued signal model might seem to be better because it avoids the use of complex-valued signal parameter, but Appendix A.1.4 shows that the separation can be performed very easily if the parameters are uncorrelated.

Equation (4.56) represents a signal-processing technique that fully accounts for signal correlations. Replica signals given by the term $\mathbf{r}(\mathbf{q}_0)$ are subtracted from the received samples before correlating them with the reference signals given by (4.49). This is equal to a parallel interference cancellation scheme [13]. Via the nondiagonal form of I, (4.56) accounts for correlations between the estimated parameters for different transmitted signals.

For a further evaluation of this LSQ problem, we need to make specific assumptions on the received signals. In Section 4.3.2, a detailed analysis for the case of $m = 1$ will be presented.

4.3.2 Single Propagation Path

For the case of a single propagation path ($m = 1$), the normal matrix takes the form

$$I = \frac{1}{2} A_1^* A_1 \tag{4.57}$$

This is a complex-valued 3 × 3 matrix.

Here, we assume that the integration interval is chosen symmetrically with respect to the received signal power such that all assumptions on $c(t)$ made in Section 1.8 are fulfilled. Because of the definition in (4.47), we write for the centered sampling epochs t_μ'

$$t_\mu' = \left(\mu - \frac{L+1}{2} \right) \Big/ f_s \tag{4.58}$$

Starting from (4.55) and using the assumptions of Section 1.8, the elements of I read as

$$\begin{aligned}
2I_{1,1} &= \sum_{\mu=1}^{L} \overline{\frac{\partial r_\mu}{\partial a_1}} \frac{\partial r_\mu}{\partial a_1} = \sum_{\mu=1}^{L} \left| c_1(t_\mu - \tau_{1,0}) \right|^2 \approx L \\
2I_{1,2} = 2\overline{I}_{2,1} &= \sum_{\mu=1}^{L} \overline{\frac{\partial r_\mu}{\partial a_1}} \frac{\partial r_\mu}{\partial \tau_1} = -a_{1,0} \sum_{\mu=1}^{L} \overline{c_1(t_\mu - \tau_{1,0})} c_1'(t_\mu - \tau_{1,0}) \approx 0 \\
2I_{1,3} = 2\overline{I}_{3,1} &= \sum_{\mu=1}^{L} \overline{\frac{\partial r_\mu}{\partial a_1}} \frac{\partial r_\mu}{\partial \omega_1} = -ia_{1,0} \sum_{\mu=1}^{L} t_\mu' \left| c_1(t_\mu - \tau_{1,0}) \right|^2 \approx 0 \\
2I_{2,2} &= \sum_{\mu=1}^{L} \overline{\frac{\partial r_\mu}{\partial \tau_1}} \frac{\partial r_\mu}{\partial \tau_1} = \left| a_{1,0} \right|^2 \sum_{\mu=1}^{L} \left| c_1'(t_\mu - \tau_{1,0}) \right|^2 \approx \left| a_{1,0} \right|^2 L R_{\bar{c}',c'}(0) = -\left| a_{1,0} \right|^2 L R_{\bar{c}_1,c_1}''(0) \\
2I_{3,2} = 2\overline{I}_{2,3} &= \sum_{\mu=1}^{L} \overline{\frac{\partial r_\mu}{\partial \omega_1}} \frac{\partial r_\mu}{\partial \tau_1} = i \left| a_{1,0} \right|^2 \sum_{\mu=1}^{L} t_\mu' \overline{c_1(t_\mu - \tau_{1,0})} c_1'(t_\mu - \tau_{1,0}) \approx 0 \\
2I_{3,3} &= \sum_{\mu=1}^{L} \overline{\frac{\partial r_\mu}{\partial \omega_1}} \frac{\partial r_\mu}{\partial \omega_1} = \left| a_{1,0} \right|^2 \sum_{\mu=1}^{L} (t_\mu')^2 \left| c_1(t_\mu - \tau_{1,0}) \right|^2 \\
&\approx \left| a_{1,0} \right|^2 \chi_{freq} \frac{L^3 - L}{12 f_s^2} \approx \frac{\chi_{freq} L^3 \left| a_{1,0} \right|^2}{12 f_s^2}
\end{aligned} \tag{4.59}$$

or in matrix notation as

$$2I \approx L \begin{pmatrix} 1 & 0 & 0 \\ 0 & -R_{\bar{c}_1,c_1}{''}(0)\left|a_{1,0}\right|^2 & 0 \\ 0 & 0 & \chi_{freq}\left|a_{1,0}\right|^2 L^2/(12f_s^2) \end{pmatrix} \tag{4.60}$$

The inverse normal matrix is given by

$$I^{-1} \approx \begin{pmatrix} \frac{2}{L} & 0 & 0 \\ 0 & \frac{2}{-LR_{\bar{c},c}{''}(0)\left|a_{1,0}\right|^2} & 0 \\ 0 & 0 & \frac{24f_s^2}{\chi_{freq}\left|a_{1,0}\right|^2 L^3} \end{pmatrix} \tag{4.61}$$

Because I is a diagonal matrix, amplitude, code phase, and frequency estimates are independent of each other. This is a consequence of the assumptions of Section 1.8.

From this point forward, all approximate equal signs ≈ will be replaced by equal signs '=' for the sake of simplicity.

4.3.2.1 Estimated Parameter Improvements

After setting up the design matrix, the LSQ adjustment can be started, taking (4.55) as the starting point. In each step, parameter improvements are calculated that are added to the a priori values. The following equations are formulated for the first iteration but can be extended for subsequent iterations.

The estimated improvement in the complex amplitude is given as

$$\begin{aligned} \Delta\hat{a}_{1,0} &= \frac{1}{L}\sum_{\mu=1}^{L}\overline{\frac{\partial r_\mu}{\partial a_1}}(S_\mu - r_\mu(\mathbf{q}_0)) \\ &= \frac{1}{L}\sum_{\mu=1}^{L}\overline{c_1(t_\mu - \tau_{1,0})}\exp\{-i\omega_{1,0}t'_\mu\}(S_\mu - a_{1,0}c_1(t_\mu - \tau_{1,0})\exp\{i\omega_{1,0}t'_\mu\}) \\ &= \frac{1}{L}\sum_{\mu=1}^{L}\overline{c_1(t_\mu - \tau_{1,0})}\exp\{-i\omega_{1,0}t'_\mu\}S_\mu - a_{1,0} \end{aligned} \tag{4.62}$$

and the estimated amplitude after the first iteration is given by

$$\hat{a}_{1,0} = a_{1,0} + \Delta\hat{a}_{1,0} = \frac{1}{L}\sum_{\mu=1}^{L}\overline{c_1(t_\mu - \tau_{1,0})}\exp\{-i\omega_{1,0}t'_\mu\}S_\mu \tag{4.63}$$

For the estimated code-phase improvement, we obtain

$$
\begin{aligned}
\Delta\hat{\tau}_{1,0} &= \frac{-1}{LR_{\bar{c}_1,c_1}''(0)\left|a_{1,0}\right|^2}\sum_{\mu=1}^{L}\overline{\frac{\partial r_\mu}{\partial \tau_1}}(S_\mu - r_\mu(\mathbf{q}_0)) \\
&= \frac{\bar{a}_{1,0}}{LR_{\bar{c}_1,c_1}''(0)\left|a_{1,0}\right|^2} \\
&\times \sum_{\mu=1}^{L}\overline{c_1'(t_\mu-\tau_{1,0})\exp\{-i\omega_{1,0}t'_\mu\}}(S_\mu - a_{1,0}c_1(t_\mu-\tau_{1,0})\exp\{i\omega_{1,0}t'_\mu\}) \\
&= \frac{1}{LR_{\bar{c}_1,c_1}''(0)a_{1,0}}\sum_{\mu=1}^{L}\overline{c_1'(t_\mu-\tau_{1,0})\exp\{-i\omega_{1,0}t'_\mu\}}S_\mu - \frac{1}{LR_{\bar{c},c}''(0)}R_{\bar{c}_1',c_1}(0) \\
&= \frac{1}{LR_{\bar{c}_1,c_1}''(0)a_{1,0}}\sum_{\mu=1}^{L}\overline{c_1'(t_\mu-\tau_{1,0})\exp\{-i\omega_{1,0}t'_\mu\}}S_\mu
\end{aligned}
\tag{4.64}
$$

and for the estimated frequency improvement:

$$
\begin{aligned}
\Delta\hat{\omega}_{1,0} &= \frac{12f_s}{\chi_{freq}\left|a_{1,0}\right|^2 L^3}\sum_{\mu=1}^{L}\overline{\frac{\partial r_\mu}{\partial \omega_1}}(S_\mu - r_\mu(\mathbf{q}_0)) \\
&= \frac{-i12f_s}{\chi_{freq}a_{1,0}L^3}\sum_{\mu=1}^{L}t'_\mu\overline{c_1(t_\mu-\tau_{1,0})}\exp\{-i\omega_{1,0}t'_\mu\}(S_\mu - a_{1,0}c_1(t_\mu-\tau_{1,0})\exp\{i\omega_{1,0}t'_\mu\}) \\
&= -\frac{i12f_s}{\chi_{freq}a_{1,0}L^3}\sum_{\mu=1}^{L}t'_\mu\overline{c_1(t_\mu-\tau_{1,0})}\exp\{-i\omega_{1,0}t'_\mu\}S_\mu + \\
&+ \frac{-i12f_s}{\chi_{freq}L^3}\sum_{\mu=1}^{L}t'_\mu\overline{c_1(t_\mu-\tau_{1,0})}c_1(t_\mu-\tau_{1,0}) \\
&= -\frac{i12f_s}{\chi_{freq}a_{1,0}L^3}\sum_{\mu=1}^{L}t'_\mu\overline{c_1(t_\mu-\tau_{1,0})}\exp\{-i\omega_{1,0}t'_\mu\}S_\mu
\end{aligned}
\tag{4.65}
$$

The estimated frequency and code phase after the first iteration are given by

$$
\begin{aligned}
\hat{\tau}_{1,0} &= \tau_{1,0} + \Delta\hat{\tau}_{1,0} \\
\hat{\omega}_{1,0} &= \omega_{1,0} + \Delta\hat{\omega}_{1,0}
\end{aligned}
\tag{4.66}
$$

4.3.2.2 Iteration

Due to the nonlinear signal model (4.46), it is, in general, necessary to repeat a LSQ procedure until the estimated parameters converge. The iteration can be written for $k > 0$ as

$$\begin{aligned}\hat{\tau}_{1,k+1} &= \hat{\tau}_{1,k} + \Delta\hat{\tau}_{1,k} \\ \hat{\omega}_{1,k+1} &= \hat{\omega}_{1,k} + \Delta\hat{\omega}_{1,k} \\ \hat{a}_{1,k+1} &= \hat{a}_{1,k} + \Delta\hat{a}_{1,k}\end{aligned} \tag{4.67}$$

At this point it should be noted that, after the first iteration, the a priori values become themselves random variables because they depend on the signal samples. This will finally result in what is referred to as "squaring loss" and will be discussed in detail below.

Formally, the parameter iteration is written as

$$\hat{\tau}_{1,*} = \lim_{k \to \infty} \hat{\tau}_{1,k} \tag{4.68}$$

and is similar for Doppler and amplitude.

4.3.2.3 How to Evaluate the LSQ Parameter Variance

Within the framework of a LSQ adjustment, the variance of the estimated parameters is given by the inverse of the normal matrix. The normal matrix (4.54) equals the Fisher information matrix, so in a first approximation we would expect that the estimated LSQ parameters achieve the CRLB.

However, this is strictly true only if the linearization point $\mathbf{q}_0$ is a deterministic quantity and if the linearization point is sufficiently near the true values. During the iterations in the LSQ adjustment, the linearization point $\mathbf{q}_0$ becomes dependent on the estimated parameters (which themselves depend of the signal samples) and has to be regarded as a random variable. Therefore, the LSQ estimates generally have an *increased variance* when iterations are required.

In a standard LSQ adjustment, the influence of stochastic parameters used to setup the design matrix is ignored. This can be justified if the observation equations can be linearized and the linearization is a good approximation to the true nonlinear observation equations. For a linear problem, the iteration converges after one step and (4.17) holds. In the case of the nonlinear high-rate pseudorange parameter-estimation problem discussed here, this linearity assumption is true if the signal power is high and the accuracy of the estimated parameters is high. Then the design matrix can be considered as nonrandom. This is in accordance with the asymptotic properties of any MLE, as discussed in Section 4.2.3. If the signal power becomes low, or errors in the estimated parameters are in a range that they cannot be ignored when building the design matrix, the standard LSQ error propagation method [see (A.29)] should not be used. This is especially important for the signal amplitude estimate. In that case, the formulas for the estimated parameter variance have to be modified and the squaring loss appears. The squaring loss generally increases the variance and is interpreted as a contribution to the parameter variance because of signal-amplitude errors when setting up the design matrix.

It should be noted that the squaring loss is usually interpreted from an engineering point of view. As noted elsewhere [14, 15], the squaring loss originates from a nonlinear operation (e.g., multiplication) of a coherent discriminator with the punctual correlator to remove the carrier-phase dependency of the discriminator [16].

4.3.2.4 Cramér–Rao Lower Bound Without Nuisance Parameters

In the next step, the variance of the estimated parameters assuming a *nonrandom* design matrix shall be evaluated. The obtained variances correspond to the CRLB if we treat all four parameters (code phase, Doppler, carrier phase, and amplitude) as useful nonrandom parameters.

Bounds can only be computed for a Gaussian noise amplitude distribution, which shall be assumed here. Fisher's information matrix is therefore given by the normal matrix (4.60). Equivalently, we treat the design matrix as a nonrandom quantity that is set up using the true parameter values. Therefore, it is sufficient to analyze the first iteration only.

The definition of the stochastic noise properties in (4.53) relates the signal amplitude to the C/N_0 value by (see also Section 1.8.1)

$$\left|a_1\right|^2 = \frac{2(C/N_0)_1}{f_s} \tag{4.69}$$

Recall that the index $m = 1$ enumerates the signal.

The variances of the estimated signal parameters are the diagonal elements of the inverse normal matrix given by (4.61).

The variance of the estimated (complex-valued) signal amplitude is given by

$$\mathrm{var}\left\langle \Delta\hat{a}_{1,0} \right\rangle_{\mathrm{N}} = \frac{2}{L} \tag{4.70}$$

It should be recalled that the complex amplitude contains information on the C/N_0 value and the carrier phase. Below, the complex amplitude will be separated into these two components and their variances will be discussed. First, however, the variance of the code phase and Doppler estimates will be calculated.

The variance of the estimated code phase is given by

$$\begin{aligned} \mathrm{var}\left\langle \Delta\hat{\tau}_{1,0} \right\rangle_{\mathrm{N}} &= \frac{2}{-LR_{\bar{c}_1,c_1}{}''(0)\left|a_{1,0}\right|^2} = \frac{f_s}{-LR_{\bar{c}_1,c_1}{}''(0)(C/N_0)_1} \\ &= \frac{1}{-T_{coh}R_{\bar{c}_1,c_1}{}''(0)(C/N_0)_1} \end{aligned} \tag{4.71}$$

where the coherent integration time T_{coh} is the ratio of the number of samples L divided by the sample rate f_s. Note that this variance applies to the complex-valued code-phase estimate. According to Appendix A.1, the variance of the real-valued code phase estimated is half of (4.71), which is the CRLB for the code-phase estimate in square seconds

$$\operatorname{var}\left\langle \operatorname{Re}\{\Delta\hat{\tau}_{1,0}\}\right\rangle_{N} = \frac{1}{-2T_{coh}R_{\bar{c}_1,c_1}{}''(0)(C/N_0)_1} \tag{4.72}$$

Similarly, we obtain for the Doppler variance in square radians per second squared

$$\operatorname{var}\left\langle \operatorname{Re}\{\Delta\hat{\omega}_{1,0}\}^2\right\rangle_{N} = \frac{6}{T_{coh}{}^3(C/N_0)_1\chi_{freq}} \tag{4.73}$$

4.3.2.5 Cramér–Rao Lower Bound with Nuisance Parameters

Fisher's information matrix (4.60) is diagonal and depends on the signal power but not on the carrier phase. Therefore, if we consider the carrier phase as a nuisance parameter, the MCRLB, ACRLB, and the JCRLB for code phase and Doppler are also given by (4.72) and (4.73). Note, however, that the true CRLB may be different from those expressions if the carrier phase is considered as a nuisance parameter.

If the signal power is also considered as a nuisance parameter, the MCRLB and the ACRLB equal each other because the amplitude is uncoupled with Doppler and code phase. Furthermore, the Fisher information matrix depends linearly on the signal power, such that the MCRLB is given by (4.72) and (4.73) if the C/N_0 value is replaced by its expected value with respect to its a priori distribution. Considering the JCRLB, it is easy to recognize that, for example, the code phase JCRLB is written as

$$JCRLB\left(\operatorname{Re}\{\Delta\hat{\tau}_1\}\right) = \frac{1}{-2T_{coh}R_{\bar{c}_1,c_1}{}''(0)}\int \frac{1}{(C/N_0)_1} p((C/N_0)_1)d((C/N_0)_1) \tag{4.74}$$

where $p((C/N_0)_1)$ is the a priori probability distribution for the signal-to-noise ratio. In contrast to the MCRLB or the ACRLB, the JCRLB depends not only on the mean signal-to-noise ratio, but also on its distribution. The less stringent the distribution, the larger the difference between the JCRLB and the MCRLB. If $(C/N_0)_1$ is from $N(C, \sigma^2)$, then for small values of σ^2,

$$JCRLB(\operatorname{Re}\{\Delta\hat{\tau}_1\}) \approx \frac{1}{-2T_{coh}R_{\bar{c}_1,c_1}{}''(0)C}\left(1+\frac{\sigma^2}{C^2}\right) \tag{4.75}$$

4.3.2.6 Squaring Loss Affecting the Code-Phase LSQ Estimate

To assess the influence of the estimated parameters on the LSQ design matrix in the sense of Section 4.3.2.3, a simplified iteration procedure will be assumed. The

LSQ adjustment is iterated and in each step the design matrix is recalculated using the signal amplitude estimate from the last step. The influence of code-phase errors and Doppler errors on the design matrix is not considered, and the initial values $\tau_{1,0}$ and $\omega_{1,0}$ are retained. Their influence will be discussed later in Section 4.3.2.10. The iteration starts at any nonzero value of the amplitude.

The simplified iteration procedure converges after the first iteration because the observation equation for the complex signal amplitude is linear.

Let us examine (4.64), which describes the code-phase improvement. After the iteration has converged, this equation will take the form

$$\Delta\hat{\tau}_{1,*} = \frac{1}{LR_{\bar{c}_1,c_1}{}''(0)\hat{a}_{1,*}} \sum_{\mu=1}^{L} \overline{c_1{}'(t_\mu - \tau_{1,0})\exp\{-i\omega_{1,0}t'_\mu\}S_\mu} \tag{4.76}$$

where the subscript "*" denotes the converged values. The initial value for the complex amplitude $a_{1,0}$ has been replaced by the converged estimated amplitude value $\hat{a}_{1,*}$.

To calculate the variance of the code-phase estimate obtained with the converged amplitude estimate, we rewrite (4.76) as

$$\Delta\hat{\tau}_{1,*} = \frac{a_1}{\hat{a}_{1,*}} \frac{1}{LR_{\bar{c}_1,c_1}{}''(0)a_1} \sum_{\mu=1}^{L} \overline{c'(t_\mu - \tau_{1,0})\exp\{-i\omega_{1,0}t'_\mu\}S_\mu} = \frac{a_1}{\hat{a}_{1,*}}\Delta\hat{\tau}_{1,0} \tag{4.77}$$

The estimated signal amplitude and the code-phase improvement are uncorrelated, as can be seen from (4.60). Because the expected value of the code-phase improvement vanishes, the following equation holds

$$\left\langle \Delta\hat{\tau}_{1,*} \right\rangle = \left\langle \frac{a_1}{\hat{a}_{1,*}}\Delta\hat{\tau}_{1,0} \right\rangle = \left\langle \frac{a_1}{\hat{a}_{1,*}} \right\rangle \left\langle \Delta\hat{\tau}_{1,0} \right\rangle = 0 \tag{4.78}$$

Note that now and for the rest of the squaring loss discussion, the subscript **N** is omitted from the expectation values and all expectation values are understood with respect to the thermal noise.

For the variance we obtain

$$\left\langle \left|\Delta\hat{\tau}_{1,*}\right|^2 \right\rangle = \left\langle \left| \frac{a_1}{\hat{a}_{1,*}}\Delta\hat{\tau}_{1,0} \right|^2 \right\rangle = \left\langle \left| \frac{a_1}{\hat{a}_{1,*}} \right|^2 \right\rangle \left\langle \left|\Delta\hat{\tau}_{1,0}\right|^2 \right\rangle \tag{4.79}$$

The first term of the product in the above equation is the squaring loss. If the estimated value of the signal amplitude is precise (i.e., $\hat{a}_{1,*} = a_1$), the squaring loss will be 1 (= 0 dB). This stands in analogy to the JCRLB of Section 4.3.2.5.

For the general case (i.e., $|\hat{a}_{1,*}| \neq |a_1|$), we perform a Taylor series expansion as

$$\begin{aligned}\left\langle \left|\frac{a_1}{\hat{a}_{1,*}}\right|^2 \right\rangle &= \left\langle \left|\frac{a_1}{(\hat{a}_{1,*}-a_1)+a_1}\right|^2 \right\rangle = \left\langle \left|\frac{1}{\frac{(\hat{a}_{1,*}-a_1)}{a_1}+1}\right|^2 \right\rangle \approx \left\langle \left|1-\frac{(\hat{a}_{1,*}-a_1)}{a_1}\right|^2 \right\rangle \\ &= 1-2\,\mathrm{Re}\left\langle \frac{(\hat{a}_{1,*}-a_1)}{a_1} \right\rangle + \left\langle \left|\frac{(\hat{a}_{1,*}-a_1)}{a_1}\right|^2 \right\rangle = 1+\frac{\mathrm{var}\langle \hat{a}_{1,*} \rangle}{|a_1|^2}\end{aligned} \tag{4.80}$$

Using (4.61) and (4.69), we obtain the squaring loss,

$$\left\langle \left|\frac{a_1}{\hat{a}_{1,*}}\right|^2 \right\rangle = 1+\frac{f_s}{L(C/N_0)_1} = 1+\frac{1}{T_{coh}(C/N_0)_1} \tag{4.81}$$

The squaring loss decreases with increasing C/N_0 and with an increasing coherent integration time T_{coh}.

Finally, the variance of the LSQ code-phase estimate (i.e., the real part of the complex-valued random variable) evaluates to

$$\begin{aligned}\left\langle \mathrm{Re}\{\Delta\hat{\tau}_{1,*}\}^2 \right\rangle &= \frac{1}{2}\left\langle |\Delta\hat{\tau}_{1,0}|^2 \right\rangle \left\langle \left|\frac{a_1}{\hat{a}_{1,*}}\right|^2 \right\rangle \\ &= \frac{1}{-2T_{coh}R_{\bar{c}_1,c_1}{}''(0)(C/N_0)_1}\left(1+\frac{1}{T_{coh}(C/N_0)_1}\right)\end{aligned} \tag{4.82}$$

The first part of this equation is equal to the code-phase CRLB (without nuisance parameters), the second is the squaring loss. Similar equations are derived elsewhere [16, 17].

4.3.2.7 Squaring Loss Affecting the Doppler LSQ Estimate

Comparison of (4.65) with (4.64) shows that the squaring loss affecting the Doppler frequency estimate has the same form as (4.81):

$$\left\langle |\Delta\hat{\omega}_{1,*}|^2 \right\rangle = \left\langle \left|\frac{a_1}{\hat{a}_{1,*}}\right|^2 \right\rangle \left\langle |\Delta\hat{\omega}_{1,0}|^2 \right\rangle \tag{4.83}$$

Taking the real component of the complex-valued random variable yields

$$\left\langle \mathrm{Re}\{\Delta\hat{\omega}_{1,*}\}^2 \right\rangle = \frac{12 f_s^{\,2}}{L^3 \left|a_{1,0}\right|^2 \chi_{freq}} \left(1 + \frac{1}{T_{coh}(C/N_0)_1}\right)$$
$$= \frac{6 f_s^{\,3}}{L^3 (C/N_0)_1 \chi_{freq}} \left(1 + \frac{1}{T_{coh}(C/N_0)_1}\right) = \frac{6}{T_{coh}^{\,3} (C/N_0)_1 \chi_{freq}} \left(1 + \frac{1}{T_{coh}(C/N_0)_1}\right) \tag{4.84}$$

Again, the Doppler variance is given by the product of the CRLB multiplied by the squaring loss. The factor χ_{freq} measures the nonuniformity of the signal-power distribution in time and is defined in Section 1.8.3. The CRLB obtained here corresponds to equation (16) of the article by Rife and Boorstyn under the assumption that the phase is unknown [18].

4.3.2.8 Squaring Loss Affecting the Carrier-Phase LSQ Estimate

What remains is to analyze the squaring loss on the signal-amplitude LSQ estimate itself. First, we have to note that the real and imaginary components of the complex signal amplitude itself are not affected by a squaring loss because the estimated complex amplitude is independent on the assumed amplitude a priori value. However, the squaring loss enters when the magnitude and phase are evaluated.

To calculate the estimated phase we assume (without a loss of generality) that the imaginary part of the true signal amplitude vanishes. A second-order Taylor series expansion is performed in the difference of the estimated minus the true values. To formalize this series expansion, the temporary parameter ε is introduced. The expansion is given as

$$\hat{\varphi}_{1,*} = \tan^{-1}\left(\frac{\hat{a}_{1,*,im}}{\hat{a}_{1,*,re}}\right)$$
$$= \tan^{-1}\left(\frac{\varepsilon \hat{a}_{1,*,im}}{\varepsilon(\hat{a}_{1,*,re} - a_{1,re}) + a_{1,re}}\right) \approx \varepsilon \frac{\hat{a}_{1,*,im}}{a_{1,re}} + \varepsilon^2 \frac{\hat{a}_{1,*,im}(\hat{a}_{1,*,re} - a_{1,re})}{a_{1,re}^{\,2}} \tag{4.85}$$

The parameter ε is a pure auxiliary parameter for the Taylor series expansion. It tags the signal-amplitude estimation errors and is used to collect terms that are proportional to the estimation errors, terms that are proportional to the squared estimation errors, and so on. After performing the series expansion, the parameter ε is ignored (set to 1) and we obtain

$$\hat{\varphi}_{1,*} \approx \frac{\hat{a}_{1,*,im}}{a_{1,re}} + \frac{\hat{a}_{1,*,im}(\hat{a}_{1,*,re} - a_{1,re})}{a_{1,re}^{\,2}} \tag{4.86}$$

Real and imaginary signal amplitudes are unbiased and uncorrelated random variables, thus

$$\left\langle \hat{\varphi}_{1,*} \right\rangle \approx 0 \tag{4.87}$$

The variance is given as

$$\begin{aligned}
\left\langle \hat{\varphi}_{1,*}^{\ 2} \right\rangle &\approx \left\langle \left(\frac{\hat{a}_{1,*,im}}{a_{1,re}} + \frac{\hat{a}_{1,*,im}(\hat{a}_{1,*,re} - a_{1,re})}{a_{1,re}^{\ 2}} \right)^2 \right\rangle \\
&= \left\langle \left(\frac{\hat{a}_{1,*,im}}{a_{1,re}} \right)^2 + 2\frac{\hat{a}_{1,*,im}^{\ 2}(\hat{a}_{1,*,re} - a_{1,re})}{a_{1,re}^{\ 3}} + \frac{\hat{a}_{1,*,im}^{\ 2}(\hat{a}_{1,*,re} - a_{1,re})^2}{a_{1,re}^{\ 4}} \right\rangle \\
&= \frac{\operatorname{var}\left\langle \hat{a}_{1,*} \right\rangle}{2|a_1|^2} + \frac{\operatorname{var}\left\langle \hat{a}_{1,*} \right\rangle^2}{4|a_1|^4} = \frac{\operatorname{var}\left\langle \hat{a}_{1,*} \right\rangle}{2|a_1|^2}\left(1 + \frac{\operatorname{var}\left\langle \hat{a}_{1,*} \right\rangle}{2|a_1|^2} \right) \\
&= \frac{1}{L|a_1|^2}\left(1 + \frac{1}{L|a_1|^2} \right) = \frac{f_s}{2L(C/N_0)_1}\left(1 + \frac{f_s}{2L(C/N_0)_1} \right) \\
&= \frac{1}{2T_{coh}(C/N_0)_1}\left(1 + \frac{1}{2T_{coh}(C/N_0)_1} \right)
\end{aligned} \tag{4.88}$$

Note that the following identify holds for a complex-valued random variable

$$\left\langle \hat{a}_{1,*,im}^{\ 2} \right\rangle = \left\langle (\hat{a}_{1,*,re} - a_{1,re})^2 \right\rangle = \frac{\operatorname{var}\left\langle \hat{a}_{1,*} \right\rangle}{2} \tag{4.89}$$

whose expected imaginary part vanishes. Furthermore, all random variables involved in this expression are Gaussian due to the large number of involved samples. Uncorrelated Gaussian random variables are stochastically independent.

Again, a term similar to the squaring loss enters because, in general, $\hat{a}_{1,*,re} \neq a_{1,re}$. Otherwise, the second-order term in (4.85) would vanish.

Equation (4.88) gives the variance of a carrier-phase estimate, being the product of the carrier-phase CRLB and the squaring loss. This equation also shows that the carrier-phase variance expressed in square radians is independent of the carrier frequency or the waveform $c(t)$. An identical expression to (4.88) was derived in [16] using a different methodology and using a Taylor series expansion up to the third order.

4.3.2.9 Estimating the Signal Power

The estimated squared signal magnitude is related to the C/N_0 value and given by

$$\begin{aligned}
\hat{b}_{1,*}^{\ 2} &= \left|\hat{a}_{1,*}\right|^2 = \hat{a}_{1,*,im}^{\ 2} + ((\hat{a}_{1,*,re} - a_{1,re}) + a_{1,re})^2 \\
&= \hat{a}_{1,*,im}^{\ 2} + 2a_{1,re}(\hat{a}_{1,*,re} - a_{1,re}) + a_{1,re}^{\ 2} + (\hat{a}_{1,*,re} - a_{1,re})^2
\end{aligned} \tag{4.90}$$

We assume (without loss of generality) that the imaginary true value $a_{1,im}$ of $\hat{a}_{1,*}$ vanishes. The squared signal-magnitude estimate evaluates to

$$\begin{aligned}\left\langle \hat{b}_{1,*}{}^2 \right\rangle &= |a_1|^2 + \operatorname{var}\left\langle \hat{a}_{1,*} \right\rangle = |a_1|^2 \left(1 + \frac{\operatorname{var}\left\langle \hat{a}_{1,*} \right\rangle}{|a_1|^2}\right) \\ &= |a_1|^2 \left(1 + \frac{1}{T_{coh}(C/N_0)_1}\right)\end{aligned} \tag{4.91}$$

A C/N_0 estimator is obtained by scaling this expression with the sample rate:

$$\begin{aligned}&\left\langle \hat{b}_{1,*}{}^2 \right\rangle = \frac{2(C/N_0)_1}{f_s}\left(1 + \frac{1}{T_{coh}(C/N_0)_1}\right) = \frac{2(C/N_0)_1}{f_s} + \frac{2}{f_s T_{coh}} \\ &\Leftrightarrow \frac{f_s}{2}\left(\left\langle \hat{b}_{1,*}{}^2 \right\rangle - \frac{2}{f_s T_{coh}}\right) = (C/N_0)_1\end{aligned} \tag{4.92}$$

An unbiased C/N_0 estimator is therefore given as

$$\widehat{(C/N_0)}_1 = \frac{f_s}{2}\left(|\hat{a}_{1,*}|^2 - \frac{2}{f_s T_{coh}}\right) \tag{4.93}$$

The variance of the C/N_0 estimator is given as

$$\operatorname{var}\left\langle \widehat{(C/N_0)}_1 \right\rangle = \frac{f_s^2}{4}\operatorname{var}\left\langle |\hat{a}_{1,*}|^2 \right\rangle = \frac{f_s^2}{4}\operatorname{var}\left\langle \hat{b}_{1,*}{}^2 \right\rangle \tag{4.94}$$

The variance of $\hat{b}_{1,*}{}^2$ shall be evaluated in the following by expanding the complex-valued random variable into its real- and imaginary-valued components

$$\begin{aligned}\operatorname{var}\left\langle \hat{b}_{1,*}{}^2 \right\rangle &= \left\langle \left(\hat{a}_{1,*,re}{}^2 + \hat{a}_{1,*,im}{}^2 - \left\langle \hat{a}_{1,*,re}{}^2 + \hat{a}_{1,*,im}{}^2 \right\rangle\right)^2 \right\rangle \\ &= \left\langle \left((\hat{a}_{1,*,re} - a_{1,re})^2 + 2\hat{a}_{1,*,re}a_{1,re} + \hat{a}_{1,*,im}{}^2 - \left\langle (\hat{a}_{1,*,re} - a_{1,re})^2 + 2\hat{a}_{1,*,re}a_{1,re} + \hat{a}_{1,*,im}{}^2 \right\rangle\right)^2 \right\rangle \\ &= \left\langle \left((\hat{a}_{1,*,re} - a_{1,re})^2 + 2a_{1,re}(\hat{a}_{1,*,re} - a_{1,re}) + \hat{a}_{1,*,im}{}^2 - \left\langle (\hat{a}_{1,*,re} - a_{1,re})^2 + \hat{a}_{1,*,im}{}^2 \right\rangle\right)^2 \right\rangle\end{aligned} \tag{4.95}$$

Because for any zero-mean Gaussian random variable any moment of uneven order vanishes, and because the real and imaginary part of the complex-amplitude estimate $\hat{a}_{1,*}$ are uncorrelated, we can rewrite the equation as

$$
\begin{aligned}
&\operatorname{var}\left\langle \hat{b}_{1,*}{}^{2} \right\rangle \\
&= \left\langle \left((\hat{a}_{1,*,re} - a_{1,re})^2 + 2a_{1,re}(\hat{a}_{1,*,re} - a_{1,re}) + \hat{a}_{1,*,im}{}^2 - \left\langle (\hat{a}_{1,*,re} - a_{1,re})^2 \right\rangle - \left\langle \hat{a}_{1,*,im}{}^2 \right\rangle \right)^2 \right\rangle \\
&= \left\langle \left(\left[(\hat{a}_{1,*,re} - a_{1,re})^2 - \left\langle (\hat{a}_{1,*,re} - a_{1,re})^2 \right\rangle \right] + 2a_{1,re}(\hat{a}_{1,*,re} - a_{1,re}) \right.\right. \\
&\left.\left. + \left[\hat{a}_{1,*,im}{}^2 - \left\langle \hat{a}_{1,*,im}{}^2 \right\rangle \right] \right)^2 \right\rangle \\
&= \left\langle (\hat{a}_{1,*,re} - a_{1,re})^4 \right\rangle - \left\langle (\hat{a}_{1,*,re} - a_{1,re})^2 \right\rangle^2 + 4a_{1,re}{}^2 \left\langle (\hat{a}_{1,*,re} - a_{1,re})^2 \right\rangle \\
&+ \left\langle \hat{a}_{1,*,im}{}^4 \right\rangle - \left\langle \hat{a}_{1,*,im}{}^2 \right\rangle^2
\end{aligned}
\tag{4.96}
$$

Using the identity

$$
\left\langle x^4 \right\rangle = 3\left\langle x^2 \right\rangle^2 \tag{4.97}
$$

being a special case of

$$
\left\langle x^n \right\rangle = \frac{2^{\frac{n-2}{2}} (1 + (-1)^n) \left\langle x^2 \right\rangle^{n/2} \Gamma\left(\frac{1+n}{2} \right)}{\sqrt{\pi}} \tag{4.98}
$$

which holds for any real-valued zero-mean Gaussian random variable, yields

$$
\operatorname{var}\left\langle \hat{b}_{1,*}{}^2 \right\rangle = 2\left\langle (\hat{a}_{1,*,re} - a_{1,re})^2 \right\rangle^2 + 4a_{1,re}{}^2 \left\langle (\hat{a}_{1,*,re} - a_{1,re})^2 \right\rangle + 2\left\langle \hat{a}_{1,*,im}{}^2 \right\rangle^2 \tag{4.99}
$$

and by (4.61) the equation is evaluated as

$$
\operatorname{var}\left\langle \hat{b}_{1,*}{}^2 \right\rangle = \frac{2}{L^2} + 4a_{1,re}{}^2 \frac{1}{L} + \frac{2}{L^2} = \frac{4}{L^2} + \frac{8(C/N_0)_1}{Lf_s} \tag{4.100}
$$

According to (4.94), the variance of the C/N_0 estimator is obtained by multiplying (4.100) with $f_s^4/4$, yielding

$$
\begin{aligned}
\operatorname{var}\left\langle \widehat{(C/N_0)}_1 \right\rangle &= \frac{f_s{}^2}{4}\left(\frac{4}{L^2} + \frac{8(C/N_0)_1}{Lf_s} \right) \\
&= \frac{f_s{}^2}{L^2}\left(1 + \frac{2L(C/N_0)_1}{f_s} \right) = \frac{1}{T_{coh}{}^2}(1 + 2T_{coh}(C/N_0)_1)
\end{aligned}
\tag{4.101}
$$

The same expression was obtained, experimentally analyzed, and compared to a differential C / N_0 estimator in the work [19]. For the relative signal-power variance, the expression

$$\operatorname{var}\left\langle \frac{\widehat{(C/N_0)_1}}{(C/N_0)_1} \right\rangle = \frac{2}{(C/N_0)_1 T_{coh}} \left(1 + \frac{1}{2T_{coh}(C/N_0)_1} \right) \tag{4.102}$$

is derived from (4.101). It shows the same structure as for the code-phase, Doppler, and carrier-phase estimate; the term within the parenthesis could be interpreted as a squaring loss.

4.3.2.10 Doppler and Code-Phase Error Limits—Linearity Conditions

The above section showed that errors in the signal-amplitude LSQ estimate increase the variance of other-parameter LSQ estimates (code-phase and Doppler) because the signal-amplitude estimate is part of the design matrix and normal matrix. The question arises whether additional errors in the Doppler or code-phase estimate can cause an even further increase in the estimated parameter variances.

This section proves (under certain assumptions called *linearity conditions*) that this is not the case and that estimation errors in the code phase or Doppler do not increase the variance of other parameters. Therefore, there is no analog expression for the squaring loss in case Doppler or code-phase errors are considered. The reason lies in the fact that the normal matrix (4.60) depends on the signal amplitude but not on the code phase or Doppler. This shall be discussed in the following statements.

We assume that $\mathbf{q}$ denotes the true parameter values and, although the design matrix $A_\mathbf{q}$ depends on $\mathbf{q}$, the normal matrix I is independent of it. We assume that $\mathbf{q}$ contains the code-phase and the Doppler but not the signal amplitude. Thereby, we completely separate the complex amplitude estimation problem from the code-phase/Doppler estimation problem.

According to (4.17), the LSQ estimator using the true (but unknown) code-phase and Doppler values to set up the design matrix and the signal model is written as

$$\hat{\mathbf{q}} = \mathbf{q} + \Delta\hat{\mathbf{q}} = \mathbf{q} + \frac{1}{2} I^{-1} A_\mathbf{q}{}^* (\mathbf{S} - \mathbf{r}(\mathbf{q})) \tag{4.103}$$

and achieves the CRLB. However, this estimator cannot be used in practice and the normal matrix and the signal model is set up with different parameters that shall be denoted now as $\mathbf{q}_0$. The assumed parameter values differ from the true values by

$$\delta\mathbf{q}_0 = \mathbf{q}_0 - \mathbf{q} \tag{4.104}$$

The LSQ problem then yields the solution

$$\hat{\mathbf{q}}_0 = \mathbf{q}_0 + \Delta\hat{\mathbf{q}}_0 = \mathbf{q}_0 + \frac{1}{2} I^{-1} A_{\mathbf{q}_0}{}^* (\mathbf{S} - \mathbf{r}(\mathbf{q}_0)) \tag{4.105}$$

In the following, we derive a condition that needs to be fulfilled to have $\hat{\mathbf{q}}_0 = \hat{\mathbf{q}}$; that is, to ensure that the Doppler and code-phase estimates are MVUEs even if the true code-phase and Doppler values are *not* used to set up the design matrix and the signal model.

Due to the assumptions of Section 1.8 for any value of $\mathbf{q}$ and $m = 1$,

$$A_{\mathbf{q}}^{*}\mathbf{r}(\mathbf{q}) = 0 \tag{4.106}$$

The requirement of equal estimates reads as

$$0 = \hat{\mathbf{q}}_0 - \hat{\mathbf{q}} = \mathbf{q}_0 + \Delta\hat{\mathbf{q}}_0 - \mathbf{q} - \Delta\hat{\mathbf{q}} = \delta\mathbf{q}_0 + \frac{1}{2}I^{-1}(A_{\mathbf{q}_0}{}^{*}\mathbf{S} - A_{\mathbf{q}}^{*}\mathbf{S}) \tag{4.107}$$

which translates using a model for the signal samples to

$$\begin{aligned}\delta\mathbf{q}_0 &= \frac{1}{2}I^{-1}(A_{\mathbf{q}}^{*}(\mathbf{r}(\mathbf{q}) + \mathbf{N}) - A_{\mathbf{q}_0}{}^{*}(\mathbf{r}(\mathbf{q}) + \mathbf{N})) \\ &= \frac{1}{2}I^{-1}(A_{\mathbf{q}}^{*}\mathbf{N} - A_{\mathbf{q}_0}{}^{*}(\mathbf{r}(\mathbf{q}) + \mathbf{N}))\end{aligned} \tag{4.108}$$

Taking the expected value with respect to the noise, we obtain the linearity condition

$$\delta\mathbf{q}_0 = -\frac{1}{2}I^{-1}A_{\mathbf{q}_0}{}^{*}\mathbf{r}(\mathbf{q}) \tag{4.109}$$

an equation being independent of the noise terms that can rewritten as

$$-A_{\mathbf{q}_0}{}^{*}\mathbf{r}(\mathbf{q}) = 2I\delta\mathbf{q}_0 \tag{4.110}$$

Inserting (4.110) into (4.108) yields

$$0 = (A_{\mathbf{q}}^{*}\mathbf{N} - A_{\mathbf{q}_0}{}^{*}\mathbf{N}) \tag{4.111}$$

Now we will prove that (4.111) is also fulfilled if (4.110) is fulfilled. We show that the variance of (4.111) vanishes because the derivative of (4.110) with respect to $\mathbf{q}$ yields

$$\begin{aligned}&\frac{\partial}{\partial\mathbf{q}}(-A_{\mathbf{q}_0}{}^{*}\mathbf{r}(\mathbf{q})) = \frac{\partial}{\partial\mathbf{q}}(2I\delta\mathbf{q}_0) \\ &\Leftrightarrow A_{\mathbf{q}_0}{}^{*}A_{\mathbf{q}} = 2I\end{aligned} \tag{4.112}$$

Therefore,

$$\begin{aligned}\operatorname{var}\left\langle A_{\mathbf{q}}^{*}\mathbf{N} - A_{\mathbf{q}_0}{}^{*}\mathbf{N}\right\rangle &= \left\langle (A_{\mathbf{q}}^{*}\mathbf{N} - A_{\mathbf{q}_0}{}^{*}\mathbf{N})(\mathbf{N}^{*}A_{\mathbf{q}} - \mathbf{N}^{*}A_{\mathbf{q}_0})\right\rangle \\ &= 2(A_{\mathbf{q}}^{*}A_{\mathbf{q}} + A_{\mathbf{q}_0}{}^{*}A_{\mathbf{q}_0} - A_{\mathbf{q}_0}{}^{*}A_{\mathbf{q}} - A_{\mathbf{q}}^{*}A_{\mathbf{q}_0}) = 0\end{aligned} \tag{4.113}$$

Altogether, this shows that, as long as the linearity condition (4.110) is fulfilled, code-phase and Doppler errors in the setup of the design matrix do not degrade the estimation results and (4.107) is fulfilled. The discussion can also be seen as an interpretation of (4.17) in the context of LSQ adjustment.

An explicit expression for the linearity condition is derived by an analysis of the product of the design matrix with the deterministic signal model. Because the complex amplitude is assumed to be known, the product reads as

$$
\begin{aligned}
A_{\mathbf{q}_0}{}^*\mathbf{r}(\mathbf{q}) &= \begin{pmatrix} \sum_{\mu=1}^{L} \overline{\dfrac{\partial r_\mu(\mathbf{q})}{\partial \tau_1}\Big|_{\mathbf{q}=\mathbf{q}_0}} r_\mu(\mathbf{q}) \\ \sum_{\mu=1}^{L} \overline{\dfrac{\partial r_\mu(\mathbf{q})}{\partial \omega_1}\Big|_{\mathbf{q}=\mathbf{q}_0}} r_\mu(\mathbf{q}) \end{pmatrix} \\
&= \begin{pmatrix} -|a_1|^2 \sum_{\mu=1}^{L} \overline{c_1{}'(t_\mu - \tau_{1,0})} c_1(t_\mu - \tau_1) \exp\{it'_\mu(\omega_1 - \omega_{1,0})\} \\ -i|a_1|^2 \sum_{\mu=1}^{L} t'_\mu \overline{c_1(t_\mu - \tau_{1,0})} c_1(t_\mu - \tau_1) \exp\{it'_\mu(\omega_1 - \omega_{1,0})\} \end{pmatrix} \\
&= \begin{pmatrix} |a_1|^2 L R_{\bar{c}_1,c_1}{}'(\tau_{1,0} - \tau_1)\kappa(\omega_1 - \omega_{1,0}) \\ -i|a_1|^2 \sum_{\mu=1}^{L} t'_\mu \overline{c_1(t_\mu - \tau_{1,0})} c_1(t_\mu - \tau_1) \exp\{it'_\mu(\omega_1 - \omega_{1,0})\} \end{pmatrix}
\end{aligned}
\tag{4.114}
$$

On the other hand, (4.60) yields

$$
2I\delta\mathbf{q}_0 = \begin{pmatrix} -L R_{\bar{c}_1,c_1}{}''(0)|a_1|^2 (\tau_{1,0} - \tau_1) \\ \chi_{freq} |a_1|^2 L^3 / (12 f_s)(\omega_{1,0} - \omega_1) \end{pmatrix} \tag{4.115}
$$

and we assume that the normal matrix I is set up using the true complex signal-amplitude value. Setting both equations equal yields two equations: the code-phase linearity condition and the Doppler linearity condition.

The *code-phase linearity condition* is a linearity requirement for the first derivate of the correlation function, written as

$$
R_{\bar{c}_1,c_1}{}''(0)(\tau_{1,0} - \tau_1) = R_{\bar{c}_1,c_1}{}'(\tau_{1,0} - \tau_1)\kappa(\omega_1 - \omega_{1,0}) \tag{4.116}
$$

The *Doppler linearity condition* reads as

$$
\chi_{freq} L^3 / (12 f_s^2)(\omega_{1,0} - \omega_1) = i \sum_{\mu=1}^{L} t'_\mu \overline{c_1(t_\mu - \tau_{1,0})} c_1(t_\mu - \tau_1) \exp\{it'_\mu(\omega_1 - \omega_{1,0})\} \tag{4.117}
$$

In addition to these two conditions, a third linearity condition will be derived, which assesses the influence of the code phase and Doppler errors on the signal-amplitude estimate.

Looking at (4.61) and reviewing the discussion of Section 4.3.2.3, it might be reasonable to require that the expected estimated signal power obtained at the correlation point is equal to true signal power; that is,

$$\left\langle \left|\hat{a}_1\right|^2 \right\rangle\Big|_{\mathbf{q}_0} = \left|a_1\right|^2 \tag{4.118}$$

Using (4.63), the *amplitude linearity condition* reads as

$$\left|R_{\bar{c}_1,c_1}(\tau_{1,0} - \tau_1)\kappa(\omega_1 - \omega_{1,0})\right| = 1 \tag{4.119}$$

This is another condition on the code phase and Doppler, which needs to be sufficiently fulfilled.

Summarizing the derivations above, the three linearity conditions—(4.116), (4.117), and (4.119)—limit the range of tolerable Doppler and code-phase errors. Furthermore, if the conditions are sufficiently fulfilled, the resulting code-phase, Doppler, or amplitude estimates will be optimal. Note that a violation does not necessarily cause the estimates to be suboptimal. It should be noted that the conditions are directly derived from the assumed navigation signal. Different navigation signals may be more or less sensitive to code-phase or Doppler errors. The linearity conditions presume that the navigation receiver performs optimal processing of the signal to achieve the CRLB. If, on the contrary, the signal processing within the receiver is different (e.g., a simple early–late tracking scheme is used), then the linearity conditions do not apply anymore (at least not in this form) but the receiver performance will also be suboptimal.

4.3.3 Correlation Point

Calculating a LSQ estimate using equations like (4.63), (4.64), or (4.65) requires computation of correlation values of the signal samples with a receiver internally generated reference signal. The reference signal is computed using the code-phase and Doppler parameters given by the vector $\mathbf{q}_0$. This vector defines the *correlation point* used to obtain the estimated values.

An important requirement of the correlation point is that it is near the true code-phase and Doppler values such that (4.116), (4.117), and (4.119) are fulfilled sufficiently. Fulfilling this requirement ensures that the obtained estimates are stochastically independent of the correlation point and that, because of (4.17), the CRLB is achieved. In practice, the correlation point is generally a function of previously estimated code-phase and Doppler values. By avoiding a stochastic (nonlinear) dependence of the current estimates from the previous estimates, the total positioning problem can be treated analytically.

Furthermore, it should be pointed out that a key element for an efficient navigation software receiver implementation is the *reuse of internally generated reference signals* for multiple correlations (see Section 9.6). The reuse does not degrade the estimates as long as the correlation point is sufficiently near the true signal parameters such that the linearity conditions are fulfilled. When the difference becomes too large, the internal reference signals are refreshed, thereby shifting the correlation point back to the true parameter values.

Within a navigation receiver, several concepts may be used to choose the correlation point, depending upon which kind of information is used and how the correlation is implemented. The concepts are outlined below. It should be pointed out that the following considerations focus on the code-phase and Doppler dimensions. The signal amplitude and carrier phase are less important because the estimation problem is effectively linear for those parameters if the complex-valued amplitude is used. Carrier-phase estimation will be discussed in Section 4.3.3.4.

4.3.3.1 Tracking Loop

Tracking loops are commonly used in GNSS receivers to control the estimation process within a receiver's channel for one received signal. In the following, tracking loops are described in the framework of a LSQ adjustment. The description is slightly different (i.e., more conceptual) compared to other standard early–late tracking-loop descriptions [15, 20].

A tracking loop determines, for each integration interval, one set of correlation values and uses it to determine code-phase, Doppler, and complex-amplitude estimates. We assume that a LSQ scheme is used, but the LSQ iteration (or an approximation of it) is stopped after the first iteration. More specifically, the following steps are performed for each interval:

1. Generate reference signals for the current correlation point;
2. Single-step LSQ;
   a. Estimate signal amplitude using (4.63);
   b. Estimate code-phase and Doppler improvements using (4.64) and (4.65);
3. Filter code-phase and Doppler corrections;
4. Apply filtered corrections and update correlation point.

By continuously estimating code-phase and Doppler improvements, the tracking loop follows the received signal. Obviously, the tracking loop has to be initialized with starting values. The key element is the loop filter, which fulfills two functions simultaneously. First, the correlation point is kept near the true values. Second, the filtered code-phase and Doppler values are of reduced noise. Both functions are conceptually different and could, in general, be fulfilled using two different filters. A block diagram illustrating this modified procedure is shown in Figure 4.2. An optional smoothing filter is included.

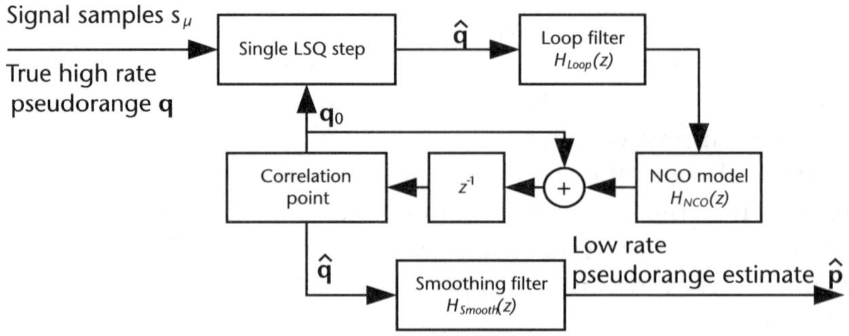

Figure 4.2 Tracking-loop block diagram.

The mathematical model for the "Single LSQ step" is based on the considerations of Section 4.3.2, which can be summarized for a single estimation step as

$$\Delta\hat{\mathbf{q}} = \mathbf{q} - \mathbf{q}_0 + \varepsilon_{\Delta\hat{\mathbf{q}}} \tag{4.120}$$

where $\varepsilon_{\Delta\hat{\mathbf{q}}}$ is the single-step high-rate pseudorange estimation error. The high-rate pseudorange estimates are used to define the correlation point for the next interval

$$\mathbf{q}_0 = \hat{\mathbf{q}} \tag{4.121}$$

The estimation is performed exactly once per integration interval. Applying the z-transform to analyze the temporal evolution of the tracking loop over multiple intervals yields

$$z\mathbf{q}_0(z) = H_{NCO}(z)H_{Loop}(z)\Delta\hat{\mathbf{q}}(z) + \mathbf{q}_0(z) \tag{4.122}$$

which can be solved for the correlation point,

$$\begin{aligned} &z\mathbf{q}_0(z) = H_{NCO}(z)H_{Loop}(z)(\mathbf{q}(z) - \mathbf{q}_0(z) + \varepsilon_{\Delta\hat{\mathbf{q}}}(z)) + \mathbf{q}_0(z) \\ &\Leftrightarrow \mathbf{q}_0(z)(z - 1 + H_{NCO}(z)H_{Loop}(z)) = H_{NCO}(z)H_{Loop}(z)(\mathbf{q}(z) + \varepsilon_{\Delta\hat{\mathbf{q}}}(z)) \\ &\Leftrightarrow \mathbf{q}_0(z) = \frac{H_{NCO}(z)H_{Loop}(z)\mathbf{q}(z)}{(z - 1 + H_{NCO}(z)H_{Loop}(z))} + \frac{H_{NCO}(z)H_{Loop}(z)\varepsilon_{\Delta\hat{\mathbf{q}}}(z)}{(z - 1 + H_{NCO}(z)H_{Loop}(z))} \end{aligned} \tag{4.123}$$

and for the final low-rate pseudorange estimate

$$\hat{\mathbf{p}}(z) = \frac{H_{Smooth}(z)H_{NCO}(z)H_{Loop}(z)\mathbf{q}(z)}{(z - 1 + H_{NCO}(z)H_{Loop}(z))} + \frac{H_{Smooth}(z)H_{NCO}(z)H_{Loop}(z)\boldsymbol{\varepsilon}_{\Delta\hat{\mathbf{q}}}(z)}{(z - 1 + H_{NCO}(z)H_{Loop}(z))} \tag{4.124}$$

For many applications, the loop filter can be chosen such that no smoothing filter is required [e.g., $H_{Smooth}(z) = 1$] because the requirement of obtaining smooth estimates is often much more stringent than the linearity condition (4.116) and (4.117). In other words, the accuracy of the measurements is much higher than the extension of the linearity region. An important exception is the case of occasional high line-of-sight dynamics. In that case, it is beneficial to design the loop filter with a wide bandwidth and the smoothing filter with a narrow bandwidth. This avoids loss-of-lock events during periods of high dynamics while maintaining accurate estimates during normal tracking. High line-of-sight dynamics may occur during acquisition-to-tracking handover, short blockage of the signals, or during specific user events. A smoothing filter is used in a patent by Thomas to improve the accuracy of the low-rate pseudorange estimates in case the low-rate pseudorange output rate is much slower (e.g., 30 seconds) than the inverse of the involved tracking-loop bandwidths (e.g., 1 Hz) [21].

The NCO-model $H_{NCO}(z)$ is introduced in Figure 4.2 because, in many practical implementations, the correlation-point update capabilities are limited. For example, in some GNSS receivers, only the rate of change of $\mathbf{q}_0$ can be controlled, but not $\mathbf{q}_0$ directly. Typically $H_{NCO}(z)$ equals the coherent integration time T_{coh}, indicating that the rate of change of $\mathbf{q}_0$ muliplied with T_{coh} is added to the previous value of $\mathbf{q}_0$ [16].

A number of different possibilities exist to realize the loop filter. For the example of a GNSS receiver, Jaffe–Rechtin filters are often used [22] and are described in many textbooks [15, 20, 23, 24]. Jaffe–Rechtin filters optimize the sum of thermal noise plus transient errors and they require only the rate of change of $\mathbf{q}_0$ to be controlled.

Kaplan describes tracking-loop *stability* conditions that are formulated based on similar linearity requirements such as (4.116) and (4.117) [20]. Those conditions transfer the linearity requirement into allowable signal parameters (e.g., maximal dynamics, signal-to-noise ratio) that must be met to keep the correlation point near the true values. Because of the feedback, a tracking loop may become unstable for an unfortunate choice of the loop filter. A detailed investigation on this topic has been done by Eissfeller and Kazemi [25, 26]. A more recent approach is to realize the loop filter via a Kalman filter. The effective transfer function stands in very close relationship with a Jaffe–Rechtin filter, but the Kalman filter is optimal under certain assumptions. The Kalman filter approach requires direct control of $\mathbf{q}_0$, not only of the rate of change of $\mathbf{q}_0$ [27].

The variance of a scalar low-rate pseudorange parameter p is given as [16]

$$\begin{aligned}\operatorname{var}\left\langle \hat{p} \right\rangle\big|_{\mathrm{N}} &= \operatorname{var}\left\langle \varepsilon_{\Delta\hat{\mathbf{q}}} \right\rangle\big|_{\mathrm{N}} \frac{1}{2\pi}\int_{\vartheta=0}^{2\pi}\left|\frac{H_{Smooth}(e^{i\vartheta})H_{\mathrm{NCO}}(e^{i\vartheta})H_{Loop}(e^{i\vartheta})}{(e^{i\vartheta}-1+H_{\mathrm{NCO}}(e^{i\vartheta})H_{Loop}(e^{i\vartheta}))}\right|^2 d\vartheta \\ &= 2T_{coh}B_L \operatorname{var}\left\langle \varepsilon_{\Delta\hat{\mathbf{q}}} \right\rangle\end{aligned} \tag{4.125}$$

where the noise equivalent bandwidth B_L in [Hz] is defined as

$$B_L = \frac{1}{4\pi T_{coh}}\int_{\vartheta=0}^{2\pi}\left|\frac{H_{Smooth}(e^{i\vartheta})H_{\mathrm{NCO}}(e^{i\vartheta})H_{Loop}(e^{i\vartheta})}{(e^{i\vartheta}-1+H_{\mathrm{NCO}}(e^{i\vartheta})H_{Loop}(e^{i\vartheta}))}\right|^2 d\vartheta \tag{4.126}$$

Note that [16] additionally considered rate-of-change variation within the correlation process.

4.3.3.2 Grid Search and LSQ Re-Iteration

The conventional tracking-loop approach of Section 4.3.3.1 relies on the assumption that the correlation point can be kept near the true values. If this assumption is not fulfilled, a different tracking scheme can be employed for high C/N_0 values that truly solves the MLE equation

$$\hat{\mathbf{q}}(\mathbf{s}) = \arg\max_{\mathbf{q}\in\Lambda} p_{\mathbf{q}}(\mathbf{s}) \tag{4.127}$$

for a given a priori range Λ of admissible values for $\mathbf{q}$. The algorithm may consist of two main steps, which are executed for each integration interval:

1. Brute force evaluation of (4.127) over a grid spanning all admissible values $\mathbf{q}$;

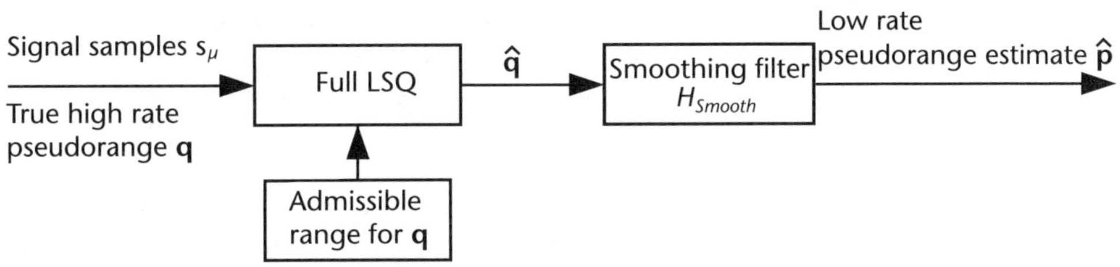

Figure 4.3 Local search and reiteration block diagram.

2. Iterative LSQ (see Section 4.3.2.2) starting from the maximum obtained from step (1).

Both steps together are called "Full LSQ" in the scheme depicted in Figure 4.3. It should be mentioned that a number of other algorithms are available to solve the maximization problem of (4.127) and might be better compared to the method presented here, depending on the specific implementation boundary conditions [28].

The admissible range Λ of high-rate pseudorange values has to be chosen in a way that it contains the true parameter values. A smoothing filter can be applied to reduce the noise of the estimated values.

The estimates (before the smoothing filter) are timely independent estimates and do not show transient errors like the approach of Section 4.3.3.1. The major disadvantage of this approach is the requirement that (4.127) needs to have a uniquely identifiable maximum located around the true values. This is only fulfilled for sufficiently high signal-to-noise ratios. Overall, the grid search and LSQ reiteration algorithm is quite similar to signal acquisition and computationally very demanding. Because no feedback is involved, it is intrinsically *perfectly stable* as long as the admissible range is chosen to be sufficiently large. Its main application lies in the field of high-dynamics applications [29].

4.3.3.3 Vector Loop

In the vector tracking approach, the correlation point is predicted based on the position solution, eventually using other sensor data [19, 30]. The vector approach uses considerably more input data to define the correlation point than the signal estimates of a single channel alone. The correlation point derived in this way is substantially more stable compared to the single channel approach, especially during periods of single-channel signal attenuation. If the position solution is aided by independent sensor data (e.g., IMU measurements), a significant improvement in stability is reached [31–33].

A block diagram of vector tracking is shown in Figure 4.4. It is conceptually simpler than Figure 4.2, but the technical realizations for the line-of-sight predictor, and also for the smoothing filter, might be cumbersome, especially if limited computational resources are available.

The low rate pseudorange estimates are given as

$$\hat{\mathbf{p}} = H_{Smooth}\{\hat{\mathbf{q}}\} = H_{Smooth}\{\mathbf{q}_0 + \Delta\hat{\mathbf{q}}\} = H_{Smooth}\{\mathbf{q} + \varepsilon_{\Delta\hat{\mathbf{q}}}\} \tag{4.128}$$

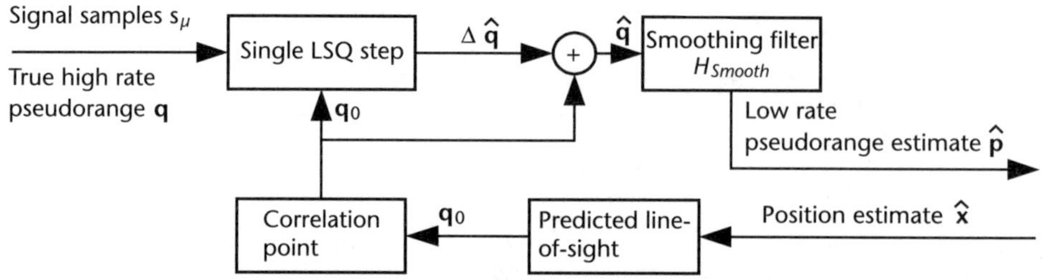

Figure 4.4 Vector tracking-loop block diagram.

To further work with this equation, we should require that the high-rate pseudorange estimates are *independent* of the correlation point and are timely uncorrelated random variables. This is achieved if the linearity conditions are fulfilled, because in that case identical estimates are obtained as if taking the true parameters as the correlation point.

The vector tracking-loop *stability* analysis can be performed like that for the single-channel filter, but now the stability analysis includes the positioning filter. The stability equations derive from the requirement that the projected line-of-sight pseudorange and Doppler (derived from the position solution) fall within the linear region centered at the true values. This also ensures that the high-rate pseudorange estimates are stochastically independent from the position solution.

If a Kalman positioning filter is used, thermal-noise stability can be derived by steady-state covariance matrices for a static GPS receiver [34]. Thermal-noise stability increases with an increasing number of satellites. In the same work, dynamic stability is investigated for various jerk values. Overall, the vector tracking stability analysis, including user dynamics, can only be performed if a specific satellite constellation is assumed and specific assumptions on the user trajectory are imposed.

4.3.3.4 Remark on Carrier-Phase Tracking

The discussion of the preceding sections showed different ways to choose the correlation point in the code-phase and Doppler dimension. In contrast, the carrier-phase estimation problem has a slightly different structure, as the carrier phase is uniquely defined by the complex-valued amplitude (4.45), which itself is a linear parameter to be estimated, as can be seen from (4.46). Therefore, the carrier-phase estimation problem can be formulated in two steps:

1. Complex amplitude estimation;
2. Wrapped phase determination and unwrapping.

Both steps work independently of the chosen carrier-phase correlation point. Carrier-phase unwrapping adds multiples of π to the wrapped phase values to obtain a smoth unwrapped phase curve.

However, at this point it should be clearly mentioned that, in many GNSS receivers, the carrier phase is tracked using a conventional tracking loop as described in Section 4.3.3.1 [15, 20]. The carrier phase of the internal reference signal is adjusted using the estimated and filtered carrier phase (4.85). Adjusting the carrier

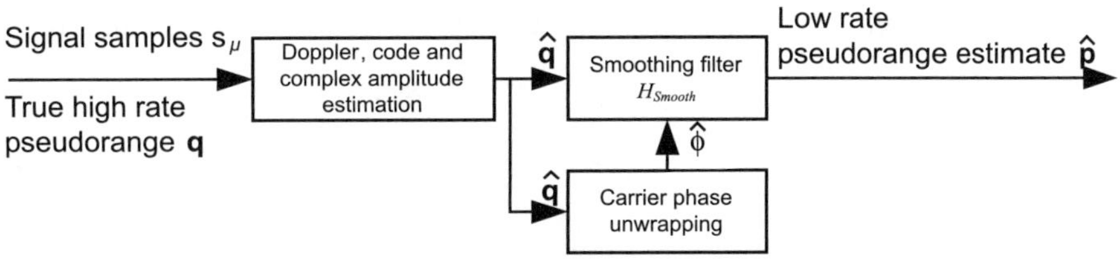

Figure 4.5 Carrier-phase tracking-loop block diagram.

phase implies adjusting the Doppler frequency. Tracking the carrier phase in this manner requires accurate Doppler initialization values, otherwise the carrier tracking loop may become unstable.

A different carrier-phase concept is presented in Figure 4.5, which is based on one of the three estimation concepts of Sections 4.3.3.1 through 4.3.3.3. Basically, only code-phase and Doppler are "tracked" and the carrier phase is estimated in a secondary block, which provides no feedback to the tracking. The estimated carrier-phase values may, however, be used in the smoothing filter to obtain the low-rate pseudorange estimates.

The presented concept has, for example, been exploited [35] such that 20 carrier-phase unwrapping blocks (each for one possible position of the data-bit transition) working on the same data are used to efficiently resolve the GPS C/A-code data-bit ambiguity under low signal-to-noise ratio conditions [36].

If the position-estimation algorithm demands carrier-phase estimates, phase unwrapping is required. On the other hand, certain carrier-phase operations (e.g., forming of carrier-phase differences between two signals) can also be expressed in terms of the complex amplitude. Unwrapping and difference forming can therefore be exchanged. Because unwrapping may perform more stably with differenced correlators due to cancellation of common mode errors, this option has been investigated in Sections 10.5.2 and 7.4 to allow carrier-phase estimation even in degraded signal environments.

4.3.4 Linearization

The discussion of Sections 4.3.2.10 and 4.3.3 showed that a necessary condition for the LSQ estimation to be optimal is to keep the correlation point near the true values. On the other hand, if this condition is true, then the signal model (4.46) can be rewritten as

$$
\begin{aligned}
r_\mu &= \sum_{m=1}^{M} a_m c_m(t_\mu - \tau_m)\exp\{i\,\omega_m t'_\mu\} \\
&= \sum_{m=1}^{M} \begin{pmatrix} a_m c_m(t_\mu - \tau_{m,0})\exp\{i\omega_{m,0}t'_\mu\} - a_m \Delta\tau_m c_m{}'(t_\mu - \tau_{m,0})\exp\{i\omega_{m,0}t'_\mu\} + \\ +it'_\mu a_m \Delta\omega_m c_m(t_\mu - \tau_{m,0})\exp\{i\omega_{m,0}t'_\mu\} \end{pmatrix} \\
&= \sum_{m=1}^{M} \begin{pmatrix} a_m c_m(t_\mu - \tau_{m,0})\exp\{i\omega_{m,0}t'_\mu\} - \Delta\tau'_m c_m{}'(t_\mu - \tau_{m,0})\exp\{i\omega_{m,0}t'_\mu\} + \\ +\Delta\omega'_m it'_\mu c_m(t_\mu - \tau_{m,0})\exp\{i\omega_{m,0}t'_\mu\} \end{pmatrix}
\end{aligned}
\tag{4.129}
$$

with the *modified* high-rate pseudorange parameters to be estimated

$$
\begin{aligned}
&a_m \\
&\Delta\tau'_m = a_m \Delta\tau_m \\
&\Delta\omega'_m = a_m \Delta\omega_m
\end{aligned}
\tag{4.130}
$$

If the estimation problem is expressed using the parameters (4.130), it becomes a truly linear problem. The relationship between the modified and unmodified high-rate pseudorange data is nonlinear. Thus, the modified high-rate pseudorange data relates to the position (or to the low-rate pseudorange data) in a nonlinear way. Only in the case when the squaring loss can be ignored, can the relationship (4.130) between the normal and the modified high-rate pseudorange parameters be considered as effectively linear because, in that case, precise complex signal-amplitude estimates are available. The linearized signal model is the basis for the Kalman filter formulation in Section 4.5.2.

4.3.5 Multiple Propagation Paths

In this section, the results obtained for a single propagation path of Section 4.3.2 are extended to the case of propagation along two or more paths.

To evaluate the normal matrix (or the Fisher information matrix) I given by (4.54), additional nondiagonal submatrices of the form $A_k^* A_l$ with $k \neq l$ have to be evaluated. The submatrices account for the coefficient's coupling between different propagation paths.

If we define

$$
I^{k,l} = \frac{1}{2} A_k^* A_l = (I^{l,k})^* \tag{4.131}
$$

then

$$
\begin{aligned}
2I_{1,1}^{k,l} &= \sum_{\mu=1}^{L} \overline{\frac{\partial r_\mu}{\partial a_k}} \frac{\partial r_\mu}{\partial a_l} = \sum_{\mu=1}^{L} \overline{c_k(t_\mu - \tau_{k,0})} c_l(t_\mu - \tau_{l,0}) \exp\{it'_\mu(\omega_{l,0} - \omega_{k,0})\} \\
&\approx L R_{\overline{c}_k, c_l}(\tau_{k,0} - \tau_{l,0}) \kappa(\omega_{l,0} - \omega_{k,0}) \\
2I_{1,2}^{k,l} &= \sum_{\mu=1}^{L} \overline{\frac{\partial r_\mu}{\partial a_k}} \frac{\partial r_\mu}{\partial \tau_l} = -a_{l,0} \sum_{\mu=1}^{L} \overline{c_k(t_\mu - \tau_{k,0})} c_l'(t_\mu - \tau_{l,0}) \exp\{it'_\mu(\omega_{l,0} - \omega_{k,0})\} \\
&\approx -a_{l,0} L R_{\overline{c}_k, c_l'}(\tau_{k,0} - \tau_{l,0}) \kappa(\omega_{l,0} - \omega_{k,0}) = -a_{l,0} L R_{\overline{c}_k, c_l}{}'(\tau_{k,0} - \tau_{l,0}) \kappa(\omega_{l,0} - \omega_{k,0}) \\
2I_{1,3}^{k,l} &= \sum_{\mu=1}^{L} \overline{\frac{\partial r_\mu}{\partial a_k}} \frac{\partial r_\mu}{\partial \omega_l} = i a_{l,0} \sum_{\mu=1}^{L} t'_\mu \overline{c_k(t_\mu - \tau_{k,0})} c_l(t_\mu - \tau_{l,0}) \exp\{it'_\mu(\omega_{l,0} - \omega_{k,0})\} \\
&\approx a_{l,0} L R_{\overline{c}_k, c_l}(\tau_{k,0} - \tau_{l,0}) \kappa'(\omega_{l,0} - \omega_{k,0}) = L\gamma_{a,\omega}
\end{aligned}
\tag{4.132}
$$

and

$$
\begin{aligned}
2I^{k,l}_{2,2} &= \sum_{\mu=1}^{L} \overline{\frac{\partial r_\mu}{\partial \tau_k}} \frac{\partial r_\mu}{\partial \tau_l} = \\
&= \bar{a}_{k,0} a_{l,0} \sum_{\mu=1}^{L} \overline{c_k{}'(t_\mu - \tau_{k,0})} c_l{}'(t_\mu - \tau_{l,0}) \exp\{i t'_\mu(\omega_{l,0} - \omega_{k,0})\} \\
&\approx -\bar{a}_{k,0} a_{l,0} L R''_{\bar{c}_k, c_l}(\tau_{k,0} - \tau_{l,0}) \kappa(\omega_{l,0} - \omega_{k,0}) \\
2I^{k,l}_{3,2} &= \sum_{\mu=1}^{L} \overline{\frac{\partial r_\mu}{\partial \omega_k}} \frac{\partial r_\mu}{\partial \tau_l} = \\
&= i\bar{a}_{k,0} a_{l,0} \sum_{\mu=1}^{L} t'_\mu \overline{c_k(t_\mu - \tau_{k,0})} c_l{}'(t_\mu - \tau_{l,0}) \exp\{i t'_\mu(\omega_{l,0} - \omega_{k,0})\} \\
&\approx \bar{a}_{k,0} a_{l,0} L R'_{\bar{c}_k, c_l}(\tau_{k,0} - \tau_{l,0}) \kappa'(\omega_{l,0} - \omega_{k,0}) = L \overline{\gamma_{\omega,\tau}} \\
2I^{k,l}_{3,3} &= \sum_{\mu=1}^{L} \overline{\frac{\partial r_\mu}{\partial \omega_k}} \frac{\partial r_\mu}{\partial \omega_l} = \\
&= \bar{a}_{k,0} a_{l,0} \sum_{\mu=1}^{L} (t'_\mu)^2 \overline{c_k(t_\mu - \tau_{k,0})} c_l(t_\mu - \tau_{l,0}) \exp\{i t'_\mu(\omega_{l,0} - \omega_{k,0})\} \\
&\approx -\bar{a}_{k,0} a_{l,0} L R_{\bar{c}_k, c_l}(\tau_{k,0} - \tau_{l,0}) \kappa''(\omega_{l,0} - \omega_{k,0})
\end{aligned}
\tag{4.133}
$$

In this case, the nondiagonal submatrices $I^{k,l}$ of the Fisher information matrix evaluate to

$$
2I^{k,l} = L \times \begin{pmatrix}
R_{\bar{c}_k,c_l}(\Delta\tau)\kappa(\Delta\omega) & -a_{l,0} R'_{\bar{c}_k,c_l}(\Delta\tau)\kappa(\Delta\omega) & \gamma_{a,\omega} \\
\overline{a_{k,0} R'_{\bar{c}_k,c_l}(\Delta\tau)\kappa(\Delta\omega)} & -\bar{a}_{k,0} a_{l,0} R''_{\bar{c}_k,c_l}(\Delta\tau)\kappa(\Delta\omega) & \gamma_{\omega,\tau} \\
\overline{\gamma_{a,\omega}} & \overline{\gamma_{\omega,\tau}} & -\bar{a}_{k,0} a_{l,0} R_{\bar{c}_k,c_l}(\Delta\tau)\kappa''(\Delta\omega)
\end{pmatrix}
\tag{4.134}
$$

with

$$
\begin{aligned}
\Delta\omega &= \omega_{l,0} - \omega_{k,0} \\
\Delta\tau &= \tau_{k,0} - \tau_{l,0}
\end{aligned}
\tag{4.135}
$$

The matrix $I^{k,l}$ can be simplified if Doppler estimates decouple from code phase and amplitude estimates. This is the case if we assume

$$
\gamma_{a,\omega} = \gamma_{\omega,\tau} = 0 \tag{4.136}
$$

which implies that the matrix $I^{k,l}$ is block diagonal. The Doppler block is decoupled from the code-phase and amplitude block. As a consequence, the Doppler parameter estimates are independent of the code-phase and amplitude estimates, because the submatrices (4.60) of the Fisher information matrix are diagonal.

The complete Fisher information matrix can be written in block diagonal form as

$$I = \begin{pmatrix} I_{a,\tau} & 0 \\ 0 & I_{\omega} \end{pmatrix} \tag{4.137}$$

The order of estimated parameters is a_k, τ_k, a_l, τ_l, ω_k, ω_l if two propagation paths are considered.

For many applications, (4.136) is a reasonable assumption because the Doppler differences between the direct and the reflected signals are much smaller than the inverse of the coherent integration time [this implies $\kappa'(\Delta\omega) = 0$]. For example, Doppler differences of signals received by a static GNSS antenna are caused by the satellite motion and are of the order of $(5\text{–}10 \text{ min})^{-1}$. For a moving receiver, the Doppler difference is bounded by $2\, f_{RF}\, v_{USER} \,/\, c$, which evaluates to 10.5 Hz on GPS L1 for a pedestrian moving at 1 m/s. On the other hand, typical integration times range from 1 to 20 ms.

4.3.6 Two Propagation Paths: Code-Phase CRLB

In this section, the case of two propagation paths shall be evaluated in detail. The two propagation paths are denoted by the index k and the index l. The signal at baseband $c(t)$ is identical for both propagation paths. It should be mentioned that it is *not* possible to, for example, assign the index k to the line-of-sight signal and the index l to the multipath signal. No estimator can distinguish between these signals because both signal models are identical. Instead, some other algorithm should be used for signal identification. For example, the signal with the shorter estimated delay or with the highest signal power can be identified as the line-of-sight signal.

We assume that Doppler estimation decouples from code-phase/amplitude estimation. Under (4.136), the Fisher information matrix for amplitude and code phase $I_{a,\tau}$ is expressed as

$$I_{a,\tau} = \frac{L}{2} \times \begin{pmatrix} 1 & 0 & \vartheta R_{\bar{c},c}(\Delta\tau) & -\vartheta a_{l,0} R'_{\bar{c},c}(\Delta\tau) \\ 0 & -R''_{\bar{c},c}(0)\left|a_{k,0}\right|^2 & \overline{\vartheta a_{k,0} R'_{\bar{c},c}(\Delta\tau)} & -\vartheta \bar{a}_{k,0} a_{l,0} R''_{\bar{c},c}(\Delta\tau) \\ \vartheta \overline{R_{\bar{c},c}(\Delta\tau)} & \vartheta a_{k,0} R'_{\bar{c},c}(\Delta\tau) & 1 & 0 \\ \overline{-\vartheta a_{l,0} R'_{\bar{c},c}(\Delta\tau)} & -\vartheta a_{k,0} \bar{a}_{l,0} \overline{R''_{\bar{c},c}(\Delta\tau)} & 0 & -R''_{\bar{c},c}(0)\left|a_{l,0}\right|^2 \end{pmatrix} \tag{4.138}$$

where for notational simplicity we define a Doppler correlation coefficient as

$$\vartheta = \kappa(\omega_{l,0} - \omega_{k,0}) \tag{4.139}$$

which assumes a value of 1 if both signals have the identical Doppler frequency.

The determinant D of $I_{a,\tau}$ divided by the squared magnitude of both signals is obtained via a symbolic mathematics program as

$$D = \begin{pmatrix} 2\vartheta^2 \left|R_{\bar{c},c}{}'(\Delta\tau)\right|^2 R_{\bar{c},c}{}''(0) \\ +\vartheta^4 \left|R_{\bar{c},c}{}'(\Delta\tau)\right|^2 \left(\left|R_{\bar{c},c}{}'(\Delta\tau)\right|^2 + 2R_{\bar{c},c}(\Delta\tau)R_{\bar{c},c}{}''(\Delta\tau) \right) \\ +\theta\left(R_{\bar{c},c}{}''(0)^2 - \vartheta^2 \left|R_{\bar{c},c}{}''(\Delta\tau)\right|^2 \right) \end{pmatrix} \tag{4.140}$$

and the submatrix of the inverse Fisher information matrix $I_{a,\tau}$ corresponding to the parameters (a_k, τ_k) of the signal k is given as

$$(I_{a,\tau}{}^{-1})^k = \frac{2}{LD} \times \begin{pmatrix} R_{\bar{c},c}{}''(0)^2 + \vartheta^2 R_{\bar{c},c}{}''(0)\left|R_{\bar{c},c}{}'(\Delta\tau)\right|^2 - \vartheta^2 R_{\bar{c},c}{}''(\Delta\tau) & \dfrac{\vartheta^2 R_{\bar{c},c}{}'(\Delta\tau)\left(\overline{R_{\bar{c},c}{}''(\Delta\tau)} - R_{\bar{c},c}{}''(0)R_{\bar{c},c}(\Delta\tau)\right)}{\bar{a}_{k,0}} \\ \dfrac{\vartheta^2 \overline{R_{\bar{c},c}{}'(\Delta\tau)}\left(R_{\bar{c},c}{}''(\Delta\tau) - R_{\bar{c},c}{}''(0)\overline{R_{\bar{c},c}(\Delta\tau)}\right)}{a_{k,0}} & -\dfrac{R_{\bar{c},c}{}''(0)\eta + \vartheta^2\left|R_{\bar{c},c}{}'(\Delta\tau)\right|^2}{\left|a_{k,0}\right|^2} \end{pmatrix} \tag{4.141}$$

where the constant

$$\eta = 1 - \vartheta^2 \left|R_{\bar{c},c}(\Delta\tau)\right|^2 \tag{4.142}$$

has been introduced to simplify the notation. The submatrix corresponding to the signal parameters l is identical (apart from exchanging k and l). Couplings between parameters for the signal path k and the path l shall not be considered here.

The CRLB (without nuisance parameters) for the code-phase estimate τ_k of the first signal k is being determined by the second diagonal term of $(I_{a,\tau}{}^{-1})^k$. The CRLB reads as

$$\begin{aligned} CRLB(\mathrm{Re}\{\tau_k\}) &= -\frac{1}{L}\frac{R_{\bar{c},c}{}''(0)\eta + \vartheta^2\left|R_{\bar{c},c}{}'(\Delta\tau)\right|^2}{\left|a_{k,0}\right|^2 D} \\ &= -\frac{R_{\bar{c},c}{}''(0)\eta + \vartheta^2\left|R_{\bar{c},c}{}'(\Delta\tau)\right|^2}{2(C/N_0)_k T_{coh} D} \end{aligned} \tag{4.143}$$

It should be noted that the CRLB for the signal k is independent of the signal strength or phase of the signal l. However, it depends on the code phase difference. For a further understanding of this statement, the reader is invited to compare the multipath mitigating and the multipath estimating discriminators in Chapter 8. Only the latter one has a multipath power independent performance.

4.3.6.1 Small Code-Phase Differences

For small code-phase differences $\Delta\tau$, the correlation function can be approximated because of the assumption in Section 1.8.1 as a second-order Taylor series

$$R_{\bar{c},c}(\Delta\tau) = 1 + \frac{r\Delta\tau^2}{2} \qquad |\Delta\tau| \ll 1 \tag{4.144}$$

Then, (4.143) reads as

$$CRLB(\mathrm{Re}\{\tau_k\}) \approx \frac{1}{L}\frac{\vartheta^2(4 + r^2\Delta\tau^4) - 4}{|a_{k,0}|^2 r(\vartheta^2 - 1)(\vartheta^2(r\vartheta^2 - 2)^2 - 4)} \tag{4.145}$$

For identical Doppler ($\vartheta = 1$) this expression is singular for all code-phase differences $\Delta\tau$ as long as (4.144) holds. Thus, *no* unbiased estimator exists that jointly estimates all parameters of both signals if (4.144) holds and if both signals have the identical Doppler frequency. It should be mentioned that this does not exclude the existence of any other (not jointly estimating) unbiased delay estimator that directly estimates τ_k and treats the l-parameters as nuisance parameters.

4.3.6.2 Identical Doppler

For identical Doppler values (i.e., $\vartheta = 1$), the CRLB (4.143) for the code-phase estimate of the first signal reads as

$$CRLB(\mathrm{Re}\{\tau_k\}) = -\frac{1}{L|a_{k,0}|^2} \times \frac{R_{\bar{c},c}''(0)(1 - |R_{\bar{c},c}(\Delta\tau)|^2) + |R_{\bar{c},c}'(\Delta\tau)|^2}{\left(|R_{\bar{c},c}'(\Delta\tau)|^4 + 2|R_{\bar{c},c}'(\Delta\tau)|^2(R_{\bar{c},c}''(0) - R_{\bar{c},c}(\Delta\tau)R_{\bar{c},c}''(\Delta\tau)) + \left(1 - |R_{\bar{c},c}(\Delta\tau)|^2\right)(R_{\bar{c},c}''(0)^2 - |R_{\bar{c},c}''(\Delta\tau)|^2)\right)} \tag{4.146}$$

This equation was also derived in publication by Ávila Rodríguez and it was analyzed for different GNSS signals [37].

For further analysis, a parametric model for any correlation function in the vicinity of its origin shall be defined by

$$R_{\bar{c},c}(\Delta\tau) = \begin{cases} 1 + \dfrac{r\Delta\tau^2}{2} - \dfrac{r\Delta\tau^4}{12m^2} & \ldots \quad |\Delta\tau| < m \\ 1 - \dfrac{m^2 r}{4} + \dfrac{2rm|\Delta\tau|}{3} & \ldots \quad m \le |\Delta\tau| \le \dfrac{3m}{8} - \dfrac{3}{2mr} \end{cases} \tag{4.147}$$

The correlation function is linear for $|\Delta\tau|$ being larger than the parameter m. The peak region is modeled by an even fourth-order polynomial. The parameter r represents the second derivative of the autocorrelation function at the origin.

For $|\Delta\tau| < m$, the code-phase CRLB of the signal k is given for the correlation function model (4.147) as

$$CRLB(\mathrm{Re}\{\tau_k\}) \approx -\frac{1}{L}\frac{9m^4}{\left|a_{k,0}\right|^2 r\Delta\tau^4} \tag{4.148}$$

The CRLB is nonsingular due to the fourth-order term, but strongly diverges as $\Delta\tau$ approaches zero.

4.3.6.3 Known Amplitudes and Carrier Phases Plus Identical Doppler

In cases where both signals have identical Doppler, and if only both code phases are estimated (but not the complex-valued amplitudes), then the CRLB for the code-phase estimate of the signal k reads as

$$CRLB(\mathrm{Re}\{\tau_k\}) = -\frac{1}{L}\frac{R_{\bar{c},c}''(0)}{\left|a_{k,0}\right|^2 (R_{\bar{c},c}''(0)^2 - R_{\bar{c},c}''(\Delta\tau)^2)} \tag{4.149}$$

This equation is obtained by inverting the 2 × 2 submatrix of (4.137) for the parameters (τ_k, τ_l).

The expression (4.149) is substantially simpler than (4.146), because (4.146) has to account for the coupling of all code-phase and complex-amplitude estimates. In an article by Lohan, a similar scenario was investigated and an expression similar to (4.149) was obtained [38]. However, in Lohan's work the multipath parameters were not estimated; instead, the multipath signal was considered as a deterministic bias added to the received signal.

Using the correlation function model (4.147), (4.149) reduces to

$$CRLB(\mathrm{Re}\{\tau_k\}) = -\frac{1}{La_{k,0}^2}\begin{cases} \dfrac{m^4}{r\Delta\tau^2(2m^2 - \Delta\tau^2)} & \ldots \quad \left|\Delta\tau\right| < m \\ \dfrac{1}{r} & \ldots \quad m \le \left|\Delta\tau\right| \le \dfrac{3m}{8} - \dfrac{3}{2mr} \end{cases} \tag{4.150}$$

We see that, if both signals are separated by a delay large enough such that the correlation function becomes linear, then the delay estimates achieve the same CRLB as in the case of a single signal. If the complex amplitude is known, the singularity is less severe (second order instead of fourth order) than for joint estimation of the code-phase and complex amplitude. This suggests that the coupling between the code-phase and the complex-amplitude estimates affect (4.146) significantly.

4.3.7 Two Propagation Paths: Doppler CRLB

Under the assumption (4.136), the submatrix of the Fisher information matrix (4.137) for the Doppler parameters (ω_k, ω_l) reads as

$$I_\omega = L \begin{pmatrix} \frac{L^2 \chi_{freq}}{24 f_s^2} \left| a_{k,0} \right|^2 & \frac{-\bar{a}_{k,0} a_{l,0} R_{\bar{c},c}(\Delta\tau) \kappa''(\Delta\omega)}{2} \\ \frac{-a_{k,0} \bar{a}_{l,0} \overline{R_{\bar{c},c}(\Delta\tau) \kappa''(\Delta\omega)}}{2} & \frac{L^2 \chi_{freq}}{24 f_s^2} \left| a_{l,0} \right|^2 \end{pmatrix} \tag{4.151}$$

The Doppler CRLB for the signal k for joint estimation of both Doppler parameters reads as

$$CRLB(\omega_k) = \frac{24 f_s^2 L}{\chi_{freq} \left| a_{k,0} \right|^2} \frac{1}{\left(\chi_{freq}^2 L^4 - 144 f_s^4 \left| \kappa''(\Delta\omega) \right|^2 \left| R_{\bar{c},c}(\Delta\tau) \right|^2 \right)} \tag{4.152}$$

where the Doppler is still considered as a complex random variable. Constraining it to be real-valued gives

$$\begin{aligned} CRLB(\mathrm{Re}\{\omega_k\}) &= \frac{12 f_s^2 L}{\chi_{freq} \left| a_k \right|^2} \frac{1}{\left(\chi_{freq}^2 L^4 - 144 f_s^4 \left| \kappa''(\Delta\omega) \right|^2 \left| R_{\bar{c},c}(\Delta\tau) \right|^2 \right)} \\ &= \frac{12 L}{\chi_{freq} \left| a_k \right|^2 f_s^2} \frac{1}{\left(\chi_{freq}^2 T_{coh}^4 - 144 \left| \kappa''(\Delta\omega) \right|^2 \left| R_{\bar{c},c}(\Delta\tau) \right|^2 \right)} \\ &= \frac{6 T_{coh}}{\chi_{freq} (C/N_0)_k} \frac{1}{\left(\chi_{freq}^2 T_{coh}^4 - 144 \left| \kappa''(\Delta\omega) \right|^2 \left| R_{\bar{c},c}(\Delta\tau) \right|^2 \right)} \end{aligned} \tag{4.153}$$

For identical Doppler values (i.e., $\vartheta = 1$) and identical code-phase values, this expression diverges since $\kappa''(\Delta\omega) \underset{\Delta\omega=0}{\rightarrow} -\chi_{freq} T_{coh}^2 / 12$.

4.3.8 Two Propagation Paths: Remark on Other Bounds

Thus far, the multipath signals have been considered as additional parameters that are estimated together with the line-of-sight parameters. For the case where the multipath parameters (and the carrier phase of the line-of-sight signal) are considered as nuisance parameters in the sense of Section 4.2.2, the corresponding expression for the MCRLB, ACRLB, and JCRLB are discussed in Section 4.3.8.1.

Here we will employ a two-step approach. In the first step, the carrier phase of both signals is considered as a nuisance parameter; in the second step, the delay and amplitude of the second signal is considered as a nuisance parameter.

4.3.8.1 Modified Cramér–Rao Lower Bound

The MCRLB for the code phase and amplitude is, according to Section 4.2.2.1, determined from (4.138). Averaging over the carrier-phase values removes terms that depend on the complex amplitude. The modified Fisher information matrix is

$$\tilde{I}_{a,\tau} = \left\langle I_{a,\tau} \right\rangle_{\varphi_k,\varphi_l} = \frac{L}{2} \times \begin{pmatrix} 1 & 0 & \vartheta R_{\bar{c},c}(\Delta\tau) & 0 \\ 0 & -R''_{\bar{c},c}(0)\left|a_{k,0}\right|^2 & 0 & 0 \\ \overline{\vartheta R_{\bar{c},c}(\Delta\tau)} & 0 & 1 & 0 \\ 0 & 0 & 0 & -R''_{\bar{c},c}(0)\left|a_{l,0}\right|^2 \end{pmatrix} \tag{4.154}$$

Code-phase estimates become independent from amplitude estimates. The MCRLB for the code-phase estimate is identical to the case of only the line of sight. This result may indicate that the MCRLB is rather weak for the considered settings. A further consideration of the multipath delay as a nuisance parameter does not change the code-phase MCRLB.

4.3.8.2 Asymptotic and Joint Cramér–Rao Lower Bound

The ACRLB for the code-phase estimate of the signal k in the case where both carrier phases are considered as nuisance parameters is given by averaging the inverse of the CRLB as described in Section 4.2.2.2. Because diagonal elements of the inverse Fisher information matrix (4.141)—including the determinate, (4.140)—are independent of the carrier-phase values, the ACRLB equals the JCRLB and is given by the expression (4.143).

If, in addition, the multipath delay $\Delta\tau$ is considered as a nuisance parameter, the ACRLB and the JCRLB are different. A good starting point to analyze both bounds is (4.148). The ACRLB is the inverse of the averaged-inverse CRLB

$$ACRLB(\mathrm{Re}\{\tau_k\}) = \left(\int (CRLB(\mathrm{Re}\{\tau_k\}))^{-1} p(\Delta\tau) d(\Delta\tau)\right)^{-1} \tag{4.155}$$

and the JCRLB is the averaged CRLB

$$JCRLB(\mathrm{Re}\{\tau_k\}) = \int CRLB(\mathrm{Re}\{\tau_k\}) p(\Delta\tau) d(\Delta\tau) \tag{4.156}$$

Both use the a priori probability density function $p(\Delta\tau)$ of the multipath delay. The probability density function often has the form of

$$p(\Delta\tau) \sim \frac{1}{d}\exp\left\{-\frac{\Delta\tau}{d}\right\} \tag{4.157}$$

for delays $\Delta\tau > 0$. Negative delays do not occur.

This delay distribution is, however, insufficient to compensate for the singularity of (4.148) at $\Delta\tau = 0$. Therefore, the JCRLB diverges. By contrast, the ACRLB will assume a high but finite value.

This implies that if an unbiased line-of-sight code-phase estimator exists at all when one multipath signal is present, then it will *not* be based on *joint* estimation of the line-of-sight and multipath signal parameters.

4.4 Data Reduction

The LSQ estimator discussed in Section 4.3 is based on correlating the received signal samples with derivatives of the signal model. Provided that the correlation point is near the true values, only a few values (the correlation values) are sufficient to determine the parameter estimates. The correlation values represent compressed information and act as *data reduction* from the large amount of samples to a few values.

This particular data-reduction method generally applies to many estimation techniques (MLE and Bayesian techniques) and is based on the definition of a *sufficient statistic*. The goal of a sufficient statistics is to obtain identical estimates either from working directly with the signal samples or from working with the sufficient statistic values.

In the following section, this method will be discussed on the example of high-rate pseudorange estimation from a given set of samples s, described by the probability density function $p_{\mathbf{q}}(\mathrm{s})$. In contrast to the preceding sections, the case of nonwhite noise is considered.

4.4.1 Sufficient Statistics

A *statistics* $T(\mathrm{s})$ is defined as a (vector) function of the signal samples that does *not* directly depend on the true parameter values $\mathbf{q}$. A statistic is itself a (multidimensional) random variable with a uniquely defined probability distribution depending on $p_{\mathbf{q}}(\mathrm{s})$. The dimension of the statistics is generally lower than the dimension of the signal samples (data reduction principle).

A statistics $T(\mathrm{s})$ is said to be *sufficient* for the parameters $\mathbf{q}$ if the conditional distribution of s, under the condition $T(\mathrm{s})$ = const., does not depend on $\mathbf{q}$. If a statistics is sufficient, the parameter $\mathbf{q}$ affects the probability density function $p_{\mathbf{q}}(\mathrm{s})$ only through a function $g_{\mathbf{q}}[T(\mathrm{s})]$. Stated differently, knowledge of $g_{\mathbf{q}}(T)$ is sufficient for determining $p_{\mathbf{q}}(\mathrm{s})$.

The Neyman–Fisher factorization theorem gives a simple criterion of sufficiency (see Theorem 3.1 of [4]). A statistics $T(\mathrm{s})$ is sufficient for $\mathbf{q}$ if and only if there exists a factorization of $p_{\mathbf{q}}(\mathrm{s})$ of the form

$$p_{\mathbf{q}}(\mathrm{s}) = g_{\mathbf{q}}(T(\mathrm{s}))h(\mathrm{s}) \tag{4.158}$$

with positive functions $g_{\mathbf{q}}$ and h.

Sufficient statistics are closely related to ML estimation in the sense that the ML estimate of $\mathbf{q}$ is a function of any sufficient statistics (see Theorem 3.10 in the book by Porat [4]).

The factorization theorem shall now be applied to the high-rate pseudorange probability density function

$$p_{\mathbf{q}}(\mathbf{s}) = \frac{1}{(2\pi)^L} \exp\left\{-\frac{1}{2}\sum_{\mu=1}^{L} \left|s_\mu - r_\mu(\mathbf{q})\right|^2\right\} \tag{4.159}$$

This function is extended for the case of colored noise

$$p_{\mathbf{q}}(\mathbf{s}) = \frac{1}{\det Q(2\pi)^L} \exp\left\{-\frac{1}{2}(\mathbf{s}^* - \mathbf{r}(\mathbf{q})^*)Q^{-1}(\mathbf{s} - \mathbf{r}(\mathbf{q}))\right\} \tag{4.160}$$

where Q is the colored noise covariance matrix,

$$\left\langle N_\mu N_\upsilon \right\rangle_{\mathrm{N}} = 0, \left\langle \bar{N}_\mu \bar{N}_\upsilon \right\rangle_{\mathrm{N}} = 0, \left\langle N_\mu \bar{N}_\upsilon \right\rangle_{\mathrm{N}} = 2Q_{\mu,\upsilon} \tag{4.161}$$

The quadratic form in the exponent of (4.160) is expanded into

$$-\frac{1}{2}(\mathbf{s}^* - \mathbf{r}(\mathbf{q})^*)Q^{-1}(\mathbf{s} - \mathbf{r}(\mathbf{q})) = -\frac{1}{2}\mathbf{s}^* Q^{-1}\mathbf{s} - \frac{1}{2}\mathbf{r}(\mathbf{q})^* Q^{-1}\mathbf{r}(\mathbf{q}) + \mathrm{Re}\{\mathbf{r}(\mathbf{q})^* Q^{-1}\mathbf{s}\} \tag{4.162}$$

Only the last term of the above expression is relevant for the factorization theorem, as the first term is independent of the parameters $\mathbf{q}$ (and thus goes into h) and the second term (being independent of s) can be trivially part of any function of $g_{\mathbf{q}}$. In the following section, two approximations are discussed that are commonly used to express the last term with the help of properly defined statistics.

4.4.2 Multicorrelator Approach

The multicorrelator approach relies on the evaluation of the statistics at selected points spanning the range Λ of admissible values for $\mathbf{q}$. Between the selected points, the last term of (4.162) is obtained via interpolation. The last term of (4.162) is therefore approximated as

$$\mathbf{r}(\mathbf{q})^* Q^{-1}\mathbf{s} = \sum_b \alpha_b(\mathbf{q}) T_b(\mathbf{s}) \tag{4.163}$$

where

$$T_b(\mathbf{s}) = \mathbf{r}(\mathbf{q}_b)^* Q^{-1}\mathbf{s} \tag{4.164}$$

is the multidimensional test statistics evaluated via correlation of the signal model with the signal samples at selected control points $\mathbf{q}_b$ (plus accounting for the noise covariance). The functions $\alpha_b(\mathbf{q})$ are the interpolation coefficients.

The signal model (4.46) is linear in the complex amplitude. Therefore, the interpolation in the complex-amplitude dimension is trivial and the grid spanning all admissible values is effectively two-dimensional. It extends in the code-phase and Doppler dimension.

4.4.3 First-Derivative Approach

A second possibility to express the last term of (4.162) as a function of properly defined statistics is a Taylor series expansion around a predefined correlation point.

For the sake of simplicity, this expansion shall be carried out only in the first order, but the generalization to higher-order terms is obvious.

The last term of (4.162) is approximated as

$$\begin{aligned}\mathbf{r}(\mathbf{q})^* Q^{-1}\mathbf{s} &\approx \mathbf{r}(\mathbf{q}_0)^* Q^{-1}\mathbf{s} + (\mathbf{q}-\mathbf{q}_0)^* \frac{\partial \mathbf{r}(\mathbf{q})^*}{\partial \mathbf{q}}\Big|_{\mathbf{q}=\mathbf{q}_0} Q^{-1}\mathbf{s} \\ &= \mathbf{r}(\mathbf{q}_0)^* Q^{-1}\mathbf{s} - \mathbf{q}_0^* \frac{\partial \mathbf{r}(\mathbf{q})^*}{\partial \mathbf{q}}\Big|_{\mathbf{q}=\mathbf{q}_0} Q^{-1}\mathbf{s} + \mathbf{q}^* T(\mathbf{s})\end{aligned} \tag{4.165}$$

with the *vector* $T(\mathbf{s})$ of test statistics defined as

$$T(\mathbf{s}) = \frac{\partial \mathbf{r}(\mathbf{q})^*}{\partial \mathbf{q}}\Big|_{\mathbf{q}=\mathbf{q}_0} Q^{-1}\mathbf{s} \tag{4.166}$$

The first two terms of (4.165) are independent of $\mathbf{q}$ and thus go into h.

The test statistics of the first-derivative approach relies on the correlation of the received samples with the first derivates of the signal model (and accounts for the noise covariance). The derivative of the signal model (4.46) with respect to the complex amplitude yields a correlation with the signal model itself, which will define the P-correlator. The derivatives with respect to the code phase and with respect to the Doppler define the D- and the F-correlators. All three correlators will be discussed in Section 7.3.

4.4.4 Colored Noise

The test statistics of the preceding two sections account for the noise correlations via the matrix Q. Consequently, the test statistics can be expressed as correlations of the signal samples with a modified signal model, defined as

$$\breve{\mathbf{r}}(\mathbf{q}) = Q^{-1}\mathbf{r}(\mathbf{q}) \tag{4.167}$$

For example, the test statistics (4.166) are then given as

$$T(\mathbf{s}) = \frac{\partial \breve{\mathbf{r}}(\mathbf{q})^*}{\partial \mathbf{q}}\Big|_{\mathbf{q}=\mathbf{q}_0} \mathbf{s} \tag{4.168}$$

The modified signal model $\breve{\mathbf{r}}(\mathbf{q})$ compensates for the colored-noise effects. For example, if the signal and noise are affected by the same low-pass filter, then the modified reference signal $\breve{\mathbf{r}}(\mathbf{q})$ is obtained by applying twice the inverse filter (i.e., a high-pass filter) to the infinite-bandwidth reference signal. This and related techniques will be further explored in Section 6.4.

4.5 Bayesian Approach

In the Bayesian approach, the estimated parameters (for example, the high-rate pseudorange parameters $\mathbf{q}$) are assumed to be random quantities related statistically to the observations. The parameters are endowed with an a priori probability density function $w(\mathbf{q})$, which is defined, even with the lack of actual observations. By contrast, the nonrandom-parameter estimation approach introduced in Section 4.2 does not require any a priori distribution of the parameters.

The second important difference between nonrandom-parameter estimation and the Bayesian approach is the use of a cost function in the latter case. The cost function measures the cost of estimating the true value $\mathbf{q}$ as $\hat{\mathbf{q}}$; the more incorrect the estimate, the higher the cost. A Bayesian estimate minimizes the cost function, averaged over all admissible values for $\mathbf{q}$. Bayesian estimates are not necessarily unbiased. ML estimation can be seen as a special case of Bayesian estimation for the case of a uniform distribution $w(\mathbf{q})$ and a cost function that assumes 0 for $\hat{\mathbf{q}} = \mathbf{q}$ and otherwise one (see chapter IV.D in the book by Poor [1]).

The Bayesian approach is generally less popular than the nonrandom parameter-estimation approach for the design of navigation signal-processing algorithms, probably because of two facts. First, the a priori probability density function $w(\mathbf{q})$ is often hard to determine. Second, the coupling between $w(\mathbf{q})$ and the estimated parameters complicates the determination of the statistical distribution of the estimated parameters. An important exception to this is the case of linear observations and a linear dynamical model with Gaussian noise. The resulting Bayesian estimator—the Kalman–Bucy filter—yields explicit expressions for the estimates and their probability density functions.

In the following, minimum mean-squared error estimation will be introduced; the estimator will be expressed as a function of the sufficient statistics of Section 4.4, regardless of the form of the underlying probability density functions or the relationship between parameters and observations. The necessary assumptions to apply the Kalman–Bucy filter theory, as well as a block diagram for a Kalman filter based on the sufficient statistics of Section 4.4, will be discussed in Section 6.5.2.

4.5.1 Minimum Mean-Squared Error Estimation

It is well known that the choice of the cost function should be related to the estimation problem; in the following, we focus on a quadratic cost function $\|\hat{\mathbf{q}} - \mathbf{q}\|^2$. The Bayesian risk $R(\hat{\mathbf{q}}, \mathbf{s})$ for a particular set of observed samples $\mathbf{s}$ is defined as the averaged cost function over all admissible parameter values conditioned to $\mathbf{s}$ as

$$R(\hat{\mathbf{q}}, \mathbf{s}) = \left\langle \|\hat{\mathbf{q}} - \mathbf{q}\|^2 \right\rangle \Big|_{\mathbf{q}|\mathbf{s}} = \int_{\mathbf{q} \in \Lambda} \|\hat{\mathbf{q}} - \mathbf{q}\|^2 \, p(\mathbf{q}|\mathbf{s}) d\mathbf{q} \tag{4.169}$$

This equation includes the conditional probability $p(\mathbf{q} \mid \mathbf{s})$ of observing $\mathbf{q}$ under the assumption $\mathbf{S} = \mathbf{s}$.

The Bayes estimate for the quadratic cost function is defined as

$$\hat{\mathbf{q}}(\mathbf{s}) = \arg \min_{\hat{\mathbf{q}}} R(\hat{\mathbf{q}}, \mathbf{s}) \tag{4.170}$$

and evaluates according to Case IV.B.4 in the book by Poor to the conditional expected value of the parameters itself [1]; that is,

$$\hat{\mathbf{q}}(\mathbf{s}) = \int_{\mathbf{q}\in\Lambda} \mathbf{q}p(\mathbf{q}|\mathbf{s})d\mathbf{q} = \int_{\mathbf{q}\in\Lambda} \frac{\mathbf{q}p_{\mathbf{q}}(\mathbf{s})w(\mathbf{q})}{\int_{\mathbf{q}\in\Lambda} p_{\mathbf{q}}(\mathbf{s})w(\mathbf{q})d\mathbf{q}} d\mathbf{q} \tag{4.171}$$

In the last step, the conditional probability $p(\mathbf{q} \mid \mathbf{s})$ is expressed using the sample probability density function $p_{\mathbf{q}}(\mathbf{s})$ and the a priori probability density function $w(\mathbf{q})$ using the Bayes theorem.

From Section 4.4, it is clear that for fixed observed samples $\mathbf{s}$, the probability density function $p_{\mathbf{q}}(\mathbf{s})$ can be obtained via a sufficient statistics $T(\mathbf{s})$. Therefore, after the data reduction from the signal samples to the statistics has been performed, the evaluation of (4.171) requires averaging over all parameter values without, however, performing further correlations. The MMSE parameter estimates can, for example, be obtained from the P-, D-, and F-correlator values introduced in Chapter 7, and the integral

$$\hat{\mathbf{q}}(\mathbf{s}) = \int_{\mathbf{q}\in\Lambda} \frac{\mathbf{q}g_{\mathbf{q}}(T(\mathbf{s}))w(\mathbf{q})}{\int_{\mathbf{q}\in\Lambda} g_{\mathbf{q}}(T(\mathbf{s}))w(\mathbf{q})d\mathbf{q}} d\mathbf{q} \tag{4.172}$$

represents a discriminator that converts correlation values into parameter estimates. Typically, the integration over complex-valued amplitude parameters can be carried out in closed form.

4.5.2 Kalman–Bucy Filter

The evaluation of (4.171) can be performed using linear algebra if a number of assumptions on the probability density functions and the signal model are fulfilled. The resulting estimation scheme is called *Kalman–Bucy* (or simply Kalman) filter and shall be outlined here.

To apply the Kalman filter theory, we assume that the samples are Gaussian distributed and that the a priori probability density function of the parameters is Gaussian as well. Furthermore, we assume that the signal model is essentially linear in the estimated parameters in the sense of (4.129), with the relationship between modified and true signal parameters being sufficiently linear. The latter linearity condition requires knowing the complex-signal amplitude rather precisely. According to Example IV.B.3 in the book by Poor, the MMSE estimator of the parameters can be obtained using linear algebra expressions involving the observations, the mean values, and the covariance matrices [1]. Furthermore, for this particular setting, the MMSE coincides to the minimum mean absolute error (MMAE) estimator and to the maximum a posteriori probability (MAP) estimator.

The expressions can be further simplified if a dynamical model for the estimated parameters is assumed. The estimated parameters are time-dependent and the time dependence is realized by providing an identical set of parameters for each epoch separately. A dynamical model is a linear relationship between the true parameters of successive epochs including Gaussian process noise. Furthermore, each signal sample can be uniquely assigned to one epoch (i.e., the signal model for a particular sample depends on parameters of only one epoch). Then the whole estimation problem can be formulated as a Kalman filter, as for example described in Section V.B of Poor [1]. The Kalman filter is an MMSE (and also MMAE/MAP) estimator. The formulas governing the Kalman filter are generally well-known and shall not be repeated here. Instead, the reader is referred to the multitude of textbooks, including those of Poor, Minkler, and Minkler [1, 39].

The assumption of Gaussian statistics can be relaxed and the Kalman filter is optimum not only for the Gaussian model but also for all linear estimators for the same model with non-Gaussian statistics, provided that the second-order statistics (i.e., means and covariances) remain unchanged (see Section V.C of [1]).

In this context, the observation update step of the Kalman filter is of special consideration because this step usually involves a large number of signal samples. This large number of samples could potentially cause computational performance problems. However, the update step can be expressed as a function of a sufficient statistics. The measurement or observations update step is usually written in a form like

$$\hat{\mathbf{q}}_{k|k} = \hat{\mathbf{q}}_{k|k-1} + K_k(\mathbf{s}_k - A_k\hat{\mathbf{q}}_{k|k-1}) \tag{4.173}$$

with A_k being the design matrix used to establish the linear relationship between samples and the measurements. The Kalman gain matrix K_k is given as (see Equation (IV.B.55) in the book by Poor [1])

$$\begin{aligned} K_k &= \Sigma_{k|k-1}A_k^{\ *}(A_k\Sigma_{k|k-1}A_k^{\ *} + Q_k)^{-1} \\ &= \left(A_k^{\ *}Q_k^{\ -1}A_k + \Sigma_{k|k-1}^{\ \ -1}\right)^{-1} A_k^{\ *}Q_k^{\ -1} \end{aligned} \tag{4.174}$$

Here, Q_k denotes the sample (or measurement noise) covariance matrix for all samples belonging to the parameters of the epoch k. The predicted parameter-covariance matrix is denoted as $\Sigma_{k|k-1}$. The received samples are multiplied by the inverse measurement noise-covariance matrix and the Hermite conjugate of the design matrix to update the parameter estimates. For $\Sigma_{k|k-1} \to \infty$, this expression has a similar structure like the LSQ step (4.56). The identical operation is also performed to obtain the sufficient statistics (4.166). Therefore, the Kalman filter may use as input the sufficient statistics (4.166) instead of directly operating on the samples. The resulting block diagram is shown in Figure 4.6.

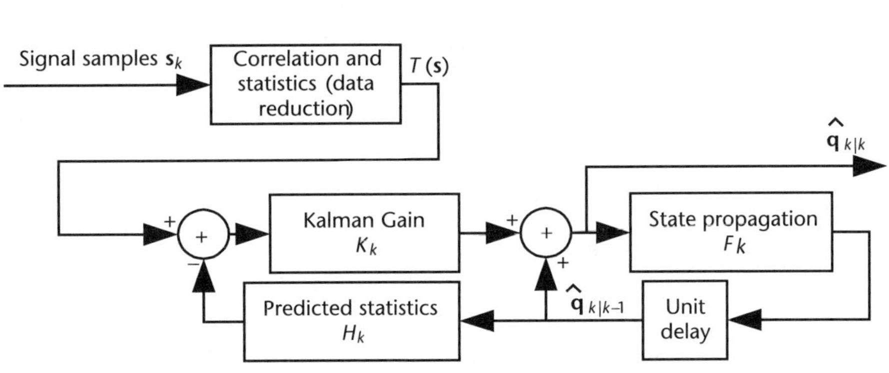

Figure 4.6 Kalman filter block diagram.

A navigation receiver realized as a Kalman filter depicted in Figure 4.6 does *not* require more correlations of the received samples with internally generated reference signals compared to a conventional receiver based on, for example, tracking loops of Section 4.3.3.1. It is, however, required that the underlying dynamical model and the observation model are correct, as is the short-period signal model of Section 1.8 that relates batches of signal samples to one set of estimated parameters. The presented scheme estimates all four fundamental signal parameters (code phase, Doppler, carrier phase, and amplitude) and uses the *predicted* complex-valued amplitude to setup the design matrix of the current epoch. By contrast, the LSQ discriminators of Section 4.3.2 use the *estimated* complex-valued amplitude of the current epoch to correct code-phase and Doppler estimates. Correct filtering and prediction of the carrier phase and amplitude is essential for this Kalman scheme. The Kalman filter intrinsically accounts for all correlations between the estimated parameters and signals, thereby causing an increased computational load.

The Kalman filter of Figure 4.6 can also be expressed in terms of the position estimates $\mathbf{x}$, instead of the high-rate pseudorange estimates $\mathbf{q}$. In this case, the design matrix not only derives from the short-period signal model of Section 1.8 but also includes the observation model (4.2). Alternatively, a cascaded Kalman filter structure can be used that first estimates $\mathbf{q}$ and then $\mathbf{x}$. Provided that all underlying models are sufficiently good and provided that the position-estimation filter correctly accounts for all correlations in the estimated high-rate pseudorange parameters $\mathbf{q}$ (also timely correlations), identical results are obtained.

The major disadvantage of the scheme in Figure 4.6 is that it is an all-in-one solution and a single failure (e.g., incorrect carrier-phase filtering) may corrupt the whole estimation process. Furthermore, the Kalman filter relies on a proper linearization of the signal model and its behavior is undefined if the correlation point diverges too much from the true value.

4.5.3 Other Filters

The two estimation schemes (LSQ estimation plus subsequent filtering and the Kalman filter) must fulfil a number of assumptions to be optimal:

1. Gaussian noise;
2. Gaussian a priori parameter distribution;
3. Linearity conditions fulfilled or linearized signal model.

The first assumption—Gaussian noise—can be considered to be fulfilled in most cases, especially if the navigation receiver is correlation-based. Then the law of large numbers applies. The Gaussian parameter distribution is a reasonable assumption and Markov models (e.g., for the user movement) can be provided. It is, however, more difficult to chose suitable parameters for the distributions.

In the author's opinion, the most difficult assumptions to fulfill are the linearity conditions. Although these conditions can only be violated if the signal power is weak (a high signal power allows ML estimation, thereby fulfilling the linearity conditions), reception of weak signals is a typical problem a navigation receiver has to cope with. In the case of GNSS signals, a linearized signal model of Section 4.3.4 is more difficult to obtain than fulfilling the linearity conditions of Section 4.3.2.10. If full linearization (including products with the complex-valued amplitude) of the signal model is not possible, but the linearity conditions are fulfilled, the LSQ estimation scheme or a noncoherent Kalman filter (see next Section 4.5.4) are still optimal. If the linearity conditions are also not fulfilled, nonlinear filters might be used.

Nonlinear filtering for navigation signal processing is a topic of ongoing research. A tutorial on nonlinear and non-Gaussian filters was written by Arulampalan [40]. Nonlinear filters are Bayesian filters and use Monte Carlo methods to evaluate Bayesian estimation integrals like (4.171). Particle filters have been used by Closas for static multipath scenarios [41]. Dynamic multipath scenarios have been presented by Lentmaier [42]. In the latter work, it is evident how the particle filter combines the non-Gaussian parameter probability density function of the different epochs and circumvents the linearization problem. A demonstration of the practical use of nonlinear filters or a comparison with linear filters is not known to the author.

A related problem is the transition between different discrete states in an estimation filter. Most importantly, the number of multipath signals has to be estimated correctly to set up the right signal model. Bayesian methods can be used for this purpose [42].

4.5.4 Use of Kalman Filters in GNSS Signal Processing

The use of a Kalman filter for signal tracking has become increasingly popular in recent years. Kalman filters are used to implement independent channel tracking loops [43] and may be used to improve the tracking performance during ionospheric scintillations [44]. At the independent tracking-loop level, a Kalman filter has a similar transfer-function structure as a carrier-aided DLL with an FLL-assisted PLL [27]. Kalman filters can also be found in ultratightly coupled GPS/INS systems that are based on vector tracking.

We categorize signal-processing Kalman filters according to the input data:

1. Navigation signal samples;
2. Correlator values;
3. Discriminator values.

A commonly used fourth type, working with pseudoranges, is not considered here because this type is considered to work at the navigation-processor level and not at the signal-processing level. Furthermore, the signal-processing Kalman filters are categorized according to the state vector:

A. (Low-rate) pseudorange parameters;
B. Position parameters.

Section 4.4 showed that categories 1 and 2 are equivalent if the correlator values represent a sufficient statistics. Category 1 and 2 are *coherent* Kalman filters because the input data depends on the carrier phase, whereas a category-3 (*noncoherent*) Kalman filter may operate with code-phase and Doppler (and eventually with amplitude) discriminator values.

A coherent Kalman filters needs to generate *predicted* complex-valued amplitudes, which are used to set up the Kalman filter design matrix of the current epoch. By contrast, a noncoherent Kalman filter based on LSQ discriminators uses the complex-valued amplitude estimate based on the *current* epoch's samples to set up the LSQ design matrix. From the theoretical point of view, this is the main difference among category 1/2 and 3. The extent to which the predicted complex-valued amplitude is more accurate than the estimated complex-valued amplitude depends primarily on the signal characteristics (data/pilot) and dynamics (including clock jitter), the duration of the coherent integration and on the availability of aiding (IMU) data.

A coherent Kalman filter is optimal (e.g., squaring-loss free) but difficult to realize for low signal power [45]. Note that the squaring loss is irrelevant for high signal power. For GNSS receivers, predicting the complex-valued amplitude—in situations where carrier phase estimation is not possible—requires at least a cm-accurate pseudorange prediction that needs a stable receiver oscillator (OCXO), a navigation-grade IMU, and a stable propagation channel (e.g., no atmospheric scintillations). Furthermore, predicting the carrier phase requires pilot signals or a data-bit wipe-off functionality. A more practical approach to incorporate an equivalent amount of aiding information in a noncoherent Kalman filter is to increase the coherent integration time to macroscopic lengths (a few seconds) and to use the aiding information within the discriminators only. Effectively, the aiding information compensates for nonlinear line-of-sight dynamics and allows for using the short-period signal model over longer time spans. Macroscopic long-integration times virtually eliminate the squaring loss.

The use of the aiding data is simpler within the discriminators than within a coherent Kalman filter [46]. Only the second-order line-of-sight dynamics (e.g., the *acceleration*) need to be compensated over the integration interval, and each interval is independent. By contrast, the aiding information for the Kalman filter needs to be stable at zeroth order over the whole period of operation. Both methods—a coherent Kalman filter or a long-coherent integration—predict the carrier phase of the line-of-sight signal and thereby mitigate contributions from multipath signals, which have a different carrier-phase behavior if the antenna is moving. This may involve predetection multipath suppression or a synthetic antenna array [47].

Category-A Kalman filters are used for independent channel tracking and category-B filters are used for vector tracking. A category B/1 Kalman filter would be well-suited to realize an intrasystem interference-cancellation receiver. Finally, it should be noted that Kalman filters of all categories allow carrier-phase estimation (see Section 4.3.3.4).

4.6 Squaring Loss Revisited

The squaring loss introduced in Section 4.3.2 decreases the accuracy of high-rate pseudorange LSQ estimates compared to the CRLB (considering the carrier phase as a useful parameter). The squaring loss is caused by estimation errors of the complex-valued signal amplitude, which affect the setup of the design matrix. The squaring loss is intrinsically coupled to the underlying LSQ estimation scheme, which jointly estimates the complex-valued signal amplitude together with the code phase and Doppler. The complex-valued amplitude is treated as a nonrandom parameter.

By contrast, if the carrier phase is treated as a nuisance parameter, none of the relevant bounds (e.g., the MCRLB or the ACRLB) is affected by a term that would correspond to the squaring loss, as has been shown in Sections 4.3.2.5 and 4.3.8. Therefore, further investigations shall be carried out in this section to clarify the question of whether the squaring loss is a specific artifact of LSQ estimation or if it is more fundamental. This is done in the following section by a mathematically exact treatment of the carrier phase as a nuisance parameter, which gives the *true CRLB* (TCRLB). A numerical evaluation of the TCRLB later shows that it is affected by the squaring loss, proving that the LSQ scheme is an optimal estimator (i.e., a MVUE).

To evaluate the TCRLB under the assumption that the carrier phase is a nuisance parameter, a simplified signal model shall be used that assumes zero Doppler. Based on (4.44), the following signal model shall be assumed

$$r_\mu(\mathbf{q},\xi) = ac(t_\mu - \tau)\exp\{-i\varphi\} \quad \mathbf{q} = \begin{pmatrix} a \\ \tau \end{pmatrix}, \quad \xi = (\varphi) \tag{4.175}$$

where the code phase τ and the real-valued amplitude a are considered as useful parameters $\mathbf{q}$ to be estimated, and the carrier phase φ is treated as a nuisance parameter ξ. For the carrier phase, we assume that it is uniformly distributed between $[0, 2\pi]$.

The probability density function of the signal samples is first expanded into

$$\begin{aligned} p_{\mathbf{q},\xi}(\mathbf{s}) &= \frac{1}{(2\pi)^L}\exp\left\{-\frac{1}{2}\sum_{\mu=1}^{L}\left|s_\mu - r_\mu(\mathbf{q},\xi)\right|^2\right\} \\ &= \frac{1}{(2\pi)^L}\exp\left\{-\frac{1}{2}\sum_{\mu=1}^{L}\left|s_\mu\right|^2 - \frac{1}{2}\sum_{\mu=1}^{L}\left|r_\mu(\mathbf{q},\xi)\right|^2 + \mathrm{Re}\{ae^{i\varphi+i\delta}B\}\right\} \end{aligned} \tag{4.176}$$

where the constants

$$\begin{aligned} B &= \left|\sum_{\mu=1}^{L} s_\mu \overline{c(t_\mu - \tau)}\right| \\ \delta &= \mathrm{angle}\left\{\sum_{\mu=1}^{L} s_\mu \overline{c(t_\mu - \tau)}\right\} \end{aligned} \tag{4.177}$$

represent the magnitude and phase of the correlation value of the received samples with the signal at baseband.

According to Section 4.2.2, the probability density function of the signal samples, parameterized only by the useful parameters, is given by averaging (4.176) over all carrier phase values; that is,

$$
\begin{aligned}
p_{\mathrm{q}}(\mathbf{s}) = \left\langle p_{\mathrm{q},\xi}(\mathbf{s}) \right\rangle_{\xi} &= \frac{1}{2\pi} \int_{\varphi=0}^{2\pi} p_{\mathrm{q},\xi}(\mathbf{s}) d\varphi \\
&= \frac{1}{(2\pi)^L} \exp\left\{ -\frac{1}{2} \sum_{\mu=1}^{L} \left| s_\mu \right|^2 - \frac{La^2}{2} + \log(I_0(aB)) \right\}
\end{aligned}
\tag{4.178}
$$

Here, I_m denotes the mth-order modified Bessel function of first kind, defined as

$$
\int_{\varphi=0}^{2\pi} \exp\{imx + x\cos(\varphi)\} d\varphi = 2\pi I_m(x) \tag{4.179}
$$

For large values of x, the following approximation holds

$$
\log(I_m(x)) \approx x \qquad |x| \gg \tag{4.180}
$$

To obtain the TCRLB of the code-phase estimate (under the assumption of a constant real-valued signal amplitude a), the derivative with respect to the code phase of the logarithm of (4.178) has to be computed as follows

$$
\frac{\partial}{\partial \tau} \log p_{\mathrm{q}}(\mathbf{s}) = \frac{\partial}{\partial \tau} \log(I_0(aB)) = \frac{aI_1(aB)}{I_0(aB)} \frac{\partial B}{\partial \tau} \tag{4.181}
$$

The derivative of B with respect to the code phase is given as

$$
\begin{aligned}
\frac{\partial B}{\partial \tau} &= -\frac{1}{2B} \sum_{\mu,\nu=1}^{L} \left(S_\mu \overline{c(t_\mu - \tau)} \overline{S}_\nu c'(t_\nu - \tau) + S_\mu \overline{c'(t_\mu - \tau)} \overline{S}_\nu c(t_\nu - \tau) \right) \\
&= -\frac{1}{B} \mathrm{Re}\left\{ \sum_{\mu=1}^{L} S_\mu \overline{c(t_\mu - \tau)} \sum_{\nu=1}^{L} \overline{S}_\nu c'(t_\nu - \tau) \right\}
\end{aligned}
\tag{4.182}
$$

It is useful to introduce the (simplified) P- and D-correlators (see Section 7.3) evaluated at the true parameter values as

$$
\begin{aligned}
P &= \sum_{\mu=1}^{L} S_\mu \overline{c(t_\mu - \tau)} \approx aL\exp\{-i\varphi\} + \sum_{\mu=1}^{L} N_\mu \overline{c(t_\mu - \tau)} \\
D &= \sum_{\mu=1}^{L} S_\mu \overline{c'(t_\mu - \tau)} \approx \sum_{\mu=1}^{L} N_\mu \overline{c'(t_\mu - \tau)}
\end{aligned}
\tag{4.183}
$$

and to write the derivative of the logarithm of (4.178) as

$$
\frac{\partial}{\partial \tau} \log p_{\mathrm{q}}(\mathbf{s}) = -\frac{aI_1(a|P|)}{I_0(a|P|)} \frac{\mathrm{Re}\{P\overline{D}\}}{|P|} \tag{4.184}
$$

This expression is substantially more complex than the corresponding expression (4.13), which treats the carrier phase as a useful parameter to be estimated.

Overall, the TCRLB for the code phase under the assumption that the real-valued amplitude is constant and that the carrier phase is a uniformly distributed nuisance parameter is given as

$$TCRLB(\tau) = \left[\left\langle \left(\frac{aI_1(a|P|)}{I_0(a|P|)} \frac{\text{Re}\{P\bar{D}\}}{|P|} \right)^2 \right\rangle_N \right]^{-1} \tag{4.185}$$

In case the real-valued signal amplitude is jointly estimated with the code phase, the two-dimensional Fisher information matrix has to be set up. Therefore, the derivative with respect to the amplitude of the logarithm of (4.178) has to be computed as

$$\frac{\partial}{\partial a} \log p_{\text{q}}(\text{S}) = -aL + \frac{\partial}{\partial a} \log(I_0(aB)) = -aL + \frac{BI_1(aB)}{I_0(aB)} \tag{4.186}$$

Then the Fisher information matrix is given as

$$I_{\tau,a} = \left\langle \begin{pmatrix} \left(\frac{aI_1(a|P|)}{I_0(a|P|)} \frac{\text{Re}\{P\bar{D}\}}{|P|} \right)^2 & \left(\frac{|P| I_1(a|P|)}{I_0(a|P|)} - aL \right) \frac{aI_1(a|P|)}{I_0(a|P|)} \frac{\text{Re}\{P\bar{D}\}}{|P|} \\ \left(\frac{|P| I_1(a|P|)}{I_0(a|P|)} - aL \right) \frac{aI_1(a|P|)}{I_0(a|P|)} \frac{\text{Re}\{P\bar{D}\}}{|P|} & \left(\frac{|P| I_1(a|P|)}{I_0(a|P|)} - aL \right)^2 \end{pmatrix} \right\rangle_N \tag{4.187}$$

The off-diagonal terms tend to average out due to the presence of the D-correlator, whose distribution is symmetric around zero-code delay, whereas the P-correlator values are centered around a. For large values of a, the approximation (4.180) can be used, yielding

$$I_{\tau,a} \underset{a>>}{=} \left\langle \begin{pmatrix} \left(a \frac{\text{Re}\{P\bar{D}\}}{|P|} \right)^2 & (|P| - aL)a \frac{\text{Re}\{P\bar{D}\}}{|P|} \\ (|P| - aL)a \frac{\text{Re}\{P\bar{D}\}}{|P|} & (|P| - aL)^2 \end{pmatrix} \right\rangle_N \tag{4.188}$$

The Fisher information matrix is best evaluated using a Monte Carlo simulation at either the signal level or the correlator level. A specific example will be performed in the next section. The code-phase TCRLB can then be computed from the first diagonal element of the inverse Fisher information matrix.

4.7 Numerical Simulation

By specifying a specific navigation signal structure, some formulas of preceding sections are explicitly evaluated and the obtained results discussed. Different code-phase (ranging) CRLBs for the presented navigation signal are evaluated and compared to each other. A Monte Carlo solution is used to assess the influence of the starting correlation point on the obtained estimates and single-iteration results are compared to a converging iterative solution.

The simulation is based on the Gaussian double-pulse signal of Section 1.9.4.

4.7.1 Evaluation of Bounds

In the following, a number of code-phase CRLBs shall be evaluated. Additionally, the variance of an LSQ code-phase estimate with squaring loss based on a single iteration using the true values as correlation point is also included. More specifically, the following figures shall be considered:

- CRLB: The code-phase CRLB (4.72), which is obtained by treating the carrier phase, code phase, Doppler, and amplitude as useful parameters to be estimated.
- TCRLB: The true code-phase CRLB—the first diagonal element of the inverse matrix of (4.187)—obtained by treating the carrier phase as a nuisance parameter distributed uniformly in $[0, 2\pi]$.
- TCRLB, a = const.: The true code-phase CRLB (4.185) obtained by treating the carrier phase as a nuisance parameter distributed uniformly in $[0, 2\pi]$ and assuming a known amplitude.
- LSQ: The code-phase variance from an LSQ estimation (4.82). The correlation points coincide with the true values.

All four procedures are based on an unbiased estimation scheme. The differences in the CRLBs originate from the different treatment of the individual parameters, which is summarized in Table 4.3. The TCRLB is the most realistic bound, and the CRLB is of importance if the carrier phase is known from a priori or if the carrier phase is estimated with high accuracy.

Figure 4.7 shows the different code-phase variances (actually the square root of the variances) as a function of the signal power in a double logarithmic plot. The LSQ variance and the two TCRLBs coincide very well and exhibit the squaring loss below around 32 dBHz. The CRLB does not show the squaring loss and effectively underestimates the influence of the unknown carrier phase on code-phase estimation.

Two important conclusions can be drawn from Figure 4.7 for the assumed exemplary navigation signal. First, the CRLB is too weak for low signal power levels. One of the TCRLBs should be used. The TCRLBs provide more realistic values (although they are significantly more difficult to calculate). Second, the LSQ estimate achieves the TCRLB and is thus an *optimal* estimator, even for low signal

Table 4.3 Code Phase Cramér–Rao Lower Bounds Used in the Comparisons

Parameter	*CRLB*	*TCRLB*	*TCRLB, a = const.*
Code phase	Estimated	Estimated	Estimated
Doppler	Estimated	Known	Known
Carrier phase	Estimated	Nuisance	Nuisance
Amplitude	Estimated	Estimated	Known

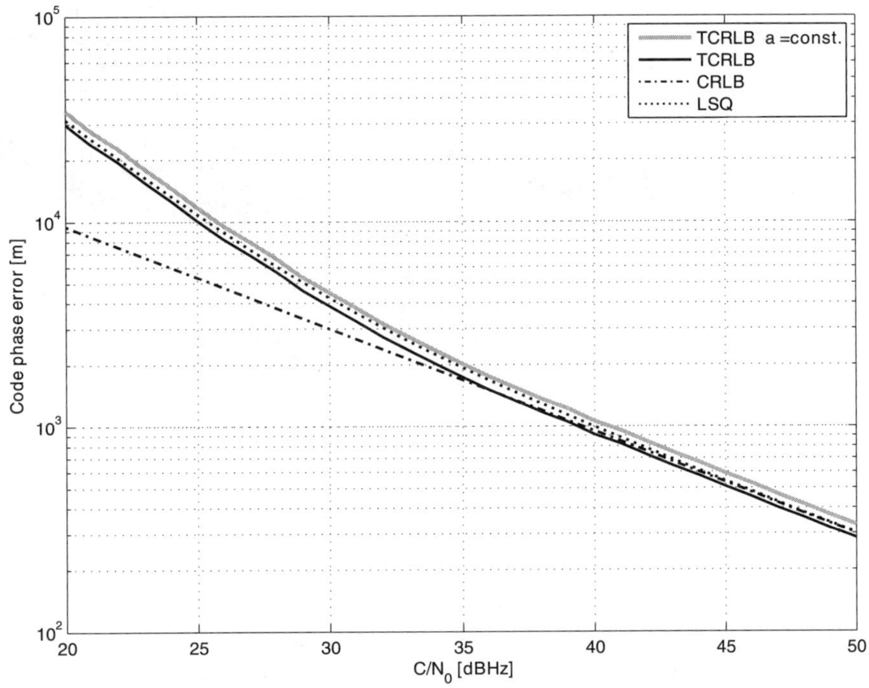

Figure 4.7 Comparison of various code-phase variance bounds as a function of C/N_0.

power. The squaring loss occurring in the LSQ estimation scheme is inevitable and no unbiased estimator exists, which is better than the LSQ estimator (provided that the correlation point is sufficiently near the true values).

4.7.2 Cost Function

The LSQ estimation of Section 4.3 minimizes a cost function defined as the sum of the squared residual signal samples. Assuming a known complex-valued amplitude, the cost function is given as

$$C(s_\mu; \hat{\tau}, \hat{\omega}) = \sum_{\mu=1}^{L} \left|(s_\mu - r_\mu(\hat{\tau}, \hat{\omega}))\right|^2 \tag{4.189}$$

and is evaluated to obtain Figures 4.8 and 4.9. Figure 4.8 refers to a C/N_0 of 42 dBHz, where the squaring loss can be ignored (see Figure 4.7). A clearly identifiable minima is visibly centered approximately around the true code-phase and Doppler values. By contrast, Figure 4.9 is obtained for a low signal power of 20 dBHz. For the left part of Figure 4.9, the minimum is centered in the upper left corner, indicating that the MLE gives useless estimation values.

Both figures are presented to demonstrate that iterating the LSQ estimation procedure is an unfavorable approach at low signal-to-noise ratios. Instead of converging near the true parameters, the iterative solution may not converge at all, thereby producing grossly erroneous estimates.

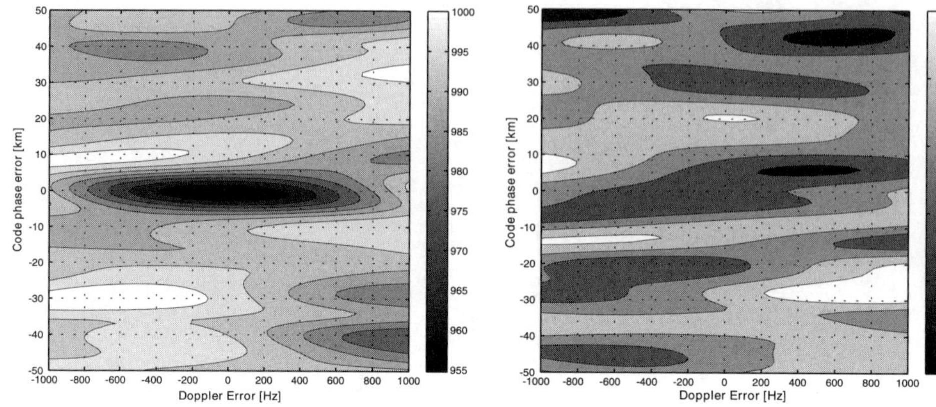

Figure 4.8 Exemplary code-phase/Doppler cost functions for $C/N_0 = 42$ dBHz.

4.7.3 LSQ Solution

We will now present a Monte Carlo simulation of the LSQ estimation scheme of Section 4.3.2 to obtain mean-values and variances of code-phase and Doppler estimates. A single iteration approach will be compared to multiple iterations. In the latter case, however, the maximum number of iterations is limited to five. The iterations start from differently chosen correlation points. Five scenarios are considered:

1. The starting correlation point coincides with the true values.
2. The starting correlation point differs in the code-phase direction by a uniformly distributed random variable in the range of –3,500 to 3,500m.
3. The starting correlation point differs in the Doppler direction by a uniformly distributed random variable in the range of –400 to 400 Hz.
4. The starting correlation point differs in code-phase direction by 3,000m.
5. The starting correlation point differs in the Doppler direction by 300 Hz.

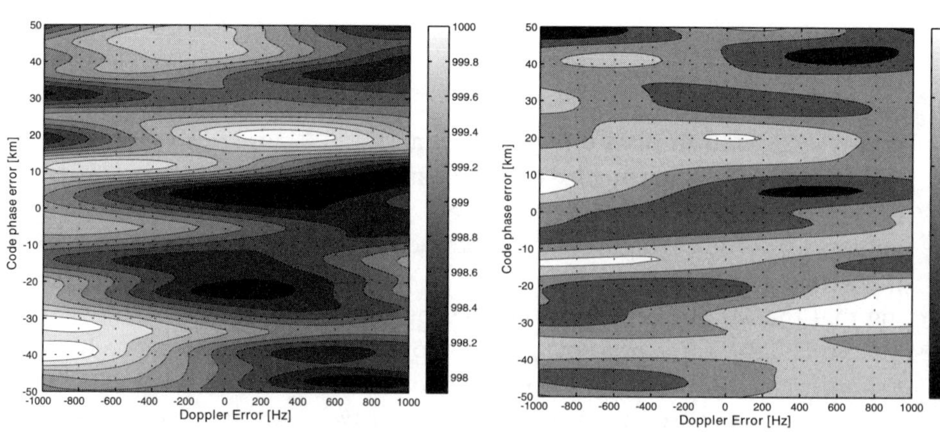

Figure 4.9 Exemplary code-phase/Doppler cost functions for $C/N_0 = 20$ dBHz.

For completeness, it should be noted that the initial carrier-phase value is taken as uniformly distributed in the range 0–2π, and the initial real-valued amplitude coincides with the true value. This information is, however, irrelevant and could be chosen differently because the complex amplitude estimation problem is linear.

The Monte Carlo simulation is performed for different signal power levels in the range of 35 to 55 dBHz. For each power level, a number of 3,000 Monte Carlo runs is performed and the mean values and variances of the estimates are calculated. As also discussed in Section 4.7.2, cases may occur where the iterative solution does not converge and gives obviously wrong estimates. Therefore, the admissible range of code-phase estimates is limited to be within ± 20 km of the true value and of the Doppler estimate to be within ± 20 kHz of the true values. The results are plotted in Figures 4.10 through 4.13 and are summarized in the following section. The figures show the mean value plus or minus the standard deviation plotted as filled polygons.

Figures 4.10 through 4.13 verify that choosing a correlation point different from the true values generally increases the variance of the obtained estimates. For zero initial errors, the variance of the code-phase estimate matches the TCLRB of Section 4.7.1. For high signal power levels (typically > 50 dBHz) the full iteration solution can compensate for an incorrectly chosen correlation point and identical estimates are obtained as for starting with the true values. For low signal power levels, the full iteration solution generally shows a worse performance than the single-iteration solution. Variances increase and biases occur.

When the offsets of the starting correlation point are chosen to be randomly distributed, the obtained estimates are still unbiased (although their variance is increased). In the case where the single-iteration solution is considered, an offset in the Doppler starting point causes a bias in the Doppler estimate, as shown in Figure

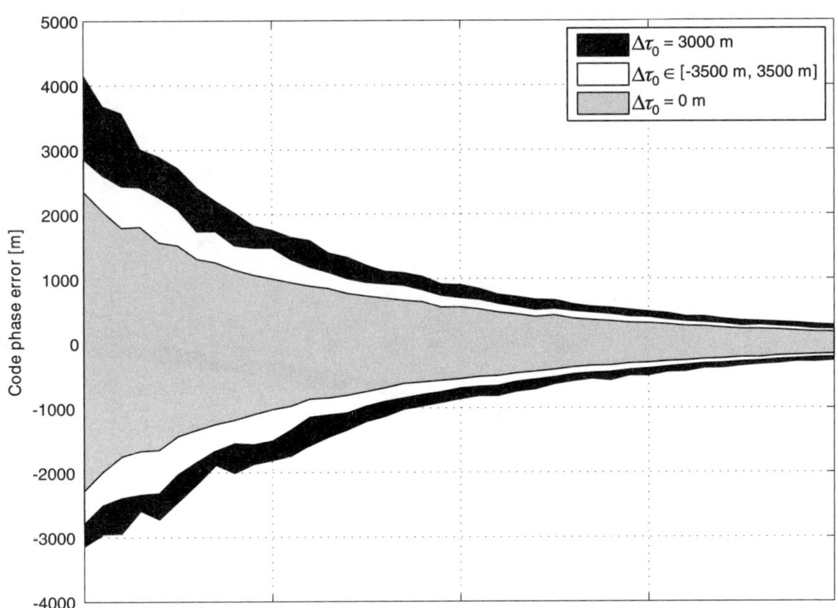

Figure 4.10 Code-phase estimation for various code-phase initial errors, one iteration.

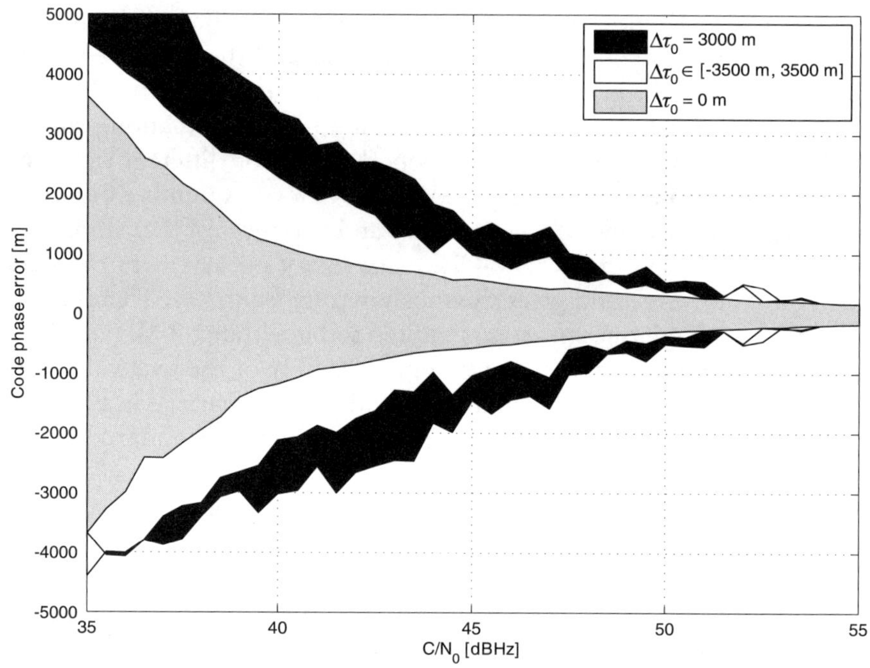

Figure 4.11 Code-phase estimation for various code-phase initial errors, max. five iterations.

4.12, even for high signal-to-noise ratios, whereas a code-phase offset in the starting point does not cause a bias in the code-phase estimate, as shown in Figure 4.10. The offsets were chosen to be outside the linearity region; therefore, the appearance of the Doppler offset is expected and understood, but the nonappearance of a code-phase offset needs further investigation, to be discussed below.

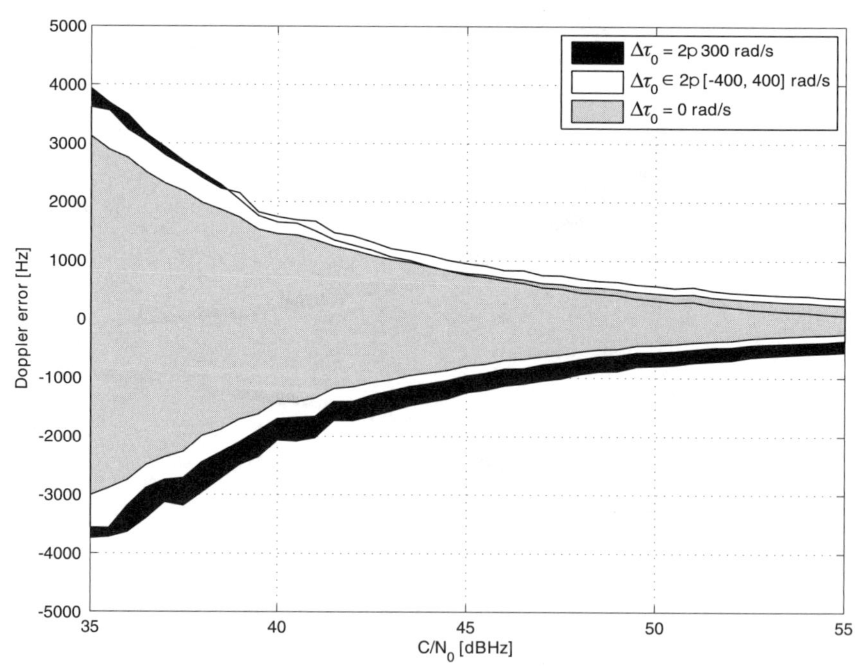

Figure 4.12 Doppler estimation for various Doppler initial errors, one iteration.

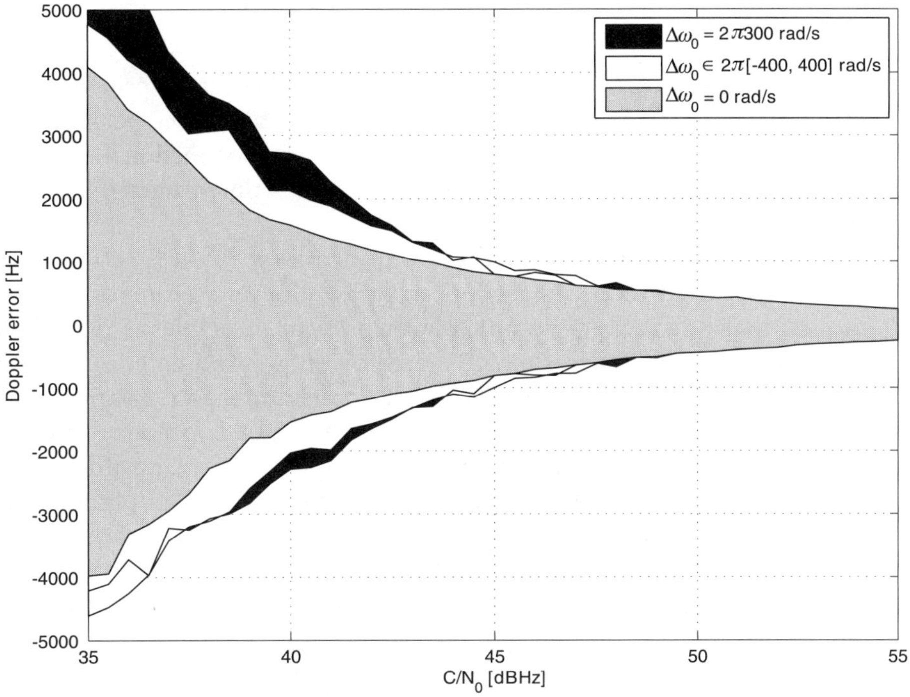

Figure 4.13 Doppler estimation for various Doppler initial errors, max. five iterations.

The nonappearance of the code-phase offset is explained by a compensation of the code-phase nonlinearities and the amplitude nonlinearites occurring during the calculation of the code-phase estimate (4.64). In fact, for the Gaussian double-pulse signal of Section 1.9.4, the ratio between the first derivative of the correlation function and the correlation function itself is linear in the code-phase error; that is,

$$\frac{R_{\bar{c},c}'(\Delta\tau)}{R_{\bar{c},c}(\Delta\tau)} = \frac{\partial}{\partial\Delta\tau}\log R_{\bar{c},c}(\Delta\tau) = g\Delta\tau \tag{4.190}$$

because the correlation function has a Gaussian bell shape of the form $\exp(g\Delta\tau^2/2)$. Here g is a constant value. Thus, the code-phase estimate is independent of the code-phase correlation point.

An investigation of the figures presented above can impose restrictions on the correlation points. For example, by requiring that biases must be smaller than a certain amount, the admissible range for the correlation point is limited. It should be mentioned that this kind of pre-evaluation is important in cases where the chosen tracking method might imply a correlation point that is biased with respect to the true values. Such tracking schemes are vector tracking (see Section 4.3.3.3) or may result from computational constraints of a GNSS SDR. For example, the internally generated reference signal may be calculated for a fixed Doppler and code-phase value and is *reused* to obtain multiple correlation values over a longer time period (see Section 9.6).

4.8 Discussion

This chapter presented and contrasted different theoretical principles to obtain position estimates from a received navigation signal: direct position estimation of Section 4.2.3, MLE or LSQ pseudorange estimation of Section 4.3, minimum mean-squared error estimation of Section 4.5.1, or a Kalman filter operating on a sufficient statistic of Section 4.5.2.

When comparing those theoretical principles to what is actually done in many GNSS navigation receivers for high-rate pseudorange estimation, it is interesting to note that normally none of these theoretical principles is directly applied. Instead, the receiver design is mostly based on an approximated single-iteration LSQ approach and the first derivative of the baseband signal is approximated by an early-minus-late replica difference. It is also common practice to determine multiple correlation values centered near the true values (e.g., a sufficient statistic) and then to construct code-phase, Doppler, carrier-phase, and amplitude estimators that are insensitive to certain effects, e.g., multipath, false-locks, carrier-phase loss, and many more. All those engineering approximations considerably improve the overall receiver performance, at the cost of an increased variance and the possible introduction of biases.

If, by contrast, truly minimum variance and unbiased estimates are desired, the receiver should estimate high-rate pseudorange parameters as follows. First, the correlation point should be within the linearity region. Second, the LSQ estimation scheme of Section 4.3 should be used with a single iteration. The LSQ estimation scheme itself is well-implementable for a GNSS SDR because the LSQ computational processing demands are negligible compared to the signal correlation (at least for the CPUs considered in chapter 3). The numerical evaluation of the TCRLB showed that the LSQ discriminator is optimal. Ensuring that the correlation point follows the true values requires a sophisticated receiver design, especially if the targeted operation scenario of the receiver does not allow independent channel tracking (see Section 4.3.3.1). Vector tracking, along with aiding information (e.g., IMU, stable clock, map-matching) is useful. Furthermore, the number and the approximate delays of possibly present multipath signals is valuable to overcome the singularity for near-range multipath mentioned in Sections 4.3.6 and 4.3.7. Section 8.3 introduces a multipath-estimating discriminator and gives practical implementation aspects.

If the relation between position parameters and pseudorange parameters is sufficiently linear (as is the case for GNSS signals), the optimality property of the LSQ pseudorange estimates transfers to the position estimates if all parameter correlations are properly accounted for.

As long as no navigation problem with high dynamics and high signal power is considered, the LSQ equations should *not* be iterated and the LSQ estimates are based on one iteration for the chosen correlation point. The code phase and Doppler of the correlation point typically derive from a predicted value based on the previous estimates. By contrast, the complex-valued amplitude is irrelevant for the correlation point, but is needed to set up the design matrix. A predicted complex-valued amplitude from previous estimates gives a coherent estimation scheme; estimating the complex-valued amplitude from the current epoch's samples gives a noncoherent scheme.

If the correlation point cannot be expected to be within the linear region, the proposed LSQ scheme produces suboptimal estimates. Suboptimal estimates will also be produced if the number of multipath signals is incorrectly estimated. Both situations should not occur for high signal-to-noise ratios. If they occur, Bayesian techniques, such as (sequential) Monte Carlo methods, may still be optimal because they do not require fulfilled linearity conditions. An alternative to Bayesian techniques is the increase of the coherent integration time. The increased integration time results in a synthetic antenna aperture, thereby already mitigating multipath signals. The long integration increases the energy of the received signals and the LSQ scheme might again be optimal. Pilot signals, short-term stable oscillators, a MEMS IMU, or difference correlators may help to increase the integration time.

References

[1] Poor, H. V., *An Introduction to Signal Detection and Estimation*, New York: Springer, 1988.

[2] Zhou, G., and G. B. Giannakis, "Harmonics in Gaussian Multiplicative and Additive Noise: Cramér–Rao Bounds," *IEEE Trans. Signal Processing*, Vol. 43, 1995, pp. 1217–1231.

[3] Lehmann, E. L., *Testing Statistical Hypothesis,* 2nd ed., New York: Wiley, 1986.

[4] Porat, B., *Digital Processing of Random Signals, Theory & Methods*, Englewood Cliffs, NJ: Prentice-Hall, 1994.

[5] D'Andrea, A. N., U. Mengali, and R. Reggiannini, "The Modified Cramér–Rao Bound and Its Application to Synchronization Problems," *IEEE Trans. Commun.*, Vol. 42, 1994, pp. 1391–1399.

[6] Gini, F., R. Reggiannini, and U. Mengali, "The Modified Cramér–Rao Bound in Vector Parameter Estimation," *IEEE Trans. Commun.*, Vol. 46, 1998, pp. 52–60.

[7] Moeneclaey, M., "On the True and the Modified Cramer-Rao Bounds for the Estimation of a Scalar Parameter in the Presence of Nuisance Parameters," *IEEE Trans. Commun.*, Vol. 46, 1998, pp. 1536–1544.

[8] Closas, P., et al., "Bayesian Direct Position Estimation," *Proc. 21st International Technical Meeting of the Satellite Division of the Institute of Navigation (ION-GNNS) 2008,* Savannah, GA, September 16–19, 2008, pp. 183–190.

[9] NXP Software, "NXP SnapSpot GPS Technology and JOBO photoGPS Capture a Location in an Instant," http://www.software.nxp.com/?pageid=139, 2007.

[10] Brown, A., P. Brown, and J. Griesbach, "GeoZigBee: A Wireless GPS Wristwatch Tracking Solution," *Proc. 19th Int. Technical Meeting of the Satellite Division of the Institute of Navigation (ION-GNNS) 2006*, Fort Worth, TX, September 26–29, 2006, pp. 2883–2888.

[11] Blewitt, G., "GPS Data Processing Methodology: From Theory to Applications," in *GPS for Geodesy*, pp. 233–270, Teunissen, P. J. G., and A. Kleusberg, (eds.), New York: Springer, 1998.

[12] Biberger, R., *Error Modelling of Pseudolite Signal Reception on Conducting Aircraft Surfaces*, University FAF Munich, Werner-Heisenberg-Weg 39, D-85577 Neubiberg, http://www.unibw.de/unibib/digibib/ediss/bauv, 2006.

[13] Varanasi, M. K., and B. Aazhang, "Multistage Detection in Asynchronous Code-Division Multiple-Access Communications," *IEEE Trans. Commun.*, Vol. 38, 1990, pp. 509–519.

[14] van Dierendonck, A. J., P. Fenton, and T. Ford, "Theory and Performance of Narrow Correlator Spacing in a GPS Receiver," *NAVIGATION, Journal of The Institute of Navigation*, Vol. 39, No. 3, 1992, pp. 265–283.

[15] Misra, P., and P. Enge, *Global Positioning System: Signals, Measurements, and Performance,* 2nd ed., Lincoln: Ganga-Jamuna Press, 2006.

[16] Pany, T., and B. Eissfeller, "Code and Phase Tracking of Generic PRN Signals with Sub-Nyquist Sample Rates," *NAVIGATION, Journal of The Institute of Navigation*, Vol. 51, No. 2, 2004, pp. 143–159.

[17] Spilker, J. J., Jr., "GPS Signal Structure and Theoretical Performance," in *Global Positioning System: Theory and Applications, Vol. I*, pp. 57–120, Parkinson, B. W., and J. J. Spilker, (eds.), Washington, D.C.: American Institute of Aeronautics and Astronautics Inc., 1996.

[18] Rife, D., and R. Boorstyn, "Single Tone Parameter Estimation from Discrete-Time Observations," *IEEE Trans. Information Theory*, Vol. 20, No. 5, 1974, pp. 591–598.

[19] Pany, T., and B. Eissfeller, "Use of a Vector Delay Lock Loop Receiver for GNSS Signal Power Analysis in Bad Signal Conditions," *PLANS 2006, IEEE/ION Position, Location and Navigation Symposium*, San Diego, CA, April 25–27, 2006, pp. 893–903.

[20] Kaplan, E. D., and C. J. Hegarty, (eds.), *Understanding GPS: Principles and Applications,* 2nd ed., Norwood, MA: Artech House, 2006.

[21] Thomas, J. B. Jr., inventor. California Institute of Technology, assignee, "Digital Signal Processor and Processing Method for GPS Receivers," U.S. Patent No. 4821294, 1989.

[22] Jaffe, R., and E. Rechtin, "Design and Performance of Phase-lock Circuits Capable of Near-Optimum Performance over a Wide Range of Input Signal and Noise Levels," *IEEE Trans. Information Theory*, Vol. 1, 1955, pp. 66–76.

[23] van Dierendonck, A. J., "GPS Receivers," in *Global Positioning System: Theory and Applications, volume I,* pp. 329–407, Parkinson, B. W., and J. J. Spilker, (eds.), Washington, D.C.: American Institute of Aeronautics and Astronautics Inc., 1996.

[24] Tsui, J. B. Y., *Fundamentals of Global Positioning System Receivers: A Software Approach,* 2nd ed., New York: Wiley, 2005.

[25] Eissfeller, B., *Schriftenreihe der Universität der Bundeswehr (55): Ein dynamisches Fehlermodell für GPS Autokorrelationsempfänger*. University of Federal Armed Forces Munich, Werner-Heisenberg-Weg 39, D-85577 Neubiberg, 1997.

[26] Kazemi, P. L., "Optimum Digital Filters for GNSS Tracking Loops," *Proc. 21st Int. Technical Meeting of the Satellite Division of the Institute of Navigation (ION-GNNS) 2008,* Savannah, GA, September 16–19, 2008, pp. 2304–2313.

[27] Won, J. H., T. Pany, and B. Eissfeller, "Design of a unified MLE tracking for GPS/Galileo software receivers," *Proc. 19th Int. Technical Meeting of the Satellite Division of the Institute of Navigation (ION-GNNS) 2006*, Fort Worth, TX, September 26–29, 2006, pp. 2396–2406.

[28] Nunes, F. D., F. M. G. Sousa, and J. M. N. Leitao, "BOC/MBOC Multicorrelator Receiver with Least-Squares Multipath Mitigation Technique," *Proc. 21st Int. Technical Meeting of the Satellite Division of the Institute of Navigation (ION-GNNS) 2008,* Savannah, GA, September 16–19, 2008, pp. 652–662.

[29] Won, J. H., T. Pany, and B. Eissfeller, "Implementation, Verification and Test Results of a MLE-Based F-Correlator Method for Multi-Frequency GNSS Signal Tracking," *Proc. 20th Int. Technical Meeting of the Satellite Division of the Institute of Navigation (ION-GNNS) 2007*, Fort Worth, TX, September 25–28, 2007, pp. 2237–2249.

[30] Spilker, J. J. Jr., "Fundamentals of Signal Tracking Theory," in *Global Positioning System: Theory and Applications, Vol. I*, pp. 245–328, Parkinson, B. W., and J. J. Spilker, (eds.), Washington, D.C.: American Institute of Aeronautics and Astronautics Inc., 1996.

[31] Alban, S., D. M. Akos, and S. M. Rock, "Performance Analysis and Architectures for INS-Aided GPS Tracking Loops," *Proc. Institute of Navigation National Technical Meeting (ION-NTM) 2003*, San Diego, CA, January 22–24, 2003, pp. 611–622.

[32] Landis, D., et al., "A Deep Integration Estimator for Urban Ground Navigation," *PLANS 2006, IEEE/ION Position, Location and Navigation Symposium*, San Diego, CA, April 25–27, 2006, pp. 927–932.

[33] Groves, P. D., C. J. Mather, and A. A. Macaulay, "Demonstration of Non-Coherent Deep INS/GPS Integration for Optimised Signal-to-Noise Performance," *Proc. 20th Int. Technical Meeting of the Satellite Division of the Institute of Navigation (ION-GNNS) 2007*, Fort Worth, TX, September 25–28, 2007, pp. 2627–2638.

[34] Lashley, M., D. M. Bevly, and J. Y. Hung, "Performance Analysis of Vector Tracking Algorithms for Weak GPS Signals in High Dynamics," *Journal of Selected Topics in Signal Processing (J-STSP) (GNSS and Robust Navigation)*, 2009.

[35] Anghileri, M., et al., "An Algorithm for Bit Synchronization And Signal Tracking In Software GNSS Receivers," *Proc. 19th Int. Technical Meeting of the Satellite Division of the Institute of Navigation 2006*, Fort Worth, TX, September 26–29, 2006, pp. 1836–1848.

[36] Ziedan, N. I. and J. L. Garrison, "Bit Synchronization and Doppler Frequency Removal at Very Low Carrier to Noise Ratio Using a Combination of the Viterbi Algorithm with an Extended Kalman Filter," *Proc. 16th Int. Technical Meeting of the Satellite Division of the Institute of Navigation (ION-GPS) 2003*, Portland, OR, September 9–12, 2003, pp. 616–627.

[37] Ávila Rodríguez, J. Á., T. Pany, and G. Hein, "Bounds on Signal Performance Regarding Multipath-Estimating Discriminators," *Proc. 19th Int. Technical Meeting of the Satellite Division of the Institute of Navigation (ION-GNNS) 2006*, Fort Worth, TX, September 26–29, 2006, pp. 1710–1722.

[38] Lohan, E.-S., R. Hamila, and M. Renfors, "Cramer Rao Bound for Multipath Time Delays in a DS-CDMA System," *Proc. 4th Int. Symp. Wireless Personal Multimedia Communications (WPMC'01)*, Aalborg, September 2001, pp. 1043–1047.

[39] Minkler, G., and J. Minkler, *Theory and Application of Kalman Filtering*, Palm Bay: Magellan Book Company, 1993.

[40] Arulampam, M. S., et al., "Tutorial on Particle Filters for Online Nonlinear/Non-Gaussian Bayesian Tracking," *IEEE Trans. Signal Processing*, Vol. 50, 2002, pp. 174–188.

[41] Closas, P., et al., "Multipath Mitigation Using Particle Filtering," *Proc. 19th Int. Technical Meeting of the Satellite Division of the Institute of Navigation (ION-GNSS) 2006*, Fort Worth, TX, September 26–29, 2006, pp. 1733–1740.

[42] Lentmaier, M., et al., "Dynamic Multipath Estimation by Sequential Monte Carlo Methods," *Proc. 20th Int. Technical Meeting of the Satellite Division of the Institute of Navigation (ION-GNSS) 2007*, Fort Worth, TX, September 25–28, 2007, pp. 1712–1721.

[43] Psiaki, M. L., "Block Acquisition of Weak GPS Signals in a Software Receiver," *Proc. 14th Int. Technical Meeting of the Satellite Division of the Institute of Navigation (ION-GPS) 2001*, Salt Lake City, UT, September 11–14, 2001, pp. 2838–2850.

[44] Humphreys, T. E., et al., "GPS Carrier Tracking Loop Performance in the Presence of Ionospheric Scintillations," *Proc. 18th Int. Technical Meeting of the Satellite Division of the Institute of Navigation (ION-GNNS) 2005*, Long Beach, CA, September 13–16, 2005, pp. 156–167.

[45] Petovello, M. G., C. O'Driscoll, and G. Lachapelle, "Weak Signal Carrier Tracking Using Extended Coherent Integration With an Ultra-Tight GNSS/IMU Receiver," *Proc. European Navigation Conference (ENC-GNSS) 2008*, Toulouse, April 22–25, 2008.

[46] Niedermeier, H., et al., "Reproduction of User Motion and GNSS Signal Phase Signatures Using MEMS INS and a Pedestrian Navigation System for HS-GNSS Applications," *Proc. 4th ESA Workshop on Satellite Navigation User Equipment Technologies, NAVITEC*, Noordwijk, the Netherlands, December 10–12, 2008.

[47] Pany, T., M. Paonni, and B. Eissfeller, "Synthetic Phased Array Antenna for Carrier/Code Multipath Mitigation," *Proc. European Navigation Conference (ENC-GNNS) 2008*, Toulouse, April 22–25, 2008.

CHAPTER 5

Signal Detection

Position determination is essentially an estimation problem, using the received signal samples to obtain a good position estimate, as described in Chapter 4. Position estimation also includes the estimation of signal power values (see Section 4.3.2.9) and one may use this information to decide if the received navigation signals are sufficient in number and strength so that the obtained position estimate can be considered valid. On the other hand, this argument can be seen as a signal-detection problem: deciding between different hypotheses whether one or more navigation signals are received at all or if only noise is received.

In the following discussion, a statistical test (more specifically, a binary-hypothesis-testing problem) is set up, where a hypothesis H_0 is contrasted to a hypothesis H_1. They are defined as:

- H_0: Only noise is received
- H_1: A navigation signal is received

A definition of the term *navigation signal* in the context of signal detection is given in Section 5.2. Mathematically, a statistical test is a function $\Phi(\mathbf{S})$ of the signal samples $\mathbf{S}$ that may take two values: zero or one. If $\Phi(\mathbf{S})$ assumes the value zero, the test decides that the signal samples stem from a distribution belonging to the hypothesis H_0 and vice versa. The probability of correctly identifying that H_1 is true shall be called *probability of detection* p_d (or power) and is given as

$$p_d = \left\langle \Phi(\mathbf{S}) \right\rangle_{\mathbf{S}} \qquad \mathbf{S}: \text{signal parameters comply to } H_1 \tag{5.1}$$

The probability to incorrectly decide for H_1 although H_0 is true is called *false alarm rate* p_{fa} (or significance level or size) and is given as

$$p_{fa} = \left\langle \Phi(\mathbf{S}) \right\rangle_{\mathbf{S}} \qquad \mathbf{S}: \text{signal parameters comply to } H_0 \tag{5.2}$$

Different detection schemes can be chosen that are either optimized according to certain criteria described in Section 5.1.1 or follow an engineering approach based on the ML estimation of Section 5.7. The latter approach also gives estimated signal parameter values that can be used as starting values for an estimation algorithm.

5.1 Detection Principles

The basic statistical groundwork for the design of detectors of a signal-in-noise problem follows directly from the theory of hypothesis testing. One approach, which will be addressed, is simple hypothesis testing in which the distribution of the signal samples S is assumed to be completely known. By contrast, for composite hypothesis testing, the distribution of the signal samples depends on one or more unknown parameters, such as the code phase or Doppler.

5.1.1 Simple Hypothesis Testing

Simple hypothesis testing is a less-commonly used approach in a navigation receiver because the basic signal sample distribution is parameter-dependent, as can be seen, for example, in (1.18). However, a parameter-independent distribution can be obtained by integrating the parameters out. This requires knowledge of the a priori distribution of the unknown parameters. The resulting sample distribution function is parameter-independent and cannot be used for any parameter estimation technique. However, the simple hypothesis-testing approach gives theoretical insight into the detection problem and may have practical applications for signal detection verification.

A probability density function under H_1 that is suitable for simple hypothesis testing is obtained from the parameterized distribution $p_{\mathbf{x}}(\mathbf{S})$ of (4.11) by treating all parameters as nuisance parameters (see Section 4.1.2) and integrating them out as described in Section II.E of the work by Poor [1]

$$p_{H_1}(\mathbf{S}) = \int\limits_{\mathbf{x} \in \Lambda} p_{\mathbf{x}}(\mathbf{S}) p(\mathbf{x}) d\mathbf{x} \tag{5.3}$$

The integral is performed over all admissible values for $\mathbf{x}$ and the a priori distribution $p(\mathbf{x})$ of the parameters $\mathbf{x}$ has to be known. The parameters can either be position parameters or high-rate pseudorange parameters, as discussed in Section 5.2. In the simple hypothesis-testing approach, we test against a distribution that is valid if no signal is present (hypothesis H_0). For both hypotheses, we assume that the noise parameters (e.g., its variance) are known; then, the sample distribution under H_0 does not depend on any parameter. The simple hypothesis approach is summarized as

$$\begin{aligned} H_1 &\leftrightarrow p_{H_1}(\mathbf{S}) \\ H_0 &\leftrightarrow p_{H_0}(\mathbf{S}) \end{aligned} \tag{5.4}$$

Bayesian and Neyman–Pearson detectors can be employed for this problem. The first detector minimizes the Bayesian risk; the second is based on the Neyman–Pearson criterion and optimizes the detection probability under the constraint of a limited false-alarm rate. Both tests are based on the likelihood ratio $L(\mathbf{s})$ and compare this ratio to a threshold γ. The thresholds are different, depending on the chosen approach (Bayesian or Neyman–Pearson) as described, for example, in Section 3 of Kay's book [2]. For a given set of measured sample $\mathbf{s}$, the hypothesis H_1 is chosen if

$$L(\mathbf{s}) = \frac{p_{H_1}(\mathbf{s})}{p_{H_0}(\mathbf{s})} > \gamma \tag{5.5}$$

otherwise H_0 is chosen. Both detectors, Bayesian and Neyman–Pearson, are optimal (minimum risk or maximum probability of detection) and are comparably simple. However, a significant computational difficulty arises because it is virtually impossible to evaluate (5.3) in a closed form. For a more detailed introduction to simple hypothesis testing, the reader is referred to the textbooks by Poor and Kay [1, 2].

5.1.2 Composite Hypothesis Testing

Composite hypothesis testing refers to a situation where the sample distribution depends on one or more parameters. In the case of a navigation signal, this is the normal situation under H_1 because the fundamental signal parameters (position or code phase, Doppler, amplitude, and carrier-phase values) are generally unknown. In the following, we use the notation

$$p_{H_1}\left(\mathbf{S}\middle|\mathbf{x}\right) \equiv p_{\mathbf{x}}(\mathbf{S}) \tag{5.6}$$

to denote the sample distribution dependency on the parameters $\mathbf{x}$. The hypothesis H_1 is represented by a family of distributions indexed by the values of the signal parameters $\mathbf{x}$. All admissible parameter values $\mathbf{x}$ for H_1 must correspond to a situation where a signal is actually present (i.e., the signal amplitude must be larger than zero). By contrast, the hypothesis H_0 may describe either a situation where only noise is received or a situation where noise plus further signals (e.g., signals from different transmitters) are received. Thus, the H_0 sample distribution may either be completely known or parameter-dependent.

5.1.2.1 Conversion from Composite to Simple Hypothesis Testing

If the a priori distributions of the signal parameters are known, then the composite hypothesis-testing problem can be converted to a simple hypothesis-testing problem, as described in Section 5.1.1, by integrating the parameters out. Therefore, simple and composite hypothesis testing are related. Elimination of signal parameters from the sample distribution may also be done for only a subset of parameters, as described in Sections 5.4, 5.7.1, 5.7.2.1, and 5.7.4.1. Finally, we understand under a *true composite* hypothesis-testing problem a situation in which the sample distribution depends on at least one parameter (i.e., a situation where not all parameters are integrated out). In Section 5.1.2.2, true composite hypothesis-testing problems are treated as Neyman–Pearson problems in an attempt to optimize the detection probability.

5.1.2.2 Nonexistence of a Uniform Most Powerful Test

For a true composite hypothesis-testing problem, it would be desirable to construct a test whose performance is optimal for all possible signal parameters. Because the signal parameters are unknown, the formula for the test statistics needs to be obviously independent of the (unknown) signal parameter's values and needs to be a function only of the signal samples. Furthermore, if the test yields a maximum detection probability (under the assumption of a limited false alarm rate) it would be called a uniform most-powerful test.

It seems impossible, however, to construct a composite hypothesis test that is optimal for all admissible signal parameters if navigation signal parameters like code phase and Doppler are unknown. A uniform most powerful test would exist only if the distribution depends on only a single parameter, the test is single-sided, and the likelihood ratio is a monotone nondecreasing function of a test statistic (see Section 4.7 of [3] or Theorem 2 in Chapter 3 of Lehmann's work [4]). Such a case, however, is not applicable here. See also Chapter 3 of Lehmann's work for a discussion of dual-sided tests. Sections 5.4 and 5.7.2.2 further discuss uniform most-powerful tests for navigation signal detection.

5.1.2.3 Uniform Most-Powerful Unbiased Test

A simple condition that one may wish to impose on a hypothesis test is that, for all admissible parameter values, the probability of detection should be larger than the false alarm rate. Unless such a condition is satisfied, there would exist specific signal parameter values where it is more likely to not detect the signal when it is present than to incorrectly detect it when it is not present. Such a test would not make sense, and this class of tests can be excluded from our considerations.

More specifically, we adopt the definition from Chapter 4 of Lehmann for an unbiased test [4]. A hypothesis test that fulfills

$$p_{fa} \leq 1-\alpha \leq p_d \tag{5.7}$$

for some constant α between zero and one is called unbiased.

If we limit ourselves to tests that are unbiased, an optimal test (i.e., a uniform most-powerful test) can be found more easily. This kind of test is then called a uniform most-powerful unbiased test (UMPUT) and is the best possible test among all unbiased tests.

5.1.2.4 Generalized Likelihood-Ratio Test

A common approach to handle a composite hypothesis-testing problem is to construct a generalized likelihood-ratio test (GLRT). A GLRT is based on the likelihood ratio (5.5) and replaces the unknown parameters with their ML estimates. Although there is no optimality associated with the GLRT, it appears to work quite well and is the commonly used basis for signal acquisition in a GNSS receiver (see Section 6.4.2. in the book by Kay [2]). For a given set of measured sample s, the GLRT decides H_1 if

$$L(\mathbf{s}) = \frac{p_{H_1}\left(\mathbf{s}\middle|\hat{\mathbf{x}}_{H_1},\hat{\boldsymbol{\xi}}_{H_1}\right)}{p_{H_0}\left(\mathbf{s}\middle|\hat{\boldsymbol{\xi}}_{H_0}\right)} > \gamma \tag{5.8}$$

Here, the parameters have been divided into $\mathbf{x}$ and ξ, where $\mathbf{x}$ refers to parameters of the signal to be detected and ξ refers to parameters related to additional signals or to noise. The symbol ^ indicates that the ML estimates are used to evaluate the sample distribution (see Section 4.2.3). The ML estimates for ξ obtained under the assumption H_0 or H_1 are, in general, different. This detection approach also provides the estimated parameter values $\hat{\mathbf{x}}$ that can, in turn, be used for starting a signal estimation algorithm.

5.2 Detection Domains

Similar to Sections 4.2.3 and 4.2.4, the detection problem can be formulated in terms of generalized position parameters or in terms of generalized pseudorange parameters. In the first case, the navigation signal is defined on the *position domain*, whereas, in the latter case, the navigation signal is defined on the *pseudorange domain*.

5.2.1 Pseudorange Domain Detection

Navigation signal detection in the pseudorange domain usually implies that the fundamental signal parameters are code phase, Doppler, carrier phase, and signal amplitude, as described in Section 1.8. If pseudorange domain detection is considered—the standard case for a GNSS receiver—a situation arises where a superposition of many navigation signals is received and their mutual influence has to be properly accounted for; this will be described in Section 5.7.3.

5.2.2 Position Domain Detection

For position domain detection, the signals of all transmitters collapse into one signal, which is the superposition of all signals. The combined signal parameters are the generalized position parameters, as introduced in Section 4.1.1. Signal detection in the position domain has similar advantages as single-step position estimation for snapshot processing, as described in Section 4.2.3. The signal power of the individual signals accumulates and the combined signal has a higher signal-to-noise ratio than the single signals.

5.3 Preprocessing

If the implementation of signal-detection (or acquisition) algorithms in a navigation receiver is considered, two important issues are considered. First, signal detection is computationally cumbersome if the signal amplitude is below or similar to the noise amplitude. The detection involves a search in the space of unknown parameters and a multitude of test parameters have to be evaluated before the presence of a navigation signal can be declared. Second, during signal detection it is generally sufficient to estimate signal parameters with a low accuracy. In the context of ML estimation, signal detection can be considered to be a coarse synchronization and signal estimation provides, via fine synchronization, precise parameter estimates.

Consequently, it proves useful to employ a number of preprocessing steps during signal detection that aim at a computational burden reduction at the cost of a decreased accuracy of the coarse synchronization parameters. Those steps may include the following points in a suitable order:

- Frequency conversion (e.g., from IF to baseband);
- Resampling (e.g., to allow FFT techniques);
- Filtering (e.g., to reduce the signal bandwidth and consequently the working sample rate).

The first two points are more technical but may introduce losses in the effective signal power due to implementation constraints. Those effects are not discussed here. By contrast, filtering is of special importance because the immense computational load can be drastically reduced if the signal bandwidth can be reduced. This allows for using a lower sample rate. For example, a GNSS signal acquisition algorithm exists that makes use of only a fraction of the full signal bandwidth. Typically this occurs when the spectral main lobe is used to acquire a BPSK signal. The BOC side-lobe acquisition algorithm effectively uses only one side lobe of a BOC signal and discards the second lobe [5]. For a BOC(n,n) signal, the sample rate is reduced by a factor of two, the signal power loss is around 3 dB, and the working correlation function has a simple triangular shape compared to the full BOC autocorrelation function. For direct GPS P(Y) code acquisition, it might be convenient to work with only a fraction μ of the bandwidth of the main lobe (the bandwidth being centered around the carrier frequency). The work sample rate can then be reduced by a factor of μ and the signal power is also reduced by a factor of μ. This signal power loss can, however, be compensated by increasing the coherent integration time by μ. The same detection sensitivity is achieved, but during a signal acquisition step a μ-times-larger portion of all code-phase values is tested [6–8].

In Section 5.4, these preprocessing steps will not be mentioned explicitly, but it should be noted that the considered navigation signal might have been subject to one or more of the above-mentioned preprocessing steps.

5.4 Clairvoyant Detector for Uniformly Distributed Phase

The term *clairvoyant detector* was introduced in Chapter 6 of Kay's book and refers to a true composite hypothesis-testing problem [2]. The detector makes use of the otherwise unknown signal parameter values. Generally, the clairvoyant detector is of a hypothetical nature and does not exist in practice.

The clairvoyant detector has theoretical importance and may serve as a reference detector in the same sense as the Cramér–Rao lower bound defines a reference for several estimation strategies. In addition, the clairvoyant detector can be used if the unknown signal parameters are available from an external source. This is of particular importance if vector tracking, as described in Section 4.3.3.3, is used.

To give an example of a clairvoyant detector, we assume that a single navigation signal is present (or not) and the signal parameters (see Section 1.8) as well as the Gaussian noise distribution parameters (see Section 1.7) are assumed to be known.

The sample distribution under H_1 is

$$p_{H_1}\left(\mathrm{s}\middle|a',\tau,\omega,\varphi\right)=\frac{1}{(2\pi)^L}\exp\left\{-\frac{1}{2}\sum_{\mu=1}^{L}\left|s_\mu-a'c(t_\mu-\tau)\exp\{i\omega t'_\mu-i\varphi\}\right|^2\right\} \tag{5.9}$$

Here, a' denotes the real-valued signal amplitude.

For this example, we assume that the carrier phase φ is uniformly distributed between 0 and 2π. The sample distribution for a uniformly distributed carrier phase is obtained by integrating out the carrier phase, yielding

$$\begin{aligned}
p_{H_1}\left(\mathrm{s}\middle|a',\tau,\omega\right) &= \frac{1}{(2\pi)^L}\frac{1}{2\pi}\int_{\varphi=0}^{2\pi}\exp\left\{-\frac{1}{2}\sum_{\mu=1}^{L}\left|s_\mu - a'c(t_\mu-\tau)\exp\{i\omega t'_\mu - i\varphi\}\right|^2\right\}d\varphi \\
&= \frac{1}{(2\pi)^{L+1}}\int_{\varphi=0}^{2\pi}\exp\left\{-\frac{1}{2}\sum_{\mu=1}^{L}\left|s_\mu\right|^2 - \frac{a'^2}{2}\sum_{\mu=1}^{L}\left|c(t_\mu-\tau)\right|^2\right. \\
&\quad \left. + a'\sum_{\mu=1}^{L}\mathrm{Re}\{\bar{s}_\mu c(t_\mu-\tau)\exp\{i\omega t'_\mu - i\varphi\}\}\right\}d\varphi \\
&= \frac{1}{(2\pi)^{L+1}}\exp\left\{-\frac{1}{2}\sum_{\mu=1}^{L}\left|s_\mu\right|^2 - \frac{La'^2}{2}\right\} \\
&\quad \times\int_{\varphi=0}^{2\pi}\exp\left\{a'\sum_{\mu=1}^{L}\mathrm{Re}\{\bar{s}_\mu c(t_\mu-\tau)\exp\{i\omega t'_\mu - i\varphi\}\}\right\}d\varphi \\
&= \frac{1}{(2\pi)^{L}}\exp\left\{-\frac{1}{2}\sum_{\mu=1}^{L}\left|s_\mu\right|^2 - \frac{La'^2}{2}\right\}I_0\left(a'\left|P(\tau,\omega)\right|\sqrt{L}\right)
\end{aligned} \tag{5.10}$$

with the definition for the correlator value $P(\tau,\omega)$ as

$$P(\tau,\omega) = \frac{1}{\sqrt{L}}\sum_{\mu=1}^{L}\bar{s}_\mu c(t_\mu-\tau)\exp\{i\omega t'_\mu\} \tag{5.11}$$

The likelihood ratio (5.5) evaluates to

$$\begin{aligned}
L(\mathbf{s}) = \frac{p_{H_1}(\mathbf{s})}{p_{H_0}(\mathbf{s})} &= \frac{\exp\left\{-\frac{1}{2}\sum_{\mu=1}^{L}\left|s_\mu\right|^2 - \frac{La'^2}{2}\right\}I_0\left(a'\left|P(\tau,\omega)\right|\sqrt{L}\right)}{\exp\left\{-\frac{1}{2}\sum_{\mu=1}^{L}\left|s_\mu\right|^2\right\}} \\
&= \exp\left\{-\frac{La'^2}{2}\right\}I_0\left(a'\left|P(\tau,\omega)\right|\sqrt{L}\right)
\end{aligned} \tag{5.12}$$

Because $a' > 0$ and I_0 (the modified-Bessel function of first kind and order zero) is a monotone increasing function, the clairvoyant detector decides H_1 if

$$\left|P(\tau,\omega)\right|^2 > \gamma \tag{5.13}$$

The threshold γ has to be chosen according to the desired optimality criterion. In the following discussion, the Neyman–Pearson principle will be employed and it will be shown with (5.15) that the threshold γ is independent of the signal amplitude a'.

Under hypothesis H_0, P is the sum of L complex zero-mean Gaussian random variables because, in this case, $S_\mu = N_\mu$. The sum is itself a zero-mean Gaussian random variable with a complex variance of two because

$$
\begin{aligned}
H_0 : \left\langle \left|P(\tau,\omega)\right|^2 \right\rangle_{\mathrm{N}} &= \left\langle \left| \frac{1}{\sqrt{L}} \sum_{\mu=1}^{L} N_\mu c(t_\mu - \tau)\exp\{i\omega t'_\mu\} \right|^2 \right\rangle_{\mathrm{N}} = \\
&= \left\langle \frac{1}{L} \sum_{\mu,\upsilon=1}^{L} N_\mu \bar{N}_\upsilon c(t_\mu - \tau)\bar{c}(t_\upsilon - \tau)\exp\{i\omega(t'_\mu - t'_\upsilon)\} \right\rangle_{\mathrm{N}} \\
&= \frac{2}{L} \sum_{\mu=1}^{L} \left|c(t_\mu - \tau)\right|^2 = 2
\end{aligned}
\tag{5.14}
$$

Here, we assume that the assumption of Section 1.8.1 holds true. According to Appendix A.4.6, $|P(\tau,\omega)|^2$ follows a central chi-squared distribution with two degrees of freedom and the false alarm rate is given as

$$
p_{fa} = P\left(\left|P(\tau,\omega)\right|^2 > \gamma \mid H_0\right) = Q_{\chi^2;2}(\gamma) = e^{-\gamma/2}
\tag{5.15}
$$

Under hypothesis H_1, $P(\tau,\omega)$ retains the same variance but the squared magnitude of its mean value equals La'^2 because

$$
\begin{aligned}
H_1 : \left|P(\tau,\omega)\right|^2 &= \frac{1}{L}\left| \sum_{\mu=1}^{L} (a'\bar{c}(t_\mu - \tau)\exp\{-i\omega t'_\mu - i\varphi\} + N_\mu) c(t_\mu - \tau)\exp\{i\omega t'_\mu\} \right|^2 \\
&= \frac{1}{L}\left| \sum_{\mu=1}^{L} (a'\bar{c}(t_\mu - \tau)c(t_\mu - \tau)\exp\{-i\varphi\} + N_\mu c(t_\mu - \tau)\exp\{i\omega t'_\mu\}) \right|^2 \\
&= \left| \sqrt{L}a'\exp\{-i\varphi\} + \frac{1}{\sqrt{L}} \sum_{\mu=1}^{L} N_\mu c(t_\mu - \tau)\exp\{i\omega t'_\mu\} \right|^2
\end{aligned}
\tag{5.16}
$$

Using the definitions of Section 2.2.3 of Kay's book summarized in Appendix A.4.6, $|P(\tau,\omega)|^2$ belongs to the noncentral chi-squared distribution with two degrees of freedom and a noncentrality parameter La'^2. Therefore, the probability of detection is

$$
\begin{aligned}
p_d &= P\left(\left|P(\tau,\omega)\right|^2 > \gamma \mid H_1\right) = Q_{\chi^2;2,La'^2}(\gamma) = Q_{\chi^2;2,\frac{2LC/N_0}{f_s}}(\gamma) \\
&= Q_{\chi^2;2,2T_{coh}C/N_0}(\gamma) = Q_{\chi^2;2,2T_{coh}C/N_0}(-2\log p_{fa})
\end{aligned}
\tag{5.17}
$$

Here, we used the relationship between the signal amplitude a' and the signal-to-noise ratio C / N_0 from Section 1.8.1.

5.5 Energy Detector

The energy detector uses the sum of squared signal samples to determine if a signal is present. The averaged sum of the squared signal samples is an estimate of the received signal energy. The energy detector is eventually the simplest detector that can be used to detect the presence of a navigation signal. The energy detector's performance is comparably low. The design of the energy detector is not related to any optimality criterion and it belongs to the class of simple hypothesis-testing detectors.

The energy detector decides for H_1 if

$$E = \sum_{\mu=1}^{L} \left| s_\mu \right|^2 > \gamma \tag{5.18}$$

for a suitable chosen threshold γ. Here, the Neyman–Pearson principle will be employed to determine the threshold.

Under the hypothesis H_0, the energy E is the sum of L complex Gaussian samples, whose real and imaginary components are each of variance one and zero mean,

$$H_0 : E = \sum_{\mu=1}^{L} \left| n_\mu \right|^2 \tag{5.19}$$

According to Appendix A.4.6, E belongs to a central chi-squared distribution with $2L$ degrees of freedom and the false alarm probability is given as

$$p_{fa} = P(E > \gamma \mid H_0) = Q_{\chi^2;2L}(\gamma) \approx Q\left(\frac{\gamma - 2L}{\sqrt{4L}} \right) \tag{5.20}$$

Equation (5.20) indicates that the chi-squared probability density function can be well approximated by the right-tail probability function of the normal distribution because of the usually large number of involved samples.

Under the hypothesis H_1, the energy E can be written as

$$\begin{aligned} H_1 : E &= \sum_{\mu=1}^{L} \left| a' c(t_\mu - \tau) \exp\{i\omega t'_\mu - i\varphi\} + n_\mu \right|^2 \\ &= \sum_{\mu=1}^{L} (\mathrm{Re}\{a' c(t_\mu - \tau) \exp\{i\omega t'_\mu - i\varphi\}\} + \mathrm{Re}\{n_\mu\})^2 + \\ &+ \sum_{\mu=1}^{L} (\mathrm{Im}\{a' c(t_\mu - \tau) \exp\{i\omega t'_\mu - i\varphi\}\} + \mathrm{Im}\{n_\mu\})^2 \end{aligned} \tag{5.21}$$

According to Appendix A.4.6, E belongs (under hypothesis H_1) to the noncentral chi-squared distribution with $2L$ degrees of freedom and a noncentrality parameter λ of

$$\begin{aligned}\lambda &= \sum_{\mu=1}^{L} \left(\mathrm{Re}\{a'c(t_\mu - \tau)\exp\{i\omega t'_\mu - i\varphi\}\}^2 + \mathrm{Im}\{a'c(t_\mu - \tau)\exp\{i\omega t'_\mu - i\varphi\}\}^2\right) \\ &= \sum_{\mu=1}^{L} \left|a'c(t_\mu - \tau)\exp\{i\omega t'_\mu - i\varphi\}\right|^2 = La'^2 = 2T_{coh}C/N_0\end{aligned} \tag{5.22}$$

The detection probability is given by (5.23) and can be well approximated by the right-tail probability function of the normal distribution due to the usually large number of involved samples

$$p_d = P(E > \gamma \mid H_1) = Q_{\chi^2;2L,2T_{coh}C/N_0}(\gamma) \approx Q\left(\frac{\gamma - 2L - 2T_{coh}C/N_0}{\sqrt{4L + 8T_{coh}C/N_0}}\right) \tag{5.23}$$

5.6 Bayesian Detector

In this section, a Bayesian detector will be derived; this requires the a priori distributions of the signal parameters. Particular assumptions will be made that allow for the derivation of a partly closed expression for the detector. The Bayesian detector integrates all parameters out and can be considered a simple hypothesis-testing detector. It will be shown that, for a uniform code-phase and Doppler a priori distribution, the Bayesian detector equals the generalized likelihood-ratio detector of Section 5.7.

We assume that the carrier phase is uniformly distributed between 0 and 2π. The sample probability density function under H_1 has already been derived for this case in Section 5.4 and is given by (5.10). The Bayesian detector is based on the likelihood ratio (5.12), but in addition to the carrier phase, the real-valued amplitude a', the code phase τ, and the Doppler ω also have to be integrated out. This integration can be directly performed with the likelihood ratio because the sample distribution under H_0 is independent of the signal parameters. The likelihood ratio of the Bayesian detector is given as

$$\begin{aligned}L(\mathbf{s}) &= \frac{p_{H_1}(\mathbf{s})}{p_{H_0}(\mathbf{s})} \\ &= \int_{a',\tau,\omega} p(a',\tau,\omega)\exp\left\{-\frac{La'^2}{2}\right\} I_0\left(a'\left|P(\tau,\omega)\right|\sqrt{L}\right) da'\,d\tau\,d\omega\end{aligned} \tag{5.24}$$

The detector is based on the same correlator value $P(\tau,\omega)$ as in Section 5.4

$$\overline{P(\tau,\omega)} = \frac{1}{\sqrt{L}} \sum_{\mu=1}^{L} \overline{s}_\mu c(t_\mu - \tau)\exp\{i\omega t'_\mu\} \tag{5.25}$$

The real-valued amplitude a', the code phase τ, and the Doppler ω are assumed to be statistically independent,

$$p(a',\tau,\omega) = p(a')p(\tau)p(\omega) \tag{5.26}$$

To further evaluate (5.24), specific expressions for the a priori distributions will be given. The distributions aim at a detector for a multipath signal that is expected to appear within a certain code phase and Doppler range. The a priori distribution for code phase $p(\tau)$ and the Doppler $p(\omega)$ within the admissible range shall be left unspecified. The multipath signal amplitude shall be Rayleigh distributed. The average power of the multipath signal is assumed to be known.

For an average received multipath signal power of $2\sigma_{a'}^2$ the probability density function of a Rayleigh fading signal is, according to Lee [9], given as

$$p(a') = \frac{a'}{2\sigma_{a'}^2}\exp\left\{-\frac{a'^2}{2\sigma_{a'}^2}\right\} \tag{5.27}$$

Using a tool for analytical mathematics, integration of (5.24) can be carried out explicitly, yielding

$$\int_{a'=0}^{\infty} p(a')\exp\left\{-\frac{La'^2}{2}\right\} I_0\left(a'\left|P(\tau,\omega)\right|\sqrt{L}\right) da' = \frac{1}{1+L\sigma_{a'}^2}\exp\left\{\frac{\left|P(\tau,\omega)\right|^2 L\sigma_{a'}^2}{2\left(1+L\sigma_{a'}^2\right)}\right\} \tag{5.28}$$

The integrations with respect to the code phase or Doppler can only be done numerically. The obtained likelihood ratio is given as

$$L(s) = \frac{1}{1+L\sigma_{a'}^2}\int p(\tau)p(\omega)\exp\left\{\frac{\left|P(\tau,\omega)\right|^2 L\sigma_{a'}^2}{2\left(1+L\sigma_{a'}^2\right)}\right\} d\tau d\omega \tag{5.29}$$

The Bayesian detector shall be denoted by B and decides for H_1 if B exceeds a certain threshold

$$B = \int p(\tau)p(\omega)\exp\left\{\frac{\left|P(\tau,\omega)\right|^2 L\sigma_{a'}^2}{2\left(1+L\sigma_{a'}^2\right)}\right\} d\tau d\omega > \gamma \tag{5.30}$$

According to Section 3.7 of Kay's book, the value for the threshold is given by [2]

$$\gamma = 2\pi\left(1+L\sigma_{a'}^2\right)\frac{(C_{10}-C_{00})p(H_0)}{(C_{01}-C_{11})p(H_1)} \tag{5.31}$$

where C_{ij} is the cost if H_i is chosen but H_j is true. The symbol $p(H_0)$ denotes the a priori probability that H_0 is true and $p(H_1)$ denotes the a priori probability that H_1 is true. Overall, the Bayesian detector minimizes the cost function

$$R = \sum_{i,j=0}^{1} C_{ij}\, p\left(H_i \middle| H_j\right) p(H_j) \tag{5.32}$$

where $p(H_i \mid H_j)$ is the probability that H_i is detected and H_j is true. The Bayesian detector is an optimum detector with respect to this criterion. Note that the Neyman-Pearson expression (5.15) was based on p_{fa} and not on any cost function.

If we assume that $C_{10} = C_{01} = 1$ and $C_{00} = C_{11} = 0$, the corresponding Bayesian detector minimizes the sum of missed detections plus false detections

$$R = p\left(H_0 \middle| H_1\right) p(H_1) + p\left(H_1 \middle| H_0\right) p(H_0) = (1 - p_d)p(H_1) + p_{fa}p(H_0) \tag{5.33}$$

It is therefore called a *minimum probability of error* detector. For a large number of samples (i.e., $L\sigma_a^2 \gg 1$), the Bayesian detector can be approximated as

$$B \underset{L\sigma_{a'}^2 \gg 1}{=} \int p(\tau)p(\omega)\exp\left\{\frac{|P(\tau,\omega)|^2}{2}\right\} d\tau d\omega \tag{5.34}$$

that is independent of the mean received multipath signal power. The Bayesian detector uses contributions from all admissible correlation values, combines them properly, and compares the result against a threshold. By contrast, generalized likelihood-ratio detectors compare only the maximum correlation value against a threshold. As a consequence, the Bayesian scheme does not provide any code-phase or Doppler estimates.

For a uniform distribution of code phase and Doppler, the Bayesian detector and a generalized likelihood detector are equivalent because the functional dependency of P as a function of Doppler and code phase is fixed. Only the value of the peak and its position in the code-phase/Doppler plane varies. Therefore, being given the value of the peak of P allows the calculation of B, as long as the whole support of P is located within the admissible code-phase and Doppler values. This dependency can be expressed as

$$B = B\left(\max_{\tau,\omega} |P(\tau,\omega)|^2\right) \tag{5.35}$$

Furthermore, B is a monotone increasing function of the peak value of P and, thus, the Bayesian test (5.30) equals, for uniform-distributed code-phase and Doppler values, the generalized likelihood-ratio test (5.42). Only in the case of a non-uniform code-phase and Doppler distribution does the Bayesian test differ from the ML scheme. By focusing the search on regions where the signal is more likely to appear, it outperforms the generalized likelihood-ratio test (5.42).

5.7 Generalized Likelihood-Ratio Detector

As mentioned in the introduction, the generalized likelihood-ratio detector is by far the most common method to detect a signal in a navigation receiver. In Section 5.7.1, different examples for this detector will be given that correspond to different signal models and different levels of the signal parameter's information availability.

5.7.1 Single Coherent Integration

The (probably) most elementary generalized likelihood-ratio detector is obtained when all four fundamental signal parameters described in Section 1.8 are treated as unknown constants. The resulting detector is a correlation detector that maximizes the correlation function and is derived in the following way.

Assume a signal model based on a complex signal amplitude a as introduced in Section 4.3. The sample distribution is given as

$$p_{H_1}\left(\mathrm{s}\middle|a_1,\tau,\omega\right) = \frac{1}{(2\pi)^L}\exp\left\{-\frac{1}{2}\sum_{\mu=1}^{L}\left|s_\mu - ac(t_\mu-\tau)\exp\{i\omega t'_\mu\}\right|^2\right\} \tag{5.36}$$

The complex signal amplitude maximum (5.36) has already been obtained in (4.63) and is given as

$$\hat{a} = \frac{1}{L}\sum_{\mu=1}^{L}\overline{c(t_\mu-\tau)\exp\{-i\omega t'_\mu\}s_\mu} \tag{5.37}$$

The complex signal amplitude estimate is related to the correlator P given in Section 5.4 via

$$\hat{a} = \frac{P(\tau,\omega)}{\sqrt{L}} \Rightarrow |\hat{a}|^2 = \frac{|P(\tau,\omega)|^2}{L} \tag{5.38}$$

Inserting this expression into (5.36) yields

$$\begin{aligned}
p_{H_1}\left(\mathrm{s}\middle|\hat{a},\tau,\omega\right) &= \frac{1}{(2\pi)^L}\exp\left\{-\frac{1}{2}\sum_{\mu=1}^{L}|s_\mu|^2 - \frac{L|\hat{a}|^2}{2} + \mathrm{Re}\left\{\hat{a}\sum_{\mu=1}^{L}\bar{s}_\mu c(t_\mu-\tau)\exp\{i\omega t'_\mu\}\right\}\right\}\\
&= \frac{1}{(2\pi)^L}\exp\left\{-\frac{1}{2}\sum_{\mu=1}^{L}|s_\mu|^2 - \frac{L|\hat{a}|^2}{2} + \mathrm{Re}\{L\hat{a}\bar{\hat{a}}\}\right\}\\
&= \frac{1}{(2\pi)^L}\exp\left\{-\frac{1}{2}\sum_{\mu=1}^{L}|s_\mu|^2 + \frac{L|\hat{a}|^2}{2}\right\} = \frac{1}{(2\pi)^L}\exp\left\{-\frac{1}{2}\sum_{\mu=1}^{L}|s_\mu|^2 + \frac{|P(\tau,\omega)|^2}{2}\right\}
\end{aligned} \tag{5.39}$$

and the ML estimates for code phase and Doppler are

$$\hat{\tau},\hat{\omega} = \arg\max_{\tau,\omega}|P(\tau,\omega)|^2 \tag{5.40}$$

The generalized likelihood ratio is formed as

$$\frac{p_{H_1}\left(\mathrm{s}\middle|\hat{a},\hat{\tau},\hat{\omega}\right)}{p_{H_0}(\mathrm{s})} = \exp\left\{\frac{|P(\hat{\tau},\hat{\omega})|^2}{2}\right\} \tag{5.41}$$

and the generalized likelihood-ratio detector decides for H_1 if

$$\left|P(\hat{\tau},\hat{\omega})\right|^2 > \gamma \tag{5.42}$$

is fulfilled. The detector is basically identical to the clairvoyant detector, apart from the difference that the clairvoyant detector uses known code-phase and Doppler values.

5.7.2 Multiple Coherent Integrations

The number of samples involved in a single coherent integration cannot be arbitrarily large because, for a large coherent integration time, the short-period signal model described in Section 1.8 will no longer hold. The assumption of a linear carrier phase (a constant Doppler) and constant amplitude are likely to fail due to variations in the propagation channel, transmitter or receiver motion, clock instabilities, or simply due to the presence of a navigation data bit.

It is therefore convenient to model a longer duration of received signal samples by a concatenation of two or more short-period signal models. This assumes that the code phase and Doppler are constant over the whole period, but the carrier phase is allowed to jump at the boundaries.

5.7.2.1 Uniformly Distributed Carrier Phase and Constant Amplitude

For the following detector, we split the received samples into two parts. The first part comprises the sample indices ranging from 1, ... ,L. The second part comprises the sample index range from L + 1, ... ,$2L$. Code phase, Doppler, and real-valued signal amplitude are assumed to be unknown, constant, and identical for both parts. The carrier phases for both parts are modeled as two independent uniformly distributed random variables between 0 and 2π.

The corresponding sample distribution is given as

$$\begin{aligned} p_{H_1}\left(\mathrm{s}\middle|a',\tau,\omega\right) &= \frac{1}{(2\pi)^{2L}} \times \frac{1}{(2\pi)^2} \\ &\times \int\limits_{\varphi_1,\varphi_2=0}^{2\pi} \exp\left\{-\frac{1}{2}\left[\left(\begin{array}{l}\sum\limits_{\mu=1}^{L}\left|s_\mu - a'c(t_\mu-\tau)\exp\{i\omega t'_\mu - i\varphi_1\}\right|^2 + \\ + \sum\limits_{\mu=L+1}^{2L}\left|s_\mu - a'c(t_\mu-\tau)\exp\{i\omega t'_\mu - i\varphi_2\}\right|^2\end{array}\right)\right]\right\} d\varphi_1 d\varphi_2 \\ &= \frac{1}{(2\pi)^{2L}}\exp\left\{-\frac{1}{2}\sum_{\mu=1}^{2L}\left|s_\mu\right|^2 - La'^2\right\} I_0\left(a'\left|P_1(\tau,\omega)\right|\sqrt{L}\right) I_0\left(a'\left|P_2(\tau,\omega)\right|\sqrt{L}\right) \end{aligned} \tag{5.43}$$

where a procedure identical to that in Section 5.4 has been used to integrate out the carrier phases.

It is convenient to define correlator values, each for one sample segment, as

$$\begin{aligned}\overline{P_1(\tau,\omega)} &= \frac{1}{\sqrt{L}}\sum_{\mu=1}^{L} \overline{s}_\mu c(t_\mu - \tau)\exp\{i\omega t'_\mu\}\\ \overline{P_2(\tau,\omega)} &= \frac{1}{\sqrt{L}}\sum_{\mu=L+1}^{2L} \overline{s}_\mu c(t_\mu - \tau)\exp\{i\omega t'_\mu\}\end{aligned} \tag{5.44}$$

The ML estimate for the real-valued signal amplitude is formally written as

$$\hat{a}'(\tau,\omega) = \arg\max_{a'}\left(-La'^2 + \log I_0\left(a'\left|P_1(\tau,\omega)\right|\sqrt{L}\right) + \log I_0\left(a'\left|P_2(\tau,\omega)\right|\sqrt{L}\right)\right) \tag{5.45}$$

and, in general, this expression has to be evaluated numerically, yielding an amplitude estimate based on the two correlator values P_1 and P_2. Code-phase and Doppler estimates are obtained by a further maximization step that is written as

$$\hat{\tau},\hat{\omega} = \arg\max_{\tau,\omega}\begin{pmatrix}-L\hat{a}'(\tau,\omega)^2 + \\ +\log I_0\left(\hat{a}'(\tau,\omega)\left|P_1(\tau,\omega)\right|\sqrt{L}\right) + \log I_0\left(\hat{a}'(\tau,\omega)\left|P_2(\tau,\omega)\right|\sqrt{L}\right)\end{pmatrix} \tag{5.46}$$

The likelihood ratio is evaluated using the ML estimates and the detector decides for H_1 if

$$\frac{p_{H_1}(\mathbf{s}|\hat{a}',\hat{\tau},\hat{\omega})}{p_{H_0}(\mathbf{s})} = \exp\left\{-L\hat{a}'(\tau,\omega)^2\right\} I_0\left(\hat{a}'\left|P_1(\hat{\tau},\hat{\omega})\right|\sqrt{L}\right) I_0\left(\hat{a}'\left|P_2(\hat{\tau},\hat{\omega})\right|\sqrt{L}\right) > \gamma \tag{5.47}$$

is fulfilled. The detector (5.47) compares to the single coherent integrator of Section 5.7.1 or to the noncoherent detector of the next section. It is based on the same correlator values but involves nonlinear operations to combine both correlator values to form a test statistic.

5.7.2.2 Uniform Most-Powerful Test

The following discussion will use the probability density function (5.43) and will assume that code phase and Doppler are known and that only the real-valued signal amplitude is unknown. Both carrier phases are assumed to be uniformly distributed between 0 and 2π. Whether the test of Section 5.7.2.1 is uniform most powerful with respect to all real-valued signal amplitude values will be investigated.

According to Corollary 2 of Section 3.3 of Lehmann, a uniform most powerful exists if the probability density function can be written in such a form as

$$p(\mathbf{s}|a') = C(a')\exp\left\{Q(a')T(\mathbf{s})\right\}h(\mathbf{s}) \tag{5.48}$$

where $C(a')$ and $Q(a')$ are functions of the amplitude and $T(\mathbf{s})$ and $h(\mathbf{s})$ are functions of the signal samples [4]. The function Q must be strictly monotone. The function $T(\mathbf{s})$ acts as a sufficient statistic, as described in Section 4.4.1. The uniform most-powerful test is then given by testing $T(\mathbf{s})$ against a threshold γ. Note that the functions C and h cannot be part of the exponential whose argument needs to be a product of a s- and an a'-expression.

Unfortunately, the density function (5.43) is not of the form (5.48) because the correlator values appear separately as arguments of the two modified-Bessel functions. However, within the limit of vanishing signal amplitudes, the density function (5.43) can be approximated via a Taylor series expansion as

$$\sqrt{L}a'\left|P_{1,2}(\tau,\omega)\right| \ll \quad \Rightarrow \quad p_{H_1}(\mathrm{s}|a',\tau,\omega) \approx$$
$$\approx \frac{1}{(2\pi)^{2L}} \exp\left\{-\frac{1}{2}\sum_{\mu=1}^{2L}\left|s_\mu\right|^2 - La'^2 + \frac{1}{4}La'^2\left(\left|P_1(\tau,\omega)\right|^2 + \left|P_2(\tau,\omega)\right|^2\right)\right\} \tag{5.49}$$

That is of the form (5.48). Based on this approximation, a uniform most-powerful test is given by

$$\sqrt{L}a'\left|P_{1,2}(\tau,\omega)\right| \ll \quad \Rightarrow \quad \left|P_1(\tau,\omega)\right|^2 + \left|P_2(\tau,\omega)\right|^2 > \gamma \tag{5.50}$$

The sum of the squared correlator values is formed similar to Section 5.7.2.3 and is compared against a threshold γ.

For an infinitely large signal amplitude, the asymptotic form of the modified Bessel function

$$I_0(x) \underset{x\gg}{\approx} \frac{1}{\sqrt{2\pi x}} \exp\{x\} \tag{5.51}$$

can be used to approximate (5.43) as

$$\sqrt{L}a'\left|P_{1,2}(\tau,\omega)\right| \gg \quad \Rightarrow \quad p_{H_1}(\mathrm{s}|a',\tau,\omega) \approx$$
$$\approx \frac{1}{(2\pi)^{2L}} \exp\left\{-\frac{1}{2}\sum_{\mu=1}^{2L}\left|s_\mu\right|^2 - La'^2\right\} \frac{\exp\left\{\sqrt{L}a'\left(\left|P_1(\tau,\omega)\right| + \left|P_2(\tau,\omega)\right|\right)\right\}}{2\pi a'\sqrt{L\left|P_1(\tau,\omega)\right|\left|P_2(\tau,\omega)\right|}} \tag{5.52}$$

This approximation fits also into (5.48) using the assignments

$$\begin{aligned}
C(a') &= \frac{1}{(2\pi)^{2L}} \exp\left\{-La'^2\right\} \frac{1}{2\pi a'} \\
Q(a') &= \sqrt{L}a' \\
T(\mathrm{s}) &= \left|P_1(\tau,\omega)\right| + \left|P_2(\tau,\omega)\right| \\
h(\mathrm{s}) &= \exp\left\{-\frac{1}{2}\sum_{\mu=1}^{2L}\left|s_\mu\right|^2\right\} \frac{1}{\sqrt{L\left|P_1(\tau,\omega)\right|\left|P_2(\tau,\omega)\right|}}
\end{aligned} \tag{5.53}$$

and a uniform most-powerful test given by $T(\mathrm{s})$ tests the sum of the magnitudes of the correlator values against a threshold γ. The test is written as

$$\sqrt{L}a'\left|P_{1,2}(\tau,\omega)\right| \gg \quad \Rightarrow \quad \left|P_1(\tau,\omega)\right| + \left|P_2(\tau,\omega)\right| > \gamma \tag{5.54}$$

To further analyze which approximation is valid, the arguments of the modified Bessel functions are rewritten as

$$\begin{aligned}\left\langle \sqrt{L}a' \left|P_{1,2}(\tau,\omega)\right| \right\rangle_N &= La'^2\left(1+\frac{1}{4T_{coh}(C/N_0)_{1,2}}\right)\\ &= 2T_{coh}(C/N_0)_{1,2}\left(1+\frac{1}{4T_{coh}(C/N_0)_{1,2}}\right)\end{aligned} \tag{5.55}$$

whose minimum value equals ½ for a vanishing signal power. For this value, the Bessel function evaluates to log $I_0(0.5) = 0.0615$ and the approximation used to obtain (5.49) evaluates to ¼ $0.5^2 = 0.0625$. Thus, for low signal power values, (5.49) seems to be a reasonable approximation.

The calculation can be loosely summarized as "if small signal amplitudes are expected, the squared correlators shall be added." If large signal amplitudes are expected, the absolute values of the correlators are added to obtain (for the respective range of amplitude values) a uniform most-powerful test.

A consequence of this investigation is that, for an unknown signal amplitude (plus the other assumptions made in Section 5.7.2.1), *no* uniform most-powerful detector exists, because without a priori knowledge of the signal amplitude we do not know how to add the correlator values for threshold comparison. It should be mentioned that, for the case of a single signal segment, a uniform most-powerful detector exists: the clairvoyant detector of Section 5.4.

5.7.2.3 Unknown Carrier-Phase Distribution and Unknown Amplitude

The following detector uses the same assumptions as in Section 5.7.2.1, the only difference is that the carrier phases are treated as unknown and their ML estimates also need to be determined. We use two independent complex signal amplitude values for the different signal segments and the amplitudes are treated as unknown constants.

Overall, the probability density function is given as

$$\begin{aligned}p_{H_1}(\mathbf{s}|a_1,a_2,\tau,\omega) &= \frac{1}{(2\pi)^{2L}}\exp\left\{-\frac{1}{2}\sum_{\mu=1}^{L}\left|s_\mu - a_1 c(t_\mu-\tau)\exp\{i\omega t'_\mu\}\right|^2\right\}\\ &\times\exp\left\{-\frac{1}{2}\sum_{\mu=L+1}^{2L}\left|s_\mu - a_2 c(t_\mu-\tau)\exp\{i\omega t'_\mu\}\right|^2\right\}\end{aligned} \tag{5.56}$$

The complex signal amplitudes maximizing (5.56) are for a fixed code-phase and Doppler value completely independent of each other and are obtained in analogy to (4.63) as

$$\begin{aligned}\hat{a}_1 &= \frac{1}{L}\sum_{\mu=1}^{L}\overline{c(t_\mu-\tau)}\exp\{-i\omega t'_\mu\}s_\mu\\ \hat{a}_2 &= \frac{1}{L}\sum_{\mu=L+1}^{2L}\overline{c(t_\mu-\tau)}\exp\{-i\omega t'_\mu\}s_\mu\end{aligned} \tag{5.57}$$

The amplitude estimates are related to the correlator definitions of (5.44) via

$$\left|\hat{a}_k\right|^2 = \frac{\left|P_k(\tau,\omega)\right|^2}{L} \qquad k = 1,2 \tag{5.58}$$

Similar to Section 5.7.1, the probability density function using the ML signal amplitude estimates is obtained as

$$p_{H_1}(\mathbf{s}|\hat{a}_1,\hat{a}_2,\tau,\omega) = \frac{1}{(2\pi)^{2L}}\exp\left\{-\frac{1}{2}\sum_{\mu=1}^{L}\left|s_\mu\right|^2 - \frac{1}{2}\left|P_1(\tau,\omega)\right|^2 - \frac{1}{2}\left|P_2(\tau,\omega)\right|^2\right\} \tag{5.59}$$

and the ML code-phase and Doppler estimates are derived by maximizing the following expression

$$\hat{\tau},\hat{\omega} = \arg\max_{\tau,\omega}\left(\left|P_1(\tau,\omega)\right|^2 + \left|P_2(\tau,\omega)\right|^2\right) \tag{5.60}$$

The likelihood ratio derives then as

$$\frac{p_{H_1}(\mathbf{s}|\hat{a}_1,\hat{a}_2,\hat{\tau},\hat{\omega})}{p_{H_0}(\mathbf{s})} = \exp\left\{-\frac{1}{2}\left|P_1(\hat{\tau},\hat{\omega})\right|^2 - \frac{1}{2}\left|P_2(\hat{\tau},\hat{\omega})\right|^2\right\} \tag{5.61}$$

and the detector decides for H_1 if the sum of the squared correlator values exceeds a properly chosen threshold; that is

$$\left|P_1(\hat{\tau},\hat{\omega})\right|^2 + \left|P_2(\hat{\tau},\hat{\omega})\right|^2 > \gamma \tag{5.62}$$

This detector can be easily extended to a multitude v of signal segments yielding a standard noncoherent detector S given as

$$S = \sum_{n=1}^{v}\left|P_n(\hat{\tau},\hat{\omega})\right|^2 > \gamma \tag{5.63}$$

For completeness, the false alarm probability and the detection probability shall be evaluated under the assumption that the true code phase and Doppler are known. The evaluation is in line with the derivation for the clairvoyant detector of Section 5.4 but the number of involved random variables is increased from 2 to $2v$. The false alarm probability is given as

$$p_{fa} = P(S > \gamma \mid H_0) = Q_{\chi^2;2v}(\gamma) \tag{5.64}$$

Under hypothesis H_1, and using the definitions in Section 2.2.3 of Kay's book summarized in Appendix A.4.6, S belong to the noncentral chi-squared distribution with $2v$ degrees of freedom and a noncentrality parameter vLa'^2. Therefore, the probability of detection is

$$p_d = P(S > \gamma \mid H_1) = Q_{\chi^2;2v,vLa'^2}(\gamma) = Q_{\chi^2;2v,2vT_{coh}C/N_0}(\gamma) \tag{5.65}$$

Both formulas can be found in many textbooks on GNSS receivers such as that of van Dierendonck [10].

It should also be mentioned that the assumption of a constant code phase for all signal segments needs to be modified if a very long overall signal time span is considered. In fact, the code phase may drift among the different segments and this has to be accounted for in the noncoherent summation of (5.63). However, the drift is uniquely determined by the Doppler and therefore, during Doppler estimation, the drift is also estimated (only if code and carrier are generated coherently). Inclusion of the code-phase drift is thus only of a technical nature and does not change the theoretical performance of the detector.

5.7.3 Considering Navigation Signal Interference

Detection of a navigation signal in the pseudorange domains is generally degraded by the presence of further signals from other transmitters [11]. These additional signals should be accounted for in the detection process.

In the following discussion, a generalized likelihood-ratio detector will be evaluated to detect the possible presence of the signal "1", while a background signal "2" is present. Consequently, under hypothesis H_1 two signals ("1" and "2") are present, while under hypothesis H_0 just one signal ("2") is present. Both signals shall be described by the short period signal model of Section 1.8 and a complex valued signal amplitude representation shall be used.

The generalized likelihood ratio is given as

$$L(\mathrm{s}) = \frac{p_{H_1}\left(\mathrm{s}\middle|a_1,\tau_1,\omega_1,a_2,\tau_2,\omega_2\right)}{p_{H_0}\left(\mathrm{s}\middle|a_2,\tau_2,\omega_2\right)} \tag{5.66}$$

where the sample probability density function under H_1 is evaluated as

$$\begin{aligned}
&p_{H_1}\left(\mathrm{s}\middle|a_1,\tau_1,\omega_1,a_2,\tau_2,\omega_2\right)\\
&=\frac{1}{(2\pi)^L}\exp\left\{-\frac{1}{2}\sum_{\mu=1}^{L}\left|s_\mu - a_1c_1(t_\mu-\tau_1)\exp\{i\omega_1t'_\mu\} - a_2c_2(t_\mu-\tau_2)\exp\{i\omega_2t'_\mu\}\right|^2\right\}\\
&=\frac{1}{(2\pi)^L}\exp\left\{-\frac{1}{2}\sum_{\mu=1}^{L}\left|s_\mu\right|^2 - \frac{L}{2}\left(\left|a_1\right|^2+\left|a_2\right|^2\right)\right\}\\
&\quad\times\exp\left\{\sum_{\mu=1}^{L}\mathrm{Re}\{a_1\bar{s}_\mu c_1(t_\mu-\tau_1)\exp\{i\omega_1t'_\mu\}\} + \sum_{\mu=1}^{L}\mathrm{Re}\{a_2\bar{s}_\mu c_2(t_\mu-\tau_2)\exp\{i\omega_2t'_\mu\}\}\right\}\\
&\quad\times\exp\left\{\sum_{\mu=1}^{L}\mathrm{Re}\{a_1\bar{a}_2c_1(t_\mu-\tau_1)\overline{c_2(t_\mu-\tau_2)}\exp\{i(\omega_1-\omega_2)t'_\mu\}\}\right\}
\end{aligned} \tag{5.67}$$

According to Section 5.7, the generalized likelihood ratio (5.66) needs to be evaluated using the ML estimates of the signal parameters. To further evaluate

(5.66) the following important assumption has to be made: The ML parameter estimates of the signal "2" are not influenced by the presence of the signal "1."

Under this assumption, the ML parameter estimates of signal "2" are identical under hypothesis H_1 and hypothesis H_0. This assumption is basically fulfilled if the cross-correlation between the two navigation signals is low and if the signal amplitude of signal "1" is *not* much larger than the amplitude of signal "2." Signal "2" is considered to be a high-power signal, which is acquired first, and signal "1" is a low-power signal, which is acquired after acquisition of signal "2."

If the navigation system under consideration has a low cross-correlation protection, or for other reasons the mutual influence of the two navigation signals cannot be ignored, the following procedure cannot be applied and other detection schemes should be employed. One of them could be detection in the positioning domain as described above. Alternatively, multiuser detectors such as a decorrelating detector or a minimum mean-squared error detector can be employed [12]. If the waveforms of all received navigation signals are known, position domain detection is preferable because it optimally combines the received signals, whereas multiuser detectors only minimize the influence of the other signals. To the author's knowledge, both techniques are seldom used.

Assuming the parameters of the signal "2" are identical under the hypothesis H_1 and hypothesis H_0, the likelihood ratio is given as

$$\frac{p_{H_1}(\mathbf{s}|a_1,\tau_1,\omega_1,a_2,\tau_2,\omega_2)}{p_{H_0}(\mathbf{s}|a_2,\tau_2,\omega_2)} = \exp\left\{-\frac{L|a_1|^2}{2} + \sum_{\mu=1}^{L} \mathrm{Re}\{a_1 \bar{s}_\mu c_1(t_\mu - \tau_1)\exp\{i\omega_1 t'_\mu\}\}\right\}$$
$$\times \exp\left\{\sum_{\mu=1}^{L} \mathrm{Re}\left\{a_1 \bar{a}_2 c_1(t_\mu - \tau_1)\overline{c_2(t_\mu - \tau_2)}\exp\{i(\omega_1 - \omega_2)t'_\mu\}\right\}\right\} \tag{5.68}$$

which can be rephrased as a signal "2" cancellation scheme expressed as

$$\frac{p_{H_1}(\mathbf{s}|a_1,\tau_1,\omega_1,a_2,\tau_2,\omega_2)}{p_{H_0}(\mathbf{s}|a_2,\tau_2,\omega_2)}$$
$$= \exp\left\{-\frac{L|a_1|^2}{2} + \sum_{\mu=1}^{L} \mathrm{Re}\{a_1 c_1(t_\mu - \tau_1)\exp\{i\omega_1 t'_\mu\}\overline{(s_\mu - a_2 c_2(t_\mu - \tau_2)\exp\{i\omega_2 t'_\mu\})}\}\right\} \tag{5.69}$$

A replica signal "2" is subtracted from the received signal samples and the difference signal is used to determine the possible presence of signal "1." Following the discussion of a single coherent integration of Section 5.7.1, an interference-corrected correlator is defined as

$$\overline{P(\tau_1,\omega_1;\tau_2,\omega_2)} = \frac{1}{\sqrt{L}}\sum_{\mu=1}^{L}\left(s_\mu - a_2 c_2(t_\mu - \tau_2)\exp\{i\omega_2 t'_\mu\}\right) c_1(t_\mu - \tau_1)\exp\{i\omega_1 t'_\mu\} \tag{5.70}$$

Interference-corrected correlators have been implemented in GNSS hardware [13] and software receivers [11].

The ML estimates for the code phase and Doppler of signal "1" are obtained via maximizing

$$\hat{\tau}_1, \hat{\omega}_1 = \arg \max_{\tau_1, \omega_1} \left| P(\tau_1, \omega_1; \tau_2, \omega_2) \right|^2 \tag{5.71}$$

The presence of signal "1" is declared if the correlator exceeds a threshold like

$$\left| P(\tau_1, \omega_1; \tau_2, \omega_2) \right|^2 > \gamma \tag{5.72}$$

Provided that the signal "2" parameter estimates are accurate, this detector achieves the same performance as the single coherent detector of Section 5.7.1.

5.7.4 Data and Pilot

Modern GNSS signals often exhibit a dual component structure composed of a data and pilot signal. This signal structure optimizes the carrier tracking performance due to the presence of the pilot signal but still allows broadcasting of the navigation data message. Within the transmitter, different multiplexing schemes may be used. For certain multiplexing schemes, the received signal may be modeled as

$$s_\mu = a(\beta c_1(t_\mu - \tau) + \alpha \psi(d) c_2(t_\mu - \tau)) \exp\{i\omega t'_\mu\} \tag{5.73}$$

where the $c_1(t)$ is used to transmit the pilot signal and $c_2(t)$ is used to transmit the data signal. Here, a complex signal amplitude representation is used. The relative power of the data component is given by α^2 and the relative power of the pilot component is given by β^2. The sum of the relative signal powers is unity

$$\alpha^2 + \beta^2 = 1 \tag{5.74}$$

The cross-correlation between c_1 and c_2 is assumed to vanish

$$\sum_{\mu=1}^{L} c_1(t_\mu - \tau) \overline{c_2(t_\mu - \tau)} = 0 \tag{5.75}$$

which can be easily fulfilled by the system designers as the relative code phase and Doppler between data and pilot is constant. The data content is coded into the phase function $\psi(d)$ that assigns a certain phase to the data signal as a function of the broadcast data bit d, which is either +1 or 0. For a QPSK signal, a phase function

$$\psi(d) = \begin{cases} i & d = 1 \\ -i & d = 0 \end{cases} \tag{5.76}$$

is typically used. For the interplex modulation (as, for example, used for the Galileo OS on E1) or for time-multiplexed signals (like the L2 civil signal), the phase function

$$\psi(d) = \begin{cases} 1 & d = 1 \\ -1 & d = 0 \end{cases} \tag{5.77}$$

can be used.

It should be noted that we use the mentioned modulation schemes in a very loose sense only to introduce the phase function. The phase function alone is sufficient to describe the relationship between the data and the pilot component. Other peculiarities, especially the intermodulation product of the interplex modulation, are ignored here, as that signal component has a low cross-correlation coefficient with either the pilot or the data signal and has a much lower power. As an alternative to using a different phase function, one could substitute $c_2(t)$ with $ic_2(t)$.

5.7.4.1 Uniformly Distributed Data-Bit Values

If we assume that a "0" data bit and a "1" data bit occur with the same probability, the data-bit dependency of the sample distribution function can be integrated out, yielding

$$\begin{aligned}
&p_{H_1}(\mathrm{s}|a,\tau,\omega)\\
&=\frac{1}{2(2\pi)^L}\sum_{d=0}^{1}\exp\left\{-\frac{1}{2}\sum_{\mu=1}^{L}\left|s_\mu-a(\beta c_1(t_\mu-\tau)+\alpha\psi(d)c_2(t_\mu-\tau))\exp\{i\omega t'_\mu\}\right|^2\right\}\\
&=\frac{1}{2(2\pi)^L}\sum_{d=0}^{1}\exp\left\{-\frac{1}{2}\sum_{\mu=1}^{L}|s_\mu|^2-\frac{1}{2}L|a|^2\right.\\
&\left.+\sum_{\mu=1}^{L}\mathrm{Re}\left\{a\bar{s}_\mu(\beta c_1(t_\mu-\tau)+\alpha\psi(d)c_2(t_\mu-\tau))\exp\{i\omega t'_\mu\}\right\}\right\}\\
&=\frac{1}{2(2\pi)^L}\exp\left\{-\frac{1}{2}\sum_{\mu=1}^{L}|s_\mu|^2-\frac{1}{2}L|a|^2+\sum_{\mu=1}^{L}\mathrm{Re}\{a\beta\bar{s}_\mu c_1(t_\mu-\tau)\exp\{i\omega t'_\mu\}\}\right\}\\
&\times\sum_{d=0}^{1}\exp\left\{\sum_{\mu=1}^{L}\mathrm{Re}\{\alpha a\psi(d)\bar{s}_\mu c_2(t_\mu-\tau)\exp\{i\omega t'_\mu\}\}\right\}\\
&=\frac{1}{2(2\pi)^L}\exp\left\{-\frac{1}{2}\sum_{\mu=1}^{L}|s_\mu|^2-\frac{1}{2}L|a|^2+\sum_{\mu=1}^{L}\mathrm{Re}\{a\beta\bar{s}_\mu c_1(t_\mu-\tau)\exp\{i\omega t'_\mu\}\}\right\}\\
&\times\left(\exp\left\{\sum_{\mu=1}^{L}\mathrm{Re}\{\alpha a\psi(0)\bar{s}_\mu c_2(t_\mu-\tau)\exp\{i\omega t'_\mu\}\}\right\}\right.\\
&\left.+\exp\left\{\sum_{\mu=1}^{L}\mathrm{Re}\{a\alpha\psi(1)\bar{s}_\mu c_2(t_\mu-\tau)\exp\{i\omega t'_\mu\}\}\right\}\right)
\end{aligned}\tag{5.78}$$

For the QPSK phase function, this expression can be further simplified to

$$
\begin{aligned}
&p_{H_1}(\mathbf{s}|a,\tau,\omega) = \\
&= \frac{1}{2(2\pi)^L}\exp\left\{-\frac{1}{2}\sum_{\mu=1}^{L}|s_\mu|^2 - \frac{1}{2}L|a|^2 + \sum_{\mu=1}^{L}\mathrm{Re}\left\{a\beta\overline{s}_\mu c_1(t_\mu-\tau)\exp\{i\omega t'_\mu\}\right\}\right\} \\
&\times 2\cosh\left(\sum_{\mu=1}^{L}\mathrm{Re}\left\{ia\alpha\overline{s}_\mu c_2(t_\mu-\tau)\exp\{i\omega t'_\mu\}\right\}\right) \\
&= \frac{1}{(2\pi)^L}\exp\left\{-\frac{1}{2}\sum_{\mu=1}^{L}|s_\mu|^2 - \frac{1}{2}L|a|^2 + \mathrm{Re}\left\{a\sqrt{L}\beta\overline{P_1(\tau,\omega)}\right\}\right\}\cosh(\mathrm{Re}\left\{ia\sqrt{L}\alpha\overline{P_2(\tau,\omega)}\right\})
\end{aligned}
\tag{5.79}
$$

and it is convenient to define two correlators, each for the data and for the pilot signal as

$$
\begin{aligned}
\overline{P_1(\tau,\omega)} &= \frac{1}{\sqrt{L}}\sum_{\mu=1}^{L}\overline{s}_\mu c_1(t_\mu-\tau)\exp\{i\omega t'_\mu\} \\
\overline{P_2(\tau,\omega)} &= \frac{1}{\sqrt{L}}\sum_{\mu=1}^{L}\overline{s}_\mu c_2(t_\mu-\tau)\exp\{i\omega t'_\mu\}
\end{aligned}
\tag{5.80}
$$

The ML estimate for the complex signal amplitude results from

$$
\hat{a}(\tau,\omega) = \arg\max_a p_{H_1}(\mathbf{s}|a,\tau,\omega) \tag{5.81}
$$

This maximization can be carried out analytically by differentiating the log-likelihood function with respect to the complex conjugate signal amplitude, yielding the equation

$$
\begin{aligned}
&\frac{d}{d\overline{a}}\log p_{H_1}(\mathbf{s}|a,\tau,\omega) = \\
&= \frac{d}{d\overline{a}}\left(-\frac{1}{2}L|a|^2 + \mathrm{Re}\{\beta\sqrt{L}a\overline{P_1(\tau,\omega)}\} + \log\cosh(\mathrm{Re}\{i\sqrt{L}a\alpha\overline{P_2(\tau,\omega)}\})\right) \\
&= \frac{d}{d\overline{a}}\left(-\frac{1}{2}La\overline{a} + \frac{\beta\sqrt{L}}{2}(a\overline{P_1(\tau,\omega)} + \overline{a}P_1(\tau,\omega)) + \log\cosh\left(\frac{\alpha\sqrt{L}}{2}(ia\overline{P_2(\tau,\omega)} - i\overline{a}P_2(\tau,\omega))\right)\right) \\
&= -\frac{1}{2}La + \frac{\beta\sqrt{L}}{2}P_1(\tau,\omega) - \frac{\sinh\left(\mathrm{Re}\left\{ia\sqrt{L}\alpha\overline{P_2(\tau,\omega)}\right\}\right)}{\cosh\left(\mathrm{Re}\left\{ia\sqrt{L}\alpha\overline{P_2(\tau,\omega)}\right\}\right)}\frac{i\alpha\sqrt{L}}{2}P_2(\tau,\omega) \overset{!}{=} 0
\end{aligned}
\tag{5.82}
$$

This derivative needs to vanish in order to achieve a maximum in the log-likelihood function. It can be rewritten as

$$
La = \sqrt{L}\left(\beta P_1(\tau,\omega) - i\alpha P_2(\tau,\omega)\tanh\left(\mathrm{Re}\left\{ia\sqrt{L}\alpha\overline{P_2(\tau,\omega)}\right\}\right)\right) \tag{5.83}
$$

The ML estimate for the complex signal amplitude is formally a function of the two correlator values being expressed as

$$\hat{a}(\tau,\omega) = \hat{a}(P_1(\tau,\omega), P_2(\tau,\omega)) \tag{5.84}$$

An approximate solution for (5.83) is obtained by iteration. For the first iteration we set $P_2 = 0$ and obtain a first amplitude estimate as

$$\tilde{a}_0(\tau,\omega) = \frac{\beta}{\sqrt{L}} P_1(\tau,\omega) \tag{5.85}$$

Inserting the first iteration into the right hand side of (5.83) yields a refined approximation expressed as

$$\begin{aligned} \tilde{a}_1(\tau,\omega) &= \frac{1}{\sqrt{L}}\left(\beta P_1(\tau,\omega) - i\alpha P_2(\tau,\omega)\tanh\left(\mathrm{Re}\left\{i\tilde{a}_0(\tau,\omega)\sqrt{L}\alpha\overline{P_2(\tau,\omega)}\right\}\right)\right) \\ &= \frac{1}{\sqrt{L}}\left(\beta P_1(\tau,\omega) - i\alpha P_2(\tau,\omega)\tanh\left(\mathrm{Re}\left\{\alpha\beta i P_1(\tau,\omega)\overline{P_2(\tau,\omega)}\right\}\right)\right) \end{aligned} \tag{5.86}$$

The equation can be further iterated and converges well. In the following example, we stop after the second iteration and use this value as a complex amplitude ML estimate:

$$\hat{a}(\tau,\omega) \approx \tilde{a}_1(\tau,\omega) \tag{5.87}$$

The likelihood ratio is given as

$$\begin{aligned} L(\mathrm{s}|a,\tau,\omega) &= \frac{p_{H_1}(\mathrm{s}|a,\tau,\omega)}{p_{H_0}(\mathrm{s})} \\ &= \exp\left\{-\frac{1}{2}L|a|^2 + \mathrm{Re}\left\{a\sqrt{L}\beta\overline{P_1(\tau,\omega)}\right\}\right\}\cosh\left(\mathrm{Re}\left\{ia\sqrt{L}\alpha\overline{P_2(\tau,\omega)}\right\}\right) \end{aligned} \tag{5.88}$$

and the log-likelihood ratio using the complex amplitude ML estimate evaluates to

$$\begin{aligned} \log L(\mathrm{s}|\tau,\omega) = &-\frac{1}{2}L|\hat{a}(\tau,\omega)|^2 + \mathrm{Re}\left\{\hat{a}(\tau,\omega)\sqrt{L}\beta\overline{P_1(\tau,\omega)}\right\} + \\ &+ \log\cosh\left(\mathrm{Re}\left\{i\hat{a}(\tau,\omega)\sqrt{L}\alpha\overline{P_2(\tau,\omega)}\right\}\right) \end{aligned} \tag{5.89}$$

The code-phase and Doppler ML estimates are obtained via maximizing

$$\hat{\tau},\hat{\omega} = \arg\max_{\tau,\omega} \log L(\mathrm{s}|\tau,\omega) \tag{5.90}$$

and the detector decides for H_1 if the detector (in this case the log-likelihood ratio) exceeds a suitable chosen threshold γ

$$\log L(\mathrm{s}|\hat{\tau},\hat{\omega}) > \gamma \tag{5.91}$$

Analyzing (5.86), the detector *softly* decides between the two data-bit possibilities based on the tanh function. For high signal power levels, the tanh evaluates to either +1 or –1, depending on the true data-bit value. Data and pilot information are then properly combined and the data bit is removed for the complex amplitude estimate. By contrast, if the signal power is low, the tanh evaluates to a small number and the complex amplitude estimate is based only on the pilot correlator value P_1.

The presented method of integrating out the data-bit distribution can also be applied to design signal tracking algorithms. The resulting code and phase discriminators are optimal and are discussed in an article by Wang [14].

5.7.4.2 Unknown Data-Bit Value Distribution

The ML principle has to be employed for the data bit, if the distribution of the data-bit values is unknown and the data-bit distribution cannot be integrated out. The data-bit value is then part of the signal sample distribution being written as

$$\begin{aligned} & p_{H_1}(\mathrm{s}|a,\tau,\omega,d) \\ &= \frac{1}{(2\pi)^L}\exp\left\{-\frac{1}{2}\sum_{\mu=1}^{L}\left|s_\mu - a(\beta c_1(t_\mu-\tau)+\alpha\psi(d)c_2(t_\mu-\tau))\exp\{i\omega t'_\mu\}\right|^2\right\} \\ &= \frac{1}{(2\pi)^L}\exp\left\{-\frac{1}{2}\sum_{\mu=1}^{L}\left|s_\mu\right|^2 - \frac{1}{2}L\left|a^2\right|\right. \\ &\left. + \operatorname{Re}\left\{\sum_{\mu=1}^{L}\bar{s}_\mu a(\beta c_1(t_\mu-\tau)+\alpha\psi(d)c_2(t_\mu-\tau))\exp\{i\omega t'_\mu\}\right\}\right\} \end{aligned} \tag{5.92}$$

The generalized likelihood ratio evaluates to

$$\begin{aligned} L\left(\mathrm{s}|a,\tau,\omega,d\right) &= \frac{p_{H_1}\left(\mathrm{s}|a,\tau,\omega,d\right)}{p_{H_0}(\mathrm{s})} \\ &= \exp\left\{-\frac{1}{2}L\left|a^2\right| + \operatorname{Re}\left\{\sum_{\mu=1}^{L}\bar{s}_\mu a(\beta c_1(t_\mu-\tau)+\alpha\psi(d)c_2(t_\mu-\tau))\exp\{i\omega t'_\mu\}\right\}\right\} \end{aligned} \tag{5.93}$$

and the log-likelihood ratio to

$$\begin{aligned} & \log L\left(\mathrm{s}|a,\tau,\omega,d\right) \\ &= -\frac{1}{2}L\left|a^2\right| + \operatorname{Re}\left\{\sum_{\mu=1}^{L}\bar{s}_\mu a(\beta c_1(t_\mu-\tau)+\alpha\psi(d)c_2(t_\mu-\tau))\exp\{i\omega t'_\mu\}\right\} \end{aligned} \tag{5.94}$$

A data-bit value dependent correlator can be defined as

$$\bar{P}(\tau,\omega,d) = \frac{1}{\sqrt{L}}\sum_{\mu=1}^{L} \bar{s}_\mu(\beta c_1(t_\mu - \tau) + \alpha\psi(d)c_2(t_\mu - \tau))\exp\{i\omega t'_\mu\} \tag{5.95}$$

and it is related to the correlators (5.80) of the previous section via

$$P(\tau,\omega,d) = (\beta P_1(\tau,\omega) + \alpha\overline{\psi(d)}P_2(\tau,\omega)) \tag{5.96}$$

Similar to a single coherent integration of Section 5.7.1, the ML estimate for the complex signal amplitude is obtained as

$$\hat{a}(\tau,\omega,d) = \frac{1}{\sqrt{L}}P(\tau,\omega,d) \tag{5.97}$$

which can be inserted into the log-likelihood function (5.94), resulting in

$$\log L(\mathrm{s}|\tau,\omega,d) = -\frac{1}{2}|P(\tau,\omega,d)|^2 + \mathrm{Re}\left\{P(\tau,\omega,d)\bar{P}(\tau,\omega,d)\right\} = \frac{1}{2}|P(\tau,\omega,d)|^2 \tag{5.98}$$

Based on this expression, the ML estimates for code phase, Doppler, and data-bit value are obtained via maximization

$$\hat{\tau},\hat{\omega},\hat{d} = \arg\max_{\tau,\omega,d} \log L(\mathrm{s}|\tau,\omega,d) = \arg\max_{\tau,\omega,d} \log|P(\tau,\omega,d)|^2 \tag{5.99}$$

and the detector decides for H_1 if the squared correlator values exceed a threshold γ, which is written as

$$\left|P(\hat{\tau},\hat{\omega},\hat{d})\right|^2 > \gamma \tag{5.100}$$

5.8 System-Detection Performance

Most implementations of generalized likelihood-ratio detectors presented in previous sections follow an identical approach. First, a grid of test-parameter values is set up and the likelihood ratio is evaluated for all grid points. The grid may extend in the code-phase, Doppler, and amplitude direction. The point corresponding to the maximum value is chosen as the reference point and its likelihood ratio is compared against a threshold. Due to computational limitations, the grid is occasionally subdivided into several subgrids and during one acquisition step, the test points in a subgrid are evaluated. The number of grid points to be tested, the size of the subgrid, and the chosen detector define the performance of the system in signal detection. This system-detection performance shall be analyzed in Section 5.8.1.

The system-detection performance is based on the detection probabilities derived in Sections 5.4 and 5.7.2.3, which assume known code-phase and Doppler values. These probabilities are called single-bin detection probabilities. In contrast, system-detection probabilities characterize the whole (false) detection performance for a given search strategy. The system-detection probabilities derive from the single-bin detection probabilities.

Another figure of merit to characterize the system-detection performance is the mean acquisition time. The mean acquisition time is based on the dwell time, which is the time needed for the search algorithm to search one subgrid. The mean acquisition time is the average time of the algorithm to detect the signal. Mean acquisition time and dwell time can be defined as signal time (i.e., the time span defined by the number of the received signal samples being processed) or as processing time (i.e., the actual processing time of the CPU or the ASIC). In certain cases, processing time and signal time may coincide.

The evaluation of the system-detection probabilities may become a complex task because a multitude of correlation values for all grid points are involved, and those correlation values are generally correlated random variables, making it difficult or impossible to derive closed-form expressions [15] for the system-detection performance [2]. Often only Monte Carlo methods can be used to evaluate the system probabilities. For simplified assumptions, which shall be outlined below, the mean acquisition time and the system-detection probabilities will be evaluated.

5.8.1 Idealized Assumptions

To evaluate the system performance, we assume that the search grid spanning all admissible parameter values contains N_G points. During one acquisition step, a number of M_G points can be evaluated. The evaluation takes D_W seconds. We assume that for all grid points the correlation values are evaluated, and for each grid point the detector has a single-bin detection probability of p_d and a false-alarm probability of p_{fa}. We assume that the true signal parameters lie exactly on one of the grid points and that the correlation values for all other grid points follow an H_0 distribution. In other words, if the signal is present on one grid point, it is not present on all others. The position of the true grid point is uniformly distributed over all grid points. Finally, we assume uncorrelated correlation values on different grid points.

The correlation function of two correlator values is discussed in Section 7.3.1. The results presented there imply that, to fulfill the last two assumptions, the grid spacing needs to be larger than half of the support of the correlation function (e.g., larger than one chip for a PRN code signal) and the Doppler grid spacing is a multiple of $1/T_{coh}$. Both requirements are generally not fulfilled in real implementations, and shorter grid spacing might be used. In this case, the results presented below overestimate the system performance slightly [15]. A further performance degradation arises because, in reality, the true code phase and Doppler do not exactly lie on the grid points. Those two simplications will be discussed in Sections 5.8.4 and 5.8.5.

5.8.2 Mean Acquisition Time

Based on the above assumptions, the mean acquisition time T_{ACQ} has been evaluated in the work by Lozow as [16]

$$T_{ACQ} = \left(\frac{2-p_d}{2p_d}\right)\left(\frac{k(1-(1-p_{fa})^{M_G})+1}{M_G}\right)N_G D_W \tag{5.101}$$

Here, k is a penalty factor to characterize the time of the system to detect a false acquisition. If a signal is falsely detected, the system is assumed to spend kD_W seconds to detect its error and then to continue with the normal acquisition procedure. The mean acquisition time decreases with an increasing p_d. However, an increased p_d is normally accompanied by a longer dwell time D_W, which compensates for the decrease.

5.8.3 System Probabilities

To derive the system-detection probabilities, we assume that N_G/M_G acquisition steps are carried out and that all search-grid values are tested. The probability that the signal is detected on the correct grid point (the system-detection probability) is denoted as π_d. The probability that the system detects a signal although none is present is denoted as π_{fa}.

According to Borio, the system false-detection probability π_{fa} is given as

$$\pi_{fa} = 1 - (1 - p_{fa})^{N_G} \tag{5.102}$$

and is independent of the size M_G of the subgrid [15].

The detection probability π_d is given as

$$\pi_d = \frac{M_G}{N_G} \frac{1-(1-p_{fa})^{N_G}}{1-(1-p_{fa})^{M_G}} \int_{\gamma'=\gamma}^{\infty} \left(1 - p_{fa}(\gamma')\right)^{M_G - 1} q_d(\gamma') d\gamma' \tag{5.103}$$

Here, $q_d(\gamma')$ is the single-bin probability density function of the detector under H_1, depending on a variable threshold γ'. The symbol γ is the detector threshold used to obtain the single-bin detection probability

$$p_d = \int_{\gamma'=\gamma}^{\infty} q_d(\gamma') d\gamma' \tag{5.104}$$

Furthermore, the single-bin false-alarm probability $p_{fa}(\gamma')$ depends also on the variable threshold γ'. For small p_{fa}, the single-bin detection probability and the system-detection probability are equal to each other, which can be shown by applying l'Hôpital's rule.

5.8.4 Independent Bin Approximation

The assumption of uncorrelated correlation values used in the previous sections is often overly simplistic because a grid compatible with this assumption needs to have a large grid spacing. In this case, however, the code-phase and Doppler losses (as discussed in Section 5.8.5) might become unacceptably large. Therefore, a practical signal-acquisition scheme using a parallel search technique is normally based on correlated correlator values. The fact that adjacent correlator values are correlated reduces the number of independent search bins that primarily affect the false-detection probability. By contrast, the probability of detection is less influenced by the number of bins to be searched.

In the following discussion, an effective number of independent bins N_G' and M_G' will be introduced; it is argued that these effective numbers are used to evaluate formulas of the mean acquisition time (5.101) and of the system false-detection probability (5.102). The detection probability remains unchanged. The effective number of bins is based on the covariance function of two correlator values, as derived in Section 7.3.1. In the following example, the covariance function is denoted by $R(\Delta\tau,\Delta\omega)$, where the arguments represent the code-phase and Doppler difference of two correlators. The suggested procedure uses Monte Carlo methods using only hypothesis H_0 signal configurations and consists of the following steps:

- Generate K representative sets of correlator values $P_k(\tau,\omega)$ on a code-phase/Doppler grid whose covariance is $R(\Delta\tau,\Delta\omega)$. The index k is used to distinguish the different runs and ranges from 1 to K. Correlator values belonging to two different sets are independent.
- Based on a specific grid point (τ_0,ω_0), determine a single-bin threshold for the squared correlator statistics $|P_k(\tau_0,\omega_0)|^2$ and for a chosen single-bin false detection probability p_{fa}.
- Using the multibin statistics $\max_{\tau,\omega} |P_k(\tau,\omega)|^2$ determine the system false-detection probability π_{fa}.
- Determine the effective bin size as

$$S' = N\frac{\log(1-p_{fa})}{\log(1-\pi_{fa})} \tag{5.105}$$

Here, N denotes the number of grid points per set involved in the above procedure. This equation is directly derived from (5.102).

The effective number of bins used to calculate the mean acquisition time or the system-detection probabilities are then given as

$$N_G' = \frac{N_G}{S'}, \quad M_G' = \frac{M_G}{S'} \tag{5.106}$$

The higher the correlation between the bins, the lower the number of effective bins and the smaller the increase in the false-detection probability.

5.8.5 Code-Phase and Doppler Losses

The assumption that the true signal parameters lie exactly on one of the grid points under consideration is used in the previous section to derive the mean acquisition time or the system-detection probability. This assumption is, however, not normally fulfilled in practical implementations because computational efforts to search the multitude of grid points (caused by the necessarily fine grid spacing) would be too high.

It is generally more convenient to work with a larger grid spacing at the cost of a decreased detection probability. The decrease in the detection probability is caused by the fact that the correlation value is reduced if the true signal parameters do not lie exactly on the grid point. The reduction of the correlation can be expressed based on the expected correlator value, which is derived in Section 7.3.1.

The code-phase and Doppler loss $L_{\tau,\omega}$ in decibels can be expressed as

$$L_{\tau,\omega} = 20\log_{10}\left(\frac{1}{\delta\tau\delta\omega\left|\left\langle P(\tau_0,\omega_0)\right\rangle_{\mathrm{N}}\right|}\int\limits_{\tau=\tau_0-\delta\tau/2}^{\tau_0+\delta\tau/2}\int\limits_{\omega=\omega_0-\delta\omega/2}^{\omega_0+\delta\omega/2}\left|\left\langle P(\tau,\omega)\right\rangle_{\mathrm{N}}\right|d\tau d\omega\right) \quad (5.107)$$

where τ_0, ω_0 denote the true code-phase and Doppler values. The symbols $\delta\tau$, $\delta\omega$ denote the code-phase and Doppler spacing. In the limit of infinitesimal spacing, the code-phase and Doppler loss is 0 dB.

The code-phase and Doppler loss can be seen as an implementation loss and effectively reduces the available power when detecting the signal.

5.9 Long Integration Times and Differential Detectors

The previous discussion clearly demonstrated that by using a long coherent integration time, nearly arbitrarily weak signals can be detected. The signal information contained in each received sample is optimally exploited. For most practical detection implementations, the use of a coherent integration implies that the short-period signal model of Section 1.8 is valid, which corresponds to a linear carrier-phase dependency on time (e.g., a constant Doppler is required). In principle, coherent integration can also be performed over nonconstant Doppler signals, but this requires external aiding information that provides the relative antenna motion—a so-called μ-trajectory—during the integration interval [17]. Long coherent integration also requires pilot signals or a data wipe-off functionality. If a linear phase model is considered, a number of physical effects limit the coherent integration time, specifically, transmitter and receiver clock jitter, nonlinear line-of-sight dynamics (e.g., user accelerations), or phase fluctuations in the propagation channel. Those effects have been assessed, for example, in the articles by Sıçramaz Ayaz, López-Risueño, and Niedermeier [17–19]. If a high-quality oscillator (e.g., OCXO) is used, if the line-of-sight signal is received (and not a reflection), and if nonlinear line-of-sight dynamics are not present or compensated for by an IMU, then the coherent integration time can be up to a few seconds. If a linear phase dependency can not be assumed (and no aiding is available), then it is useful to split up the integration into several shorter periods, assuming within each period a linear phase dependency. If the carrier phases of each period are modeled as independent random variables, the noncoherent integration scheme of Section 5.7.2 is optimal.

A more stringent requirement on the maximum duration of the coherent integration time is often imposed by computational limitations. In a first approximation the number of bins to be searched increases linearly with the coherent integration time (if the timing uncertainty is limited, the number of code phase bins is constant; the number of Doppler bins increases linearly). Furthermore, the dwell time within each bin also increases linearly with the coherent integration time. By using sophisticated algorithms, as described in Sections 9.5.5 and 9.5.6, the increase can be partly reduced but, nevertheless, the computational burden is probably the most severe constraint.

A possibility to reduce the computational burden by reducing the number of Doppler bins is to use coherent integration times that are below the maximum allowable value (i.e., multiple shorter coherent integrations are used instead of one long single coherent integration). The caveat of this method is a decreased sensitivity, usually named *squaring loss*. The squaring loss originates from the fact the correlation values are squared before they are added to remove the carrier-phase dependency. Squaring—as a nonlinear operation—reduces the overall performance compared to the (linear) coherent integration. The squaring loss is occasionally attributed to a reduced postdetection output signal-to-noise ratio value of the test statistics for the squaring detector (5.65) compared to the coherent detector (5.17). The postdetection signal-to-noise ratio value can be used to compare different acquisition strategies. The higher the signal-to-noise, the more sensitive is the detection scheme. In the article by Borio, however, it has been shown that the output postdetection signal-to-noise ratio (corresponding to the deflection coefficient) is unable to completely characterize the acquisition performance; consequently, it should not be used to compare different acquisition methods in a strict sense [20]. Instead, it is recommended to work with the complete probability distribution and to define the squaring loss as the C/N_0 difference to achieve a certain probability of detection.

If the maximum coherent integration time cannot be exploited for computational limitations, differential acquisition schemes can be used. Differential acquisition schemes use, typically, half the coherent integration time (and thus reduce the number of Doppler bins by two). They remove the carrier-phase dependency by a multiplication of two correlator values, which are obtained by correlating timely separated incoming signal sample batches; that is, an expression in the form

$$S = \left| \sum_{n=1}^{\upsilon-1} P_n(\hat{\tau},\hat{\omega})\overline{P_{n+1}(\hat{\tau},\hat{\omega})} \right| > \gamma \tag{5.108}$$

is used instead of (5.63) as test statistics. It requires that two adjacent correlator values depend similarly on the carrier phase to cancel the carrier-phase dependency via multiplication. The expected value of each product is of zero mean in contrast to a squared correlator value [used in (5.63)] which is of nonzero mean. Differential correlation was first published in Zarrabizadeh and Sousa's article and has been introduced into satellite navigation in Park's work [21, 22]. Theoretical investigations performed in articles by Ávila Rodríguez and Schmid demonstrate that the use of differential correlation brings a gain of less than 3 dB compared to the squaring scheme of Section 5.7.2.3 if the squaring scheme does not fully exploit the maximum coherent integration time and uses the same (i.e., half of the maximum) integration time as the differential scheme [23, 24]. The differential scheme can also be applied by multiplying the nonadjacent correlator values with each other, as described by Shanmugam [25]. The differential scheme is sensitive to data-bit transitions and, for best performance, it is required that neither of the multiplicands is affected by a data-bit transition. Losses caused by data-bit transitions are investigated in the works by Schmid and Shanmugam [24, 25].

5.10 Discussion

This chapter presented several methods of how signal detection may occur in a navigation receiver. It included "exotic" methods, such as the energy detector, which is easy to implement but does not yield approximate code-phase or Doppler values, and detection in the position domain, which has high implementation complexity.

Established methods are based on signal detection in the pseudorange domain. All those methods are characterized by their need to evaluate the correlation values for a multitude of code-phase and Doppler values (called grid points). This evaluation is computationally very demanding. However, especially for software receivers, there exist a number of sophisticated frequency-domain methods that partly relieve the computational demands. These will be discussed in Section 9.5.

Two different detection principles have been contrasted in this section: Bayesian techniques and generalized likelihood-ratio detectors. Bayesian techniques generally outperform generalized likelihood-ratio detectors or give at least the same performance. The sensitivity gain of using Bayesian methods is, however, small (less than 1 dB for the data/pilot example) and depends on the characteristics of the available a priori information. The more the a priori distribution deviates from a uniform distribution, the higher the expected gain.

From a theoretical point of view, an optimum detection scheme does not exist; that is, there is no formula for the test statistics, which give an optimum performance independent of the true signal-parameter values. Only the clairvoyant detector is optimal, but this detector needs to know the true code-phase, Doppler, and, eventually, amplitude values.

Overall, generalized likelihood-ratio detectors seem to give a reasonable performance and their implementation is typically simpler than that of a Bayesian detector. To design a good generalized likelihood-ratio detector the following points should be kept in mind:

- Shrink the search space to a minimum number of grid points based on all information available to the navigation receiver.
- Filter the signal to avoid fine code-phase grid spacings (see Section 5.3).
- Look for an efficient implementation to evaluate the multitude of correlators (see Section 9.5).
- Find a good compromise among coherent integration time, number of noncoherent integrations, and computational complexity (see Section 5.7.2.3).
- Find a suitable acquisition strategy that balances the computational load and simultaneously cycles optimally through the signals to be acquired.
- Allow the algorithm to do a self-calibration to determine the threshold for a given system false-detection probability.

In the author's opinion, the acquisition problem is solved satisfactorily by a generalized likelihood-ratio detector. The acquisition problem is mostly a problem of implementation that faces the difficulty of realizing a high number of correlators. Time-domain correlation, matched filters, and frequency-domain techniques are the most common solutions.

The specific signal structure imposes important boundary conditions. The signal defines the required number of code search bins and Doppler search bins. For computational purposes, short periodic signals are desirable. They allow the application of FFT-based circular correlation with Doppler preprocessing, the most effective software radio acquisition method presented in Section 9.5.6. On the contrary, those signals provide small cross-correlation protection, which could be resolved using interference cancellation methods as described in Section 5.7.3. In a software-radio approach, interference cancellation techniques are easier to implement than a huge number of correlators (as would be required by a long coherent integration time), because subtraction of two signals is a computationally easy task. It should also be noted that the used-signal modulation scheme influences only via the code-phase losses described in Section 5.8.5.

The use of Bayesian techniques, differential detectors, or other specific optimized methods is encouraged if it is useful for a specific application and the implementation cost is reasonable. However, in the author's experience, those techniques do not improve the performance by more than a few decibels compared to a generalized likelihood-ratio detector, provided they are based on the same number of correlators. For example, a high-sensitivity GPS receiver like the SiRF III copes with signal power variations of more than 20 dB and a sensitivity gain of 1–2 dB is perhaps not important [26]. More optimization potential lies in the correlator implementation. The already-mentioned Doppler preprocessing of Section 9.5.6 improves the computational FFT performance by about a factor of 10 for a GPS C/A-code signal, resulting in 10 times more noncoherent integrations and, consequently, a less than 10-dB sensitivity gain. The highest sensitivity can be achieved if long coherent integration times are achieved. This requires the use of pilot signals or data wipe-off techniques and imposes restrictions on the user dynamics and on the used oscillators. New techniques, like mobile-phone-network-assisted frequency and time stability, provision of navigation data bits via assistance data, and user dynamics compensation via an inertial sensor integrated into a mobile phone, may open new GNSS high-sensitivity applications.

References

[1] Poor, H. V., *An Introduction to Signal Detection and Estimation*, New York: Springer, 1988.

[2] Kay, S. M., *Fundamentals of Statistical Signal Processing: Detection Theory*, Englewood Cliffs, NJ: Prentice Hall, 1998.

[3] Scharf, L. L., *Statistical Signal Processing: Detection, Estimation, and Time Series Analysis*, Reading, MA: Addison-Wesley, 1991.

[4] Lehmann, E. L., *Testing Statistical Hypothesis*, 2nd ed., New York: Wiley, 1986.

[5] Betz, J. W., "Binary Offset Carrier Modulations for Radionavigation," *NAVIGATION, Journal of The Institute of Navigation*, Vol. 48, No. 4, 2001, pp. 227–246.

[6] Yang, C., J. Vasquez, and J. Chaffee, "Fast Direct P(Y)-Code Acquisition Using XFAST," *Proc. 12th Int. Technical Meeting of the Satellite Division of the Institute of Navigation (ION-GPS) 1999*, Nashville, TN, September 14–17, 1999, pp. 317–321.

[7] Yang, C., et al., "Extended Replica Folding for Direct Acquisition of GPS P-Code and Its Performance Analysis," *Proc. 13th Int. Technical Meeting of the Satellite Division of The Institute of Navigation (ION-GPS) 2000*, Salt Lake City, UT, September 19–22, 2000, pp. 2070–2078.

[8] Li, H., M. Lu, and Z. Feng, "Mathematical Modelling and Performance Analysis for Average-Based Rapid Search Method for Direct Global Position System Precision Code Acquisition," *IET Radar, Sonar & Navigation*, Vol. 3, No. 1, 2009, pp. 81–92.

[9] Lee, W. C. Y., *Mobile Communications Engineering*, New York: McGraw-Hill, 1982.

[10] van Dierendonck, A. J., "GPS Receivers," in *Global Positioning System: Theory and Applications, Vol. I,* pp. 329–407, Parkinson, B. W., and J. J. Spilker, (eds.), Washington, D.C.: American Institute of Aeronautics and Astronautics, Inc., 1996.

[11] Heinrichs, G., et al., "HIGAPS: A Large-Scale Integrated Combined Galileo/GPS Chipset for the Consumer Market," *Proc. European Navigation Conference (ENC-GNNS) 2004,* Rotterdam, May 16–19, 2004.

[12] Verdú, S., *Multiuser Detection*, Cambridge, U.K.: Cambridge University Press, 1998.

[13] Glennon, E. P., et al., "Post Correlation CWI and Cross Correlation Mitigation Using Delayed PIC," *Proc. of the 20th Int. Technical Meeting of the Satellite Division of the Institute of Navigation (ION-GNNS) 2007,* Fort Worth, TX, September 25–28, pp. 236–245.

[14] Wang, D., et al., "Optimum Tracking Loop Design for the New L2 Civil Signal Based on ML Estimation," *Proc. 17th Int. Technical Meeting of the Satellite Division of the Institute of Navigation (ION-GNNS) 2004*, Long Beach, CA, September 21–24, 2004, pp. 474–485.

[15] Borio, D., L. Camoriano, and L. Lo Presti, "Impact of Acquisition Searching Strategy on the Detection and False Alarm Probabilities in a CDMA Receiver," *PLANS 2006, IEEE/ION Position, Location and Navigation Symposium*, San Diego, April 25–28, 2006, pp. 1100–1107.

[16] Lozow, J. B., "Analysis of Direct P(Y)-Code Acquisition," *NAVIGATION, Journal of The Institute of Navigation*, Vol. 44, 1997, pp. 89–97.

[17] Niedermeier, H., et al., "Reproduction of User Motion and GNSS Signal Phase Signatures Using MEMS INS and a Pedestrian Navigation System for HS-GNSS Applications," *Proc. 4th ESA Workshop on Satellite Navigation User Equipment Technologies, NAVITEC*, Noordwijk, the Netherlands, December 10–12, 2008.

[18] Sıçramaz Ayaz, A., T. Pany, and B. Eissfeller, "Performance of Assisted Acquisition of the L2CL Code in a Multi-Frequency Software Receiver," *Proc. 20th Int. Technical Meeting of the Satellite Division of the Institute of Navigation (ION-GNNS) 2007,* Fort Worth, TX, September 25–28, 2007, pp. 1830–1838.

[19] López-Risueño, G., et al., "User Clock Impact On High Sensitivity GNSS Receivers," *Proc. European Navigation Conference (ENC-GNNS) 2008,* Toulouse, April 22–25, 2008.

[20] Borio, D., et al., "The Output SNR and Its Role in Quantifying GNSS Signal Acquisition Performance," *Proc. European Navigation Conference (ENC-GNNS) 2008*, Toulouse, April 22–25, 2008.

[21] Zarrabizadeh, M. H., and E. S. Sousa, "A Differentially Coherent PN Code Acquisition Receiver for CDMA Systems," *IEEE Trans. Commun.*, Vol. 45, 1997, pp. 1456–1465.

[22] Park, S. H., et al., "A Novel GPS Initial Synchronization Scheme Using Decomposed Differential Matched Filter," *Proc. Institute of Navigation National Technical Meeting (ION-NTM) 2002,* San Diego, CA, January 28–30, 2002, pp. 246–253.

[23] Ávila Rodríguez, J. Á., T. Pany, and B. Eissfeller, "A Theoretical Analysis of Acquisition Algorithms for Indoor Positioning," *Proc. 2nd ESA Workshop on Satellite Navigation User Equipment Technologies, NAVITEC*, Noordwijk, the Netherlands, December 8–10, 2004.

[24] Schmid, A., and A. Neubauer, "Performance Evaluation of Differential Correlation for Single Shot Measurement Positioning," *Proc. 17th Int. Technical Meeting of the Satellite Division of the Institute of Navigation (ION-GNNS) 2004*, Long Beach, CA, September 21–24, 2004, pp. 1998–2009.

[25] Shanmugam, S. K., J. Nielsen, and G. Lachapelle, "Enhanced Differential Detection Scheme for Weak GPS Signal Acquisition," *Proc. 20th Int. Technical Meeting of the Satellite Division of the Institute of Navigation (ION-GNNS) 2007*, Fort Worth, TX, September 25–28, 2007, pp. 189–202.

[26] SiRF Technology, Inc., http://www.sirf.com, 2008.

CHAPTER 6

Sample Preprocessing

Sample preprocessing summarizes a number of steps before an actual correlation takes places. Interfering pulses are eliminated, filters might be applied to ensure an optimum tracking performance, and the noise floor usually needs to be determined. Analog-to-digital conversion also requires properly chosen quantization thresholds to optimally represent the analog signal with a limited number of bits.

6.1 ADC Quantization

The signal processing algorithms of Chapters 4 and 5, as well as the correlators presented in Chapter 7, are derived under the assumption of the signal samples being floating-point numbers. In practice, however, only a limited number of bits are available to represent signal samples. The statement can be rephrased by using stochastic terms: For analytical derivations, the signal samples are assumed to be real- or complex-valued random variables, but in practice the signal samples are integer-valued random variables. This approximation is only possible because virtually all statistics derive from correlation values of the received-signal samples with some reference signals, as has been described in Section 4.4. The law of large numbers applies and the correlator outputs are sufficiently well modeled as Gaussian random variables [1]. Nevertheless, the ADC reduces the overall system performance, which is described by an effective loss in signal power called *conversion* or *quantization loss*.

Quantization errors are introduced, for example, in the work by Proakis and Manolakis and in that by van Dierendonck [2, 3]. Whereas Proakis and Manolakis consider quantization of Gaussian noise only, van Dierendonck makes use of a reference that considers quantization noise together with filtering losses plus signal-processing losses [4]. An extensive treatment of quantization errors in the presence of interference is given in Spilker and Natali's chapter [5]. A numerical simulation of bandlimiting, sampling, and quantization errors for GNSS signals has been done by Betz and investigates also the influence of the sampling rate [6, 7].

Quantization errors shall be revised here for a low-resolution (1–4 bits) ADC and a high-resolution (e.g., 16-bits) signal processing. This setting is adopted for the software-radio approach described above. Without loss of generality, only real-valued signals are considered.

6.1.1 Quantization Rule

Quantization occurring during analog-to-digital conversion of an analog sample s_μ maps the sample s_μ on a finite number of values $q(s_\mu)$

$$q(s) = \begin{cases} \alpha_0 & \beta_0 \le s \le \beta_1 \\ \alpha_1 & \beta_1 < s \le \beta_2 \\ & \vdots \\ \alpha_{B-1} & \beta_{B-1} < s \le \beta_B \end{cases} \tag{6.1}$$

The mapping is completely defined by the thresholds β_b ($b = 0 \ldots B$) and by the output values α_b ($b = 0 \ldots B - 1$). For an example of a linear symmetric 2-bit ADC, the threshold and output values are given as

$$\begin{aligned} &\alpha_b \in \left\{-\tfrac{3}{2}\delta, -\tfrac{1}{2}\delta, \tfrac{1}{2}\delta, \tfrac{3}{2}\delta\right\} \\ &\beta_b \in \{-\infty, -\delta, 0, \delta, \infty\} \end{aligned} \tag{6.2}$$

For a symmetric n-bit ADC, the threshold and output values are given as

$$\begin{aligned} \alpha_b &= \delta\frac{2b - B + 1}{2} && 0 \le b < B \\ \beta_0 &= -\infty \\ \beta_b &= \delta\frac{2b - B}{2} && 0 < b < B \\ \beta_B &= \infty \end{aligned} \tag{6.3}$$

Here, δ represents the ADC input step width, which can be adjusted to optimally fit the analog signal. The range $\pm\,\Delta$ of analog input signal levels, which are covered uniformly by the ADC, is given by half the number of output levels multiplied by the input step width:

$$\Delta = \delta\frac{B}{2} \tag{6.4}$$

The ADC output values are internally represented as integer numbers. Usually the signal processing has no direct access to δ. If the signal processing works with two-complement integers with a number of bits larger than $\log_2(2B)$ (for example, Chapter 9 describes algorithms working with 16 bits), it is reasonable to represent the values using a scheme such as

$$\alpha_{16bit,b} = 2b - B + 1 \qquad 0 \le b < B \tag{6.5}$$

Using these definitions, the expected value of a function of a signal sample s_μ and its quantized value $q(s_\mu)$ is given as

$$\left\langle f(s_\mu, q(s_\mu))\right\rangle_{s_\mu} = \sum_{b=0}^{B-1} \int_{s_\mu=\beta_b}^{\beta_{b+1}} f(s_\mu, \alpha_b) p(s_\mu) ds_\mu \tag{6.6}$$

where the expected value is computed with respect to the probability density function $p(s_\mu)$ of the sample amplitude.

6.1.2 Matched Filter

Quantization errors are evaluated with respect to a certain criterion. In the case of navigation signal processing, the output SNR of a matched filter is a reasonable criterion, because the matched filter represents a sufficient statistic (see Section 4.4). Furthermore, the (absolute) postcorrelation SNR (described below) relates to acquisition sensitivity or tracking performance. In a direct-spreading communication system, the bit error rate would be an adequate criterion [1].

To analyze quantization errors for navigation signal processing, the received-signal samples are considered to be a vector of random variables **S**. The vector is modeled as the sum of *two independent wide-sense stationary stochastic processes* **R** (navigation signal) and **N** (noise),

$$\mathbf{S} = a\mathbf{R} + \mathbf{N} \tag{6.7}$$

Note that mathematical symbols with capital letters denote random variables, whereas lower-case letters denote the realization of random variables. All signals are assumed to be real-valued. The signal process is assumed to be a Gaussian process. A Gaussian signal process is a reasonable assumption if, for example, the superposition of all transmitted GNSS signals is considered. On the other hand, it should be noted that the subsequent computations can also be carried out using other probability density functions. For example, CW interference is modeled as a $(1 - r^2)^{-1/2}$ probability density function. The statistics of the Gaussian signal process are given as

$$\begin{gathered}\left\langle R_\mu \right\rangle_{\mathbf{R}} = 0 \\ \left\langle R_\mu R_\nu \right\rangle_{\mathbf{R}} = R_{rr}(\mu - \nu) \\ R_{rr}(0) = 1, \quad R_{rr}(\mu - \nu) = R_{rr}(\nu - \mu)\end{gathered} \tag{6.8}$$

and the probability density function as

$$\begin{gathered}p(r_\mu) = \frac{1}{\sqrt{2\pi}} \exp\left\{ -\frac{r_\mu^2}{2} \right\} \\ p(r_\mu, r_\nu) = \frac{1}{2\pi\sqrt{\det Q}} \exp\left\{ -\frac{1}{2} (r_\mu \quad r_\nu) Q^{-1} \begin{pmatrix} r_\mu \\ r_\nu \end{pmatrix} \right\} \\ Q = \begin{pmatrix} 1 & R_{rr}(\mu - \nu) \\ R_{rr}(\mu - \nu) & 1 \end{pmatrix}\end{gathered} \tag{6.9}$$

The statistics of the (white) noise process are given as

$$\begin{gathered}\left\langle N_\mu \right\rangle_{\mathbf{N}} = 0 \\ \left\langle N_\mu N_\nu \right\rangle_{\mathbf{N}} = \delta_{\mu,\nu}\end{gathered} \tag{6.10}$$

and the probability density function as

$$
\begin{aligned}
p(n_\mu) &= \frac{1}{\sqrt{2\pi}} \exp\left\{-\frac{n_\mu^2}{2}\right\} \\
p(n_\mu, n_\nu) &= \frac{1}{2\pi} \exp\left\{-\frac{n_\mu^2}{2} - \frac{n_\nu^2}{2}\right\}
\end{aligned}
\tag{6.11}
$$

The signal power is given by α^2 and the corresponding C/N_0 value as (see Section A.2.2)

$$
C/N_0 = \frac{f_s a^2}{2} \tag{6.12}
$$

The matched filter output operating on the quantized signal samples is given as

$$
t = q(\mathbf{s}^T) \cdot \mathbf{r} = \sum_{\mu=1}^{L} q(s_\mu) r_\mu \tag{6.13}
$$

If the matched filter is considered to be a random variable, it is analogously written as

$$
T = q(\mathbf{S}^T) \cdot \mathbf{R} = \sum_{\mu=1}^{L} q(S_\mu) R_\mu \tag{6.14}
$$

The relative SNR of the matched filter is defined as the ratio of the squared expected value divided by the variance of the matched filter as

$$
SNR_{rel} = \frac{\langle T \rangle_{\mathbf{R,N}}^2}{\mathrm{var}\langle T \rangle_{\mathbf{R,N}}} = \frac{\left\langle q(\mathbf{S}^T) \cdot \mathbf{R} \right\rangle_{\mathbf{R,N}}^2}{\left\langle \left(q(\mathbf{S}^T) \cdot \mathbf{R}\right)^2 \right\rangle_{\mathbf{R,N}} - \left\langle q(\mathbf{S}^T) \cdot \mathbf{R} \right\rangle_{\mathbf{R,N}}^2} \tag{6.15}
$$

The relative SNR is related to the bit error rate and to the variability in time of signal power estimates. The relative SNR stands in analogy to the definition used in [7].

The absolute SNR of the matched filter is defined as the ratio of the squared expected value divided by the variance of the matched filter in the noise-only case as

$$
SNR_{abs} = \frac{\left\langle q(\mathbf{S}^T) \cdot \mathbf{R} \right\rangle_{\mathbf{R,N}}^2}{\left\langle (q(\mathbf{N}^T) \cdot \mathbf{R})^2 \right\rangle_{\mathbf{R,N}} - \left\langle q(\mathbf{N}^T) \cdot \mathbf{R} \right\rangle_{\mathbf{R,N}}^2} \tag{6.16}
$$

The absolute SNR relates to the noncentrality parameters determining the acquisition sensitivity described in Section 5.4.

Both SNR values are defined at baseband and no carrier is considered. Including the carrier is not straightforward and, eventually, Monte Carlo methods must be used [7]. As a consequence, the relationship between the SNR values and the high-rate pseudorange variances can only be given for special cases. For example, if a complex-valued baseband input signal is considered and the I-channel contains the signal plus the noise and the Q-channel contains only the noise, then the variance of the estimated carrier phase is related to the absolute SNR.

6.1.3 Evaluation of Expected Values

To evaluate the SNRs as a function of the number of output values B and an ADC range Δ, a number of expected values need to be calculated, which is illustrated in the following examples.

The expected value of the matched filter output with respect to the noise and the stochastic signal is given as

$$\begin{aligned}\left\langle q(\mathbf{S}^T)\cdot\mathbf{R}\right\rangle_{\mathbf{R,N}} &= \sum_{\mu=1}^{L}\left\langle q(S_\mu)R_\mu\right\rangle_{\mathbf{R,N}} \\ &= \sum_{\mu=1}^{L}\left\langle\left\langle q(S_\mu)\middle|R_\mu\right\rangle_{\mathbf{N}} R_\mu\right\rangle_{\mathbf{R}} = \sum_{\mu=1}^{L}\left\langle\tilde{R}_\mu R_\mu\right\rangle_{\mathbf{R}} = L\left\langle\tilde{R}_0 R_0\right\rangle_{\mathbf{R}}\end{aligned} \tag{6.17}$$

where the conditional expected value with respect to the noise of a quantized sample under the assumption of a constant signal value is defined as

$$\begin{aligned}\tilde{r}_\mu &= \left\langle q(S_\mu)\middle|r_\mu\right\rangle_{\mathbf{N}} = \left\langle q(ar_\mu + N_\mu)\right\rangle_{\mathbf{N}} = \int_{n_\mu=-\infty}^{\infty} q(ar_\mu + n_\mu)p(n_\mu)dn_\mu \\ &= \sum_{b=0}^{B-1}\int_{n_\mu=\beta_b-ar_\mu}^{\beta_{b+1}-ar_\mu}\alpha_b p(n_\mu)dn_\mu\end{aligned} \tag{6.18}$$

an expression that can be evaluated analytically for Gaussian noise. The resulting expressions are, however, lengthy and do not provide more insight; they are, consequently, not given here. For non-Gaussian noise, the expression can also be evaluated with numerical integration.

The conditional expected value with respect to noise can also be considered as a random variable (actually a function of the random variable R_μ) and is then mathematically written as

$$\tilde{R}_\mu = \left\langle q(S_\mu)\middle|R_\mu\right\rangle_{\mathbf{N}} = \left\langle q(aR_\mu + N_\mu)\right\rangle_{\mathbf{N}} = \sum_{b=0}^{B-1}\int_{n_\mu=\beta_b-aR_\mu}^{\beta_{b+1}-aR_\mu}\alpha_b p(n_\mu)dn_\mu \tag{6.19}$$

Further analogous relationships between functions of random variables and their realizations are not explicitly given here.

A second-order expected value is defined as

$$\widetilde{r_\mu^2} = \left\langle q(ar_\mu + N_\mu)^2 \right\rangle_{\mathbf{N}} = \sum_{b=0}^{B-1} \int_{n_\mu=\beta_b - ar_\mu}^{\beta_{b+1}-ar_\mu} \alpha_b^2 p(n_\mu) dn_\mu \tag{6.20}$$

In cases where no signal is present, the first- and second-order moments of the quantized noise are defined as

$$\begin{aligned} \tilde{n}_\mu &= \left\langle q(N_\mu) \right\rangle_{\mathbf{N}} = \sum_{b=0}^{B-1} \int_{n_\mu=\beta_b}^{\beta_{b+1}} \alpha_b p(n_\mu) dn_\mu \\ \widetilde{n_\mu^2} &= \left\langle q(N_\mu)^2 \right\rangle_{\mathbf{N}} = \sum_{b=0}^{B-1} \int_{n_\mu=\beta_b}^{\beta_{b+1}} \alpha_b^2 p(n_\mu) dn_\mu \end{aligned} \tag{6.21}$$

The first- and second-order moments are deterministic quantities.

The expected value of the matched filter is obtained using numerical integration of

$$\left\langle q(\mathbf{S}^T) \cdot \mathbf{R} \right\rangle_{\mathbf{R},\mathbf{N}} = L \int_{r_0=-\infty}^{\infty} \tilde{r}_0 r_0 p(r_0) dr_0 \tag{6.22}$$

For a Gaussian amplitude distribution, this expression can be evaluated analytically. It can be obtained straightforwardly by using an analytical mathematic software package.

The expected value of the squared matched filter output is given as

$$\begin{aligned} \left\langle (q(\mathbf{S}^T) \cdot \mathbf{R})^2 \right\rangle_{\mathbf{R},\mathbf{N}} &= \sum_{\mu,\nu=1}^{L} \left\langle q(S_\mu) R_\mu q(S_\nu) R_\nu \right\rangle_{\mathbf{R},\mathbf{N}} \\ &= \sum_{\substack{\mu,\nu=1 \\ \mu \neq \nu}}^{L} \left\langle \tilde{R}_\mu R_\mu \tilde{R}_\nu R_\nu \right\rangle_{\mathbf{R}} + \sum_{\mu=1}^{L} \left\langle \widetilde{R_\mu^2} R_\mu^2 \right\rangle_{\mathbf{R}} \\ &= \sum_{\substack{\mu,\nu=1 \\ \mu \neq \nu}}^{L} \left(\left\langle \tilde{R}_\mu R_\mu \right\rangle_{\mathbf{R}} \left\langle \tilde{R}_\nu R_\nu \right\rangle_{\mathbf{R}} + \left\langle R_\mu R_\nu \right\rangle_{\mathbf{R}} \left\langle \tilde{R}_\mu \tilde{R}_\nu \right\rangle_{\mathbf{R}} + \left\langle \tilde{R}_\mu R_\nu \right\rangle_{\mathbf{R}} \left\langle R_\mu \tilde{R}_\nu \right\rangle_{\mathbf{R}} \right) \\ &+ L \left\langle \widetilde{R_0^2} R_0^2 \right\rangle_{\mathbf{R}} \end{aligned} \tag{6.23}$$

The last line uses a relationship for the fourth moment of correlated Gaussian random variables. Assuming the signal samples R_μ and R_ν are uncorrelated if $|\mu - \nu| > L'$, an auxiliary index $\lambda = \mu - \nu$ is introduced; and thus,

$$
\begin{aligned}
&\left\langle (q(\mathbf{S}^T)\cdot\mathbf{R})^2 \right\rangle_{\mathbf{R},\mathbf{N}} \\
&= (L^2-L)\left\langle \tilde{R}_0 R_0 \right\rangle_{\mathbf{R}}^2 + L\left\langle \widetilde{R_0^2 R_0^2} \right\rangle_{\mathbf{R}} \\
&+\sum_{v=1}^{L}\sum_{\substack{\lambda=-L',\\ \lambda\neq 0}}^{L'} \left(\left\langle R_{v+\lambda}R_v \right\rangle_{\mathbf{R}} \left\langle \tilde{R}_{v+\lambda}\tilde{R}_v \right\rangle_{\mathbf{R}} + \left\langle \tilde{R}_{v+\lambda}R_v \right\rangle_{\mathbf{R}} \left\langle R_{v+\lambda}\tilde{R}_v \right\rangle_{\mathbf{R}} \right) \\
&= (L^2-L)\left\langle \tilde{R}_0 R_0 \right\rangle_{\mathbf{R}}^2 + L\left\langle \widetilde{R_0^2 R_0^2} \right\rangle_{\mathbf{R}} + L\sum_{\substack{\lambda=-L',\\ \lambda\neq 0}}^{L'} \left(\left\langle R_\lambda R_0 \right\rangle_{\mathbf{R}} \left\langle \tilde{R}_\lambda \tilde{R}_0 \right\rangle_{\mathbf{R}} + \left\langle \tilde{R}_\lambda R_0 \right\rangle_{\mathbf{R}} \left\langle R_\lambda \tilde{R}_0 \right\rangle_{\mathbf{R}} \right) \\
&= (L^2-L)\left\langle \tilde{R}_0 R_0 \right\rangle_{\mathbf{R}}^2 + L\left\langle \widetilde{R_0^2 R_0^2} \right\rangle_{\mathbf{R}} + 2L\sum_{\lambda=1}^{L'} \left(\left\langle R_\lambda R_0 \right\rangle_{\mathbf{R}} \left\langle \tilde{R}_\lambda \tilde{R}_0 \right\rangle_{\mathbf{R}} + \left\langle \tilde{R}_\lambda R_0 \right\rangle_{\mathbf{R}}^2 \right)
\end{aligned}
\tag{6.24}
$$

The relative SNR is given as

$$
\begin{aligned}
&SNR_{rel} = \\
&= \frac{L^2\left\langle \tilde{R}_0 R_0 \right\rangle_{\mathbf{R}}^2}{(L^2-L)\left\langle \tilde{R}_0 R_0 \right\rangle_{\mathbf{R}}^2 + L\left\langle \widetilde{R_0^2 R_0^2} \right\rangle_{\mathbf{R}} + 2L\sum_{\lambda=1}^{L'} \left(\left\langle R_\lambda R_0 \right\rangle_{\mathbf{R}} \left\langle \tilde{R}_\lambda \tilde{R}_0 \right\rangle_{\mathbf{R}} + \left\langle \tilde{R}_\lambda R_0 \right\rangle_{\mathbf{R}}^2 \right) - L^2\left\langle \tilde{R}_0 R_0 \right\rangle_{\mathbf{R}}^2} \\
&= \frac{L^2\left\langle \tilde{R}_0 R_0 \right\rangle_{\mathbf{r}}^2}{L\left\langle \widetilde{R_0^2 R_0^2} \right\rangle_{\mathbf{R}} - L\left\langle \tilde{R}_0 R_0 \right\rangle_{\mathbf{R}}^2 + 2L\sum_{\lambda=1}^{L'} \left(\left\langle R_\lambda R_0 \right\rangle_{\mathbf{R}} \left\langle \tilde{R}_\lambda \tilde{R}_0 \right\rangle_{\mathbf{R}} + \left\langle \tilde{R}_\lambda R_0 \right\rangle_{\mathbf{R}}^2 \right)}
\end{aligned}
\tag{6.25}
$$

The absolute SNR is similarly evaluated as

$$
\begin{aligned}
SNR_{abs} &= \frac{L^2\left\langle \tilde{R}_0 R_0 \right\rangle_{\mathbf{R}}^2}{L\left\langle \widetilde{n_0^2 R_0^2} \right\rangle_{\mathbf{R}} - L\left\langle \tilde{n}_0 R_0 \right\rangle_{\mathbf{R}}^2 + 2L\sum_{\lambda=1}^{L'} \left(\left\langle R_\lambda R_0 \right\rangle_{\mathbf{R}} \left\langle \tilde{n}_\lambda \tilde{n}_0 \right\rangle_{\mathbf{R}} + \left\langle \tilde{n}_\lambda R_0 \right\rangle_{\mathbf{R}}^2 \right)} \\
&= \frac{L^2\left\langle \tilde{R}_0 R_0 \right\rangle_{\mathbf{R}}^2}{L\left\langle \widetilde{n_0^2 R_0^2} \right\rangle_{\mathbf{R}}} = \frac{L^2\left\langle \tilde{R}_0 R_0 \right\rangle_{\mathbf{R}}^2}{L\widetilde{n_0^2}\left\langle R_0^2 \right\rangle_{\mathbf{R}}} = \frac{L\left\langle \tilde{R}_0 R_0 \right\rangle_{\mathbf{R}}^2}{\widetilde{n_0^2}}
\end{aligned}
\tag{6.26}
$$

because the expected value of a quantized sample vanishes if no signal is present and the noise process is of zero mean.

6.1.4 Infinite Number of Bits

In cases of loss-less quantization (i.e., the quantized sample equals the input sample), the following identities hold:

$$
\begin{aligned}
\tilde{R}_\mu &= aR_\mu \\
\widetilde{R_\mu^2} &= 1 + a^2 R_\mu^2 \\
\left\langle \tilde{R}_0 R_0 \right\rangle_{\mathbf{R}} &= a\left\langle R_0^2 \right\rangle_{\mathbf{R}} = a \\
\left\langle \widetilde{R_0^2} R_0^2 \right\rangle_{\mathbf{R}} &= \left\langle \left(1 + a^2 R_0^2\right) R_0^2 \right\rangle_{\mathbf{R}} = 1 + 3a^2 \\
a^2 \left\langle R_\lambda R_0 \right\rangle_{\mathbf{R}} = \left\langle \tilde{R}_\lambda \tilde{R}_0 \right\rangle_{\mathbf{R}} &= a\left\langle \tilde{R}_\lambda R_0 \right\rangle_{\mathbf{R}} = a^2 R_{rr}(\lambda) \\
\left\langle \widetilde{n_0^2} \right\rangle_{\mathbf{R}} &= 1
\end{aligned}
\tag{6.27}
$$

Under this assumption, idealized SNR values can be obtained, which serve as comparison values for the quantized SNR values. The difference between idealized SNR values and quantized SNR values expressed in decibels gives the quantization loss.

The idealized relative SNR is given as

$$
SNR_{rel,\,ideal} = \frac{L^2 a^2}{L(1+3a^2) - La^2 + 4La^2 \sum_{\lambda=1}^{L'} R_{rr}(\lambda)^2} = \frac{La^2}{1 + 2a^2 + 4a^2 \sum_{\lambda=1}^{L'} R_{rr}(\lambda)^2}
\tag{6.28}
$$

and the idealized absolute SNR is given as

$$
SNR_{abs,\,ideal} = La^2 \tag{6.29}
$$

6.1.5 Numerical Evaluation

Considering a scenario derived from the settings in Table 10.13, the quantization loss for the absolute and relative SNR shall be evaluated in addition to the squared expected value of the matched filter. The simulation parameters are summarized in Table 6.1. The signal correlation function approximates a GPS C/A correlation function, sampled at 4 MHz.

Table 6.1 Quantization Loss Parameters

Parameter	*Value*
C/N_0	40 dBHz, 50 dBHz
Number of bits	1–4
Number of output values B	2–16
Signal correlation function	$R_{rr}(\lambda) = 1 - \lambda/4,\ 0 \le \lambda \le 4$
Sample rate	4 MHz
Sample type	Real

First, the expected value of the absolute SNR shall be investigated. It relates to the quantized signal-to-noise ratio $\widetilde{C/N_0}$ value by

$$\widetilde{C/N_0} = SNR_{abs} \frac{f_s}{2L} \tag{6.30}$$

The quantized signal-to-noise ratio $\widetilde{C/N_0}$ is shown in Figure 6.1 for a different number of ADC bits, plotted as a function of the analog C/N_0 value. In all cases, an ADC input range $\Delta = 3$ was maintained. The figure shows that the quantized value is generally smaller than the input value. For high input C/N_0 values, the output levels off; the ADC is in saturation.

Figures 6.2 and 6.3 show the quantization loss for the absolute and the relative SNR plotted as a function of the input range Δ. Figure 6.2 considers a case of lower input signal power (40 dBHz) and Figure 6.3 considers a case of higher signal input power (50 dBHz).

The quantization loss is almost identical for the absolute and relative SNR if the 40-dBHz case is considered. The optimum Δ lies between 2 and 2.75. For the 50-dBHz case, the signal itself significantly influences the matched filter variance and the quantization losses are different for the absolute and the relative cases. In fact, the quantization loss might also become slightly negative (i.e., it becomes a gain) for the relative SNR value. This reflects the fact that the variance of the relative SNR is low if the ADC is in saturation. In this case, however, the relative SNR value becomes biased.

In the work by van Dierendonck, similar curves are presented that exhibit sharper peaks [3]. This is explained by the fact that van Dierendonck analyzes the

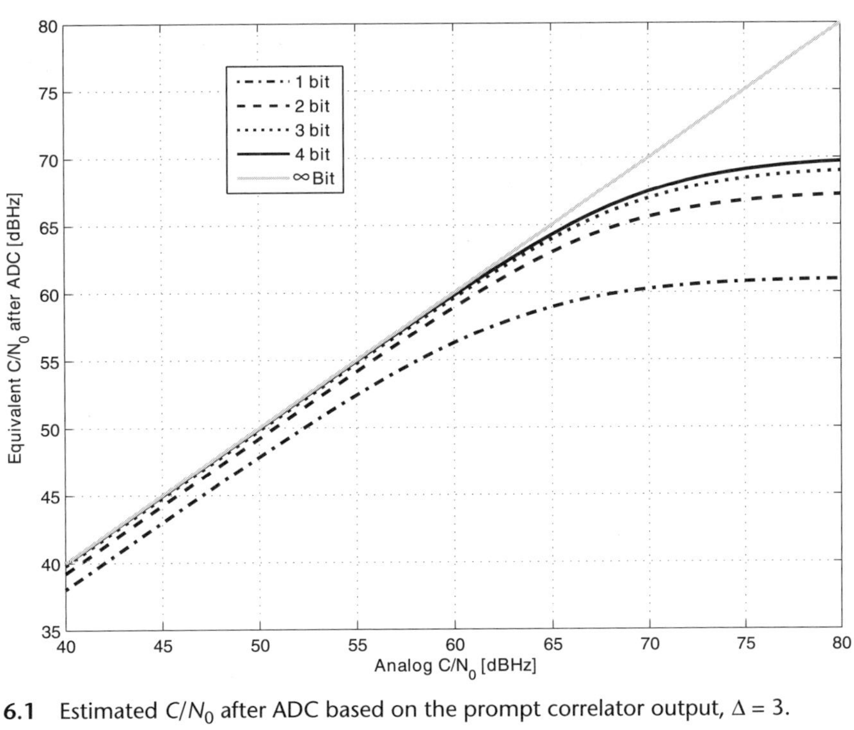

Figure 6.1 Estimated C/N_0 after ADC based on the prompt correlator output, $\Delta = 3$.

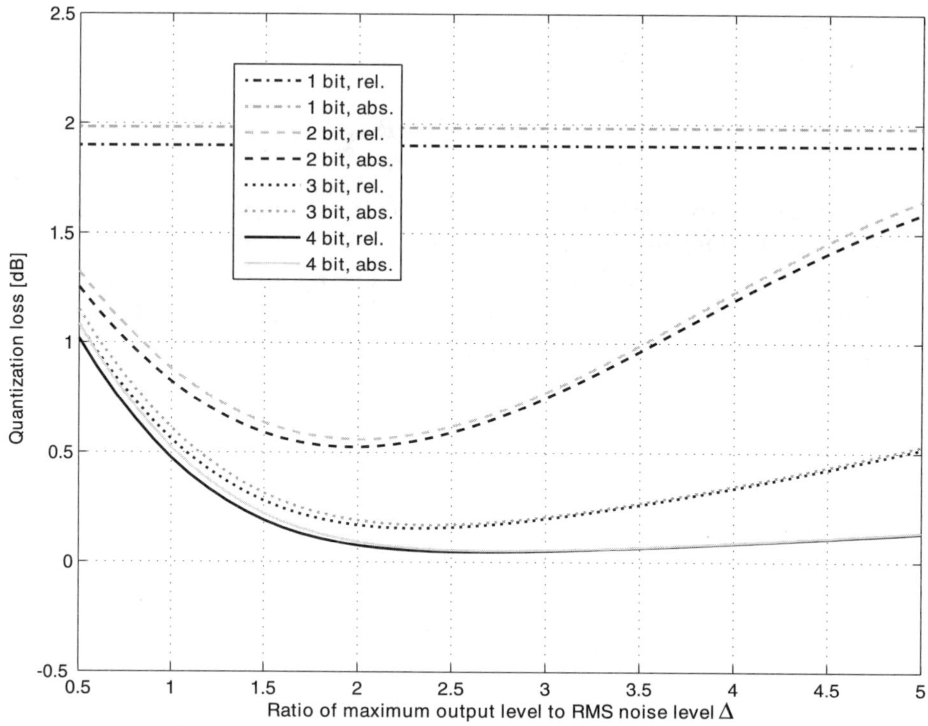

Figure 6.2 Relative and absolute quantization loss for a 1–4 bit ADC, $C/N_0 = 40$ dBHz.

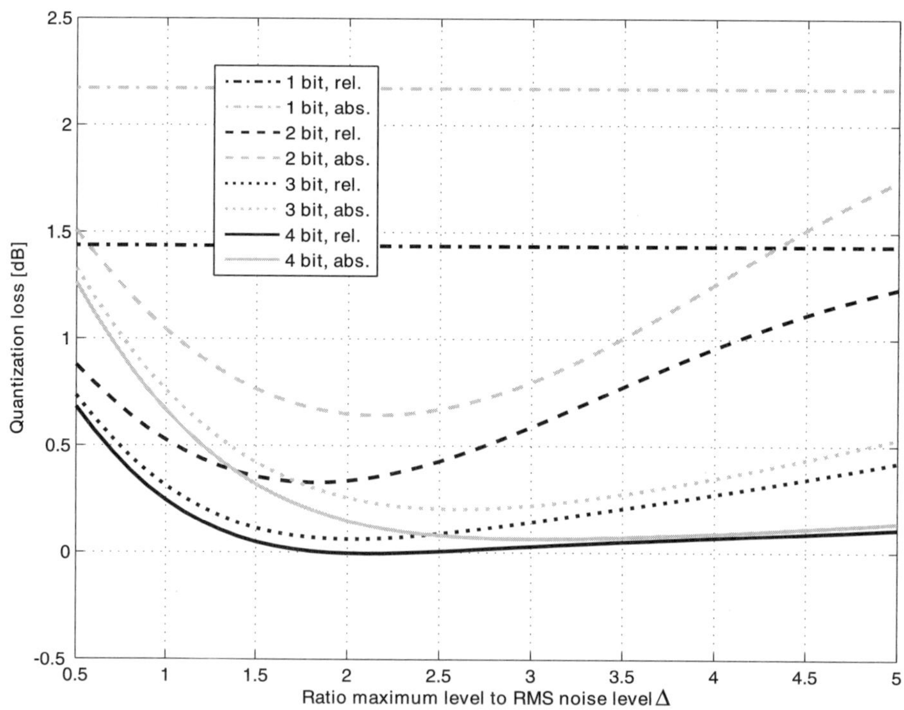

Figure 6.3 Relative and absolute quantization loss for a 1–4 bit ADC, $C/N_0 = 50$ dBHz.

effect of quantization noise *plus* low-bit signal processing, whereas in this work the quantization noise is considered isolated and the signal processing is working with a high number of bits. In van Dierendonck's work, the maximum threshold L is considered, which relates to the input range Δ via $BL = \Delta (B - 2)$ [i.e., for 2-bit ($B = 4$) we obtain $2L = \Delta$]. Generally, the Δ values for the minimum quantization loss of Figure 6.2 agree with van Dierendonck's Figure 16.

Finally, Figure 6.4 shows the normalized variance of the ADC output if only noise is applied at the input. The normalized variance measured at the ADC output, if (6.5) is used to represent the samples, can be evaluated as

$$\widetilde{n^2_{16bit;\mu}} = \frac{4}{\delta^2}\widetilde{n^2_\mu} \tag{6.31}$$

The normalized variance can be used to ensure that the ADC quantization thresholds are optimally chosen, or equivalently, that the variable gain before the ADC is set correctly. For a 2-bit ADC and the 40-dBHz case, the optimum absolute SNR value is obtained for $\Delta = 2$ (see Figure 6.2), corresponding to a normalized quantized noise sample variance of 3.53. This implies a quantized sample standard deviation of 1.88. The receiver can calculate the standard deviation of the quantized samples and compare the obtained value against the given optimum value. Eventually, the receiver may adjust the front-end gain.

It should be noted that measuring the quantized noise power is strictly possible if only noise is present at the ADC input and no signal. However, for typical GNSS applications, the signal power can be ignored with respect to the noise power be-

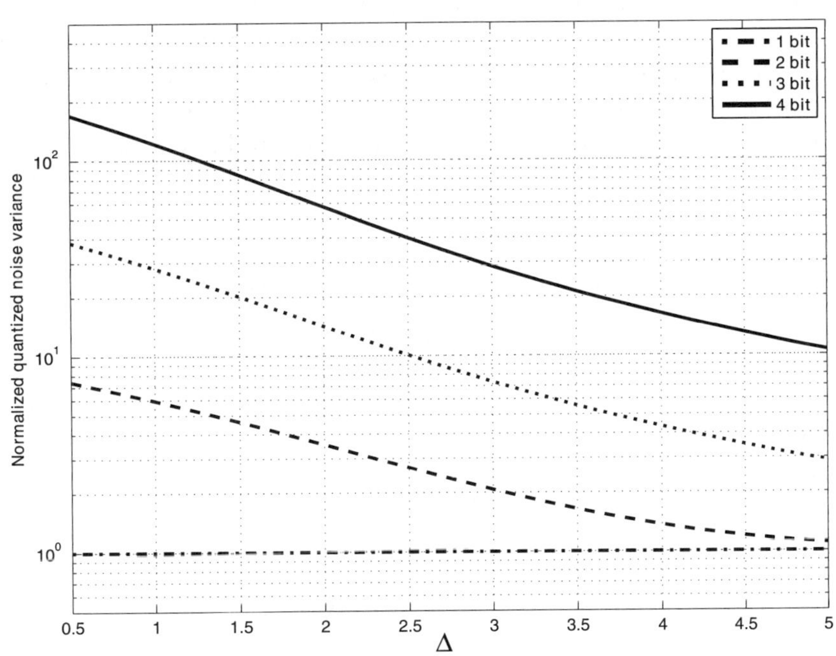

Figure 6.4 Normalized noise after quantization.

cause the total power P_{tot} is the sum of the signal power C and the noise power N and is written as

$$P_{tot} = C + N = C/N_0 \frac{N}{B} + N = N\left(1 + \frac{C/N_0}{B}\right) \tag{6.32}$$

where B is the dual-sided noise-equivalent bandwidth. C/N_0 is the cumulative signal-to-noise ratio accounting for all transmitters. As long as $C/N_0 << B$, the total quantized noise power can also be measured in the presence of one or more signals.

6.2 Noise-Floor Determination

Throughout this work, it is usually assumed that the noise variance is one for real-valued noise and two for complex-valued noise. However, Section 6.1 showed that, after ADC, the noise power is generally different from one for real-valued noise. From the example above, we could see that for a 2-bit ADC and $\Delta = 2$, the quantized noise power is 3.53 for an optimal quantization of a real-valued input signal. The difference between the true and the assumed noise variance has some (minor) consequences on the signal processing, which are discussed here. Signal processing relies on correlation values and the subsequent evaluation of them. Two cases shall be considered.

In the first case, only ratios of correlators are processed. This is done to obtain the code-phase, Doppler, or the carrier-phase estimates, as described in Section 4.3. In this case, the formulas can also be directly applied if the true noise power differs from one (or, respectively, two). The nominator and denominator scale identically and the effect of the different noise power cancels.

The second case relates to signal power estimation and determination of thresholds for signal detection. For example, a single coherent integration, as described in Section 5.7.1, compares the squared correlator value against a threshold γ. For convenience, the respective equation (5.13) shall be repeated here

$$|P(\tau, \omega)|^2 > \gamma \tag{6.33}$$

For nonunity noise power, this equation has to be modified as

$$|P(\tau, \omega)|^2 > \gamma \widetilde{n_\mu^2} \tag{6.34}$$

Other equations can be modified accordingly. For typical GNSS applications, the value of $\widetilde{n_\mu^2}$ is quite stable in time if the signals are broadcast in protected radio bands. Important exceptions only occur if (pulsed) interference is present. Pulsed interference shall be discussed in Section 6.3.

6.3 ADC Requirements for Pulse Blanking

The standard operation mode of a GNSS receiver implies that the noise floor is constant in time. For certain GNSS frequency bands (e.g., Galileo E5 or E6), this assumption is not valid and pulsed interference occurs. Pulsed interference might also occur because of the presence of pseudolite signals, which are described in Section 10.6.2.

As outlined in Section 10.6.2, pulse blanking is an effective countermeasure to mitigate the effect of the pulsed interference, as long as the pulses are received only for a small fraction of the time. In the following, some requirements for the ADC are formulated to ensure that pulse blanking works properly:

1. The front-end gain before the ADC must not react on the pulse (or least only very slowly).
2. ADC resolution and thresholds must be suitable to allow identification of pulses via an energy detector.
3. Noise-floor determination algorithms must exclude periods where pulses are present.

If these conditions are fulfilled, the algorithm of Section 10.6.2 can be applied.

6.3.1 Front-End Gain and Recovery Time

To properly handle pulsed interference by pulse blanking, it is required that the analog part of the front end not be affected by a received pulse *after* the pulse ended. In other words, it is expected that the received samples during the pulse are useless, but as soon as the pulse vanishes, the front end should return to normal operation. The time the front end needs is called *recovery time*.

Sometimes GNSS front ends include an automatic gain control (AGC) to control the ADC thresholds to obtain minimum quantization losses. AGCs are required because front ends might be used with different antennas, each having integrated LNAs with different gains. Furthermore, temperature variations of antenna/front-end internal amplifiers and mixers might require adjusting the AGC. If pulse interference is present, the AGC shall not react on this interference (i.e., the AGC time constant should be much larger than the pulse duration). An optimal solution could be to control the AGC via software, and the AGC parameters are set using noise-variance estimates obtained after pulse blanking.

6.3.2 Pulse Blanking

Pulse blanking replaces periods of the received signal contaminated with pulses by a vanishing signal

$$s_\mu = 0 \qquad \mu \in \text{pulse period} \tag{6.35}$$

Replacing the signal samples by "0" ensures that correlation values are minimally distorted. On the other hand, noise-floor determination can be corrupted if an estimation of $\widetilde{n_\mu^2}$ is carried out using blanked samples. If it is not possible for the noise-floor estimation to be aware of periods with blanked samples, the method of *pulse clipping*

$$s_\mu = \begin{cases} s_\mu & |s_\mu| < s_{\max} \\ s_{\max}\,\mathrm{sgn}(s_\mu) & |s_\mu| \geq s_{\max} \end{cases} \qquad \mu \in \text{pulse period} \tag{6.36}$$

with $s_{\max} = \sqrt{\widetilde{n_\mu^2}}$ or *sample randomization*

$$s_\mu = n_\mu \qquad n_\mu \sim \mathrm{N}(0, \widetilde{n_\mu^2}) \qquad \mu \in \text{pulse period} \tag{6.37}$$

can be applied, where either the samples are clipped if their magnitude exceeds a certain value, or the samples are replaced by simulated noise samples. Both methods further degrade the parameter estimates but allow continuous noise-floor estimation. Pulse clipping is automatically applied if a 1-bit ADC is used.

6.3.3 ADC Resolution

The ADC resolution needs to be sufficiently large to allow usage of the energy detector as described in Section 10.6.2. The variance of the energy detector output should not be affected by ADC saturation effects. Otherwise, saturation effects hinder the energy detector from detecting the presence of an interfering signal. Obviously, the number of bits needs to be larger than one (an energy detector working with 1-bit samples always gives the same output).

To further investigate the influence of saturation effects on an energy detector, the following setting is considered. Thermal noise samples n_μ with unity variance plus an interference signal ar_μ are received. The interference signal samples have a variance of α^2 and its amplitude distribution is assumed to be Gaussian. The spectral characteristics of the interfering signal (and, therefore, its autocorrelation function) are left unspecified.

An unbiased estimate of the received interference power is obtained by subtracting the noise-only energy-detector output from the noise-plus-interference energy-detector output:

$$\hat{a}^2 = \frac{1}{L'} \sum_{\mu=1}^{L'} q(s_\mu)^2 - \frac{1}{L''} \sum_{\mu=1}^{L''} q(n_\mu)^2 \tag{6.38}$$

In the case of an infinite number of bits, the expected value corresponds to the interference power, which is written as

$$\left\langle \hat{a}^2 \right\rangle_{\mathbf{R,N}} = \left\langle \widetilde{r_0^2} \right\rangle_{\mathbf{R}} - \widetilde{n_0^2} \underset{B \to \infty}{\to} a^2 \tag{6.39}$$

The ratio between interference-power and noise-power spectral density J/N_0 is defined equivalently to the C/N_0 value as

$$J / N_0 = \frac{a^2 f_s}{2} \tag{6.40}$$

and a J/N_0 estimate is obtained via

$$\widehat{J/N_0} = \frac{\hat{a}^2 f_s}{2} \tag{6.41}$$

The interference to (total) noise power ratio J / N is the J / N_0 value multiplied by the bandwidth

$$J/N = a^2 \tag{6.42}$$

and its estimate is

$$\widehat{J/N} = \hat{a}^2 \tag{6.43}$$

Using the methodology from Section 6.1, the estimated interference-to-noise density value J/N_0 and the interference-to-total-noise power J/N is plotted in Figure 6.5. For the J/N_0 plot, a bandwidth of 2 MHz was assumed. Linear ADCs were assumed to work with 2–4 bits. The ADC threshold δ was identical for all cases for ease of comparison.

The J/N graph is more universal because it is independent of the assumed bandwidth. Below a J/N value of 0 dB, all the estimates are linear; only the 2-bit case has a larger offset of a few decibels. For high J/N values, the estimates level off and the ADC is in saturation. In this case, however, the relative SNR value becomes biased.

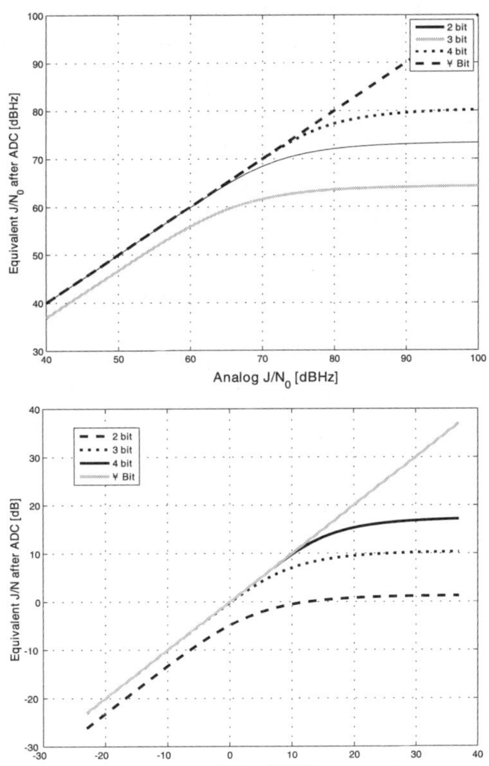

Figure 6.5 Estimated interference power after ADC based on an energy detector, $\delta = 1$.

Considering the proposed navigation system of Chapter 10, it was stated in Section 10.6.2.2 that the 1-ms energy-detector limit for pseudolite pulses was J/N_0 = 53 dBHz, a value which is clearly in the linear region for all ADC types. Therefore, pulse detection for the proposed navigation system of Chapter 10 can even be performed with a 2-bit ADC.

6.4 Handling Colored Noise

In many receiver implementations, it happens that the noise-power spectral density after ADC is nonflat. *Colored noise* appears. This case may happen because of front-end filter imperfections, which show magnitude variations in the passband. Additionally, it may happen that noise, such as interfering signals, are broadcast on the navigation signal band, effectively increasing the noise-power spectral density in certain frequency regions.

If a colored-noise signal is treated like a white-noise signal, performance degradations are the consequence. The estimation filters do not weight the spectral regions properly. Spectral regions with more noise need to be downweighted and spectral regions with less noise need to be upweighted. An estimator designed under the white-noise assumption, however, treats all spectral regions equally.

Three possibilities exist in a multibit navigation receiver to process colored-noise signals optimally. The first method (spectral whitening) consists of filtering the incoming signal such that the filtered signal has white noise. This implies a change of the reference signals. The second method consists of applying a filter on the generated reference signals and leaving the incoming signal untouched. The third method leaves the reference signals untouched and filters the incoming signal. Note that from now on a complex-valued signal representation will be used.

All methods can be directly derived from the discussion of the sufficient statistic t of Section 4.4, which is written according to (4.164) as

$$t(\mathbf{s}) = \mathbf{r}(\mathbf{q}_b)^* Q^{-1}\mathbf{s} \tag{6.44}$$

Here $\mathbf{r}(\mathbf{q}_b)$ is a vector of reference-signal samples generated at various correlation points $\mathbf{q}_b$. The vector of received-signal samples is denoted as $\mathbf{s}$ and Q is the Hermitian covariance matrix of the noise given as

$$\left\langle N_\mu N_\upsilon \right\rangle_{\mathrm{N}} = 0, \left\langle \bar{N}_\mu \bar{N}_\upsilon \right\rangle_{\mathrm{N}} = 0, \left\langle N_\mu \bar{N}_\upsilon \right\rangle_{\mathrm{N}} = 2Q_{\mu,\upsilon} \tag{6.45}$$

Assuming a wide-sense stationary noise, the covariance can be written as

$$Q_{\mu,\,\upsilon} = Q(\mu - \upsilon) \tag{6.46}$$

6.4.1 Spectral Whitening

The method of *spectral whitening* filters the incoming signal $\mathbf{s}$ and the reference signal $\mathbf{r}(\mathbf{q}_b)$ identically such that the filtered incoming signal $\mathbf{s}'$ is of white noise.

The magnitude of the filter's frequency response is determined by the colored-noise power spectral density, whereas the phase of the frequency response can be chosen arbitrarily.

Spectral whitening is written mathematically as

$$\begin{aligned} \mathbf{s}' = \Phi Q^{-1/2}\mathbf{s} \quad &\Leftrightarrow \quad Q^{1/2}\Phi^*\mathbf{s}' = \mathbf{s} \\ \mathbf{r}(\mathbf{q}_b)' = \Phi Q^{-1/2}\mathbf{r}(\mathbf{q}_b) \quad &\Leftrightarrow \quad Q^{1/2}\Phi^*\mathbf{r}(\mathbf{q}_b)' = \mathbf{r}(\mathbf{q}_b) \end{aligned} \tag{6.47}$$

where $\mathbf{\Phi}$ is an arbitrary unitary matrix $\mathbf{\Phi} * \mathbf{\Phi} = \mathbf{1}$, yielding

$$t(\mathbf{s}) = \mathbf{r}'(\mathbf{q}_b)^*\mathbf{s}' \tag{6.48}$$

The filter $\Phi Q^{-1/2}$ affects the signal and the noise part of the received-signal samples in the same way. Because the received-signal part is changed, the reference samples have to be adapted properly. The noise part N_μ' of the filtered samples is white because

$$\begin{aligned} \left\langle \mathbf{N}'\mathbf{N}'^* \right\rangle_{\mathbf{N}} &= \left\langle \Phi Q^{-1/2}\mathbf{N}\mathbf{N}^* Q^{-1/2}\Phi^* \right\rangle_{\mathbf{N}} = \Phi Q^{-1/2} \left\langle \mathbf{N}\mathbf{N}^* \right\rangle_{\mathbf{N}} Q^{-1/2}\Phi^* \\ &= \Phi Q^{-1/2} Q^* Q^{-1/2}\Phi^* = 2\Phi\Phi^* = 2\mathbf{1} \end{aligned} \tag{6.49}$$

The filter operation

$$\begin{aligned} &\mathbf{s}' = \Phi Q^{-1/2}\mathbf{s} \\ \Leftrightarrow\ &s_\mu{}' = \sum_{\lambda,\nu} \Phi(\mu-\lambda) Q^{-1/2}(\lambda-\nu) s_\nu \end{aligned} \tag{6.50}$$

can actually be realized by any digital filter. The $Q^{-1/2}$ term defines the magnitude response of the filter and $\mathbf{\Phi}$ defines the phase response. Because $\mathbf{\Phi}$ can be chosen arbitrarily and magnitude variations are usually small, a linear-phase finite impulse response (FIR) filter as described in the book by Diniz, da Silva, and Netto [8] could be an efficient way for implementation.

6.4.2 Modified Reference Signals

If colored noise is present, the reference signals can be modified to properly account for the noise correlations. This method was described in Section 4.4.4.

The modified reference signal is given by

$$\mathbf{r}(\mathbf{q}_b)' = Q^{-1}\mathbf{r}(\mathbf{q}_b) \tag{6.51}$$

and the sufficient statistic is obtained via simple correlation of the received-signal samples with the modified reference signal

$$t(\mathbf{s}) = \mathbf{r}'(\mathbf{q}_b)^*\mathbf{s} \tag{6.52}$$

The received signal itself remains unmodified and contains colored noise.

6.4.3 Overcompensation of the Incoming Signal

The method of overcompensation consists of applying a filter on the received-signal samples and leaving the reference signals unmodified. It is written as

$$\mathbf{s}' = Q^{-1}\mathbf{s} \tag{6.53}$$

The sufficient statistic is given as

$$t(\mathbf{s}) = \mathbf{r}(\mathbf{q}_b)^* \mathbf{s}' \tag{6.54}$$

The noise of the modified signal has as covariance the inverse matrix of the unmodified signal; that is,

$$\begin{aligned}\left\langle \mathbf{N}'\mathbf{N}'^* \right\rangle_{\mathbf{N}} &= \left\langle Q^{-1}\mathbf{N}\mathbf{N}^* Q^{-1} \right\rangle_{\mathbf{N}} = Q^{-1}\left\langle \mathbf{N}\mathbf{N}^* \right\rangle_{\mathbf{N}} Q^{-1} \\ &= Q^{-1} 2QQ^{-1} = 2Q^{-1}\end{aligned} \tag{6.55}$$

Effectively, the noise-power spectral density of the modified signal is the inverse of the power spectral density of the unmodified signal. The filter compensates for the nonwhite noise-power spectral density plus for the incorrectly chosen reference signal. Both effects add up and the filtered signal can be thought of as "overcompensated."

6.4.4 Implementation Issues

The presented algorithms are typically implemented after ADC. If the signal samples are represented by 16-bit values (see Chapter 9), the algorithms are expected to have negligible finite precision effects. Implementing a digital filter operating on the incoming signals is of small computational costs; a digital filter operating on all generated reference signals would need more computational power.

Two cases shall be considered: timely constant colored noise and timely variable colored noise. In the first case, the method of spectral whitening seems to be most applicable because the phase of the filter frequency response can be chosen freely (thereby allowing a linear-phase FIR filter) and the resulting signal is of white noise. White noise is favorable for spectral monitoring issues or for visualization purposes. If the power spectral density varies with time, the method of overcompensation could be used because this method requires only modification of the incoming signal. An important aspect of this method is, however, the requirement of determination of the timely variable spectrum.

6.5 Sub-Nyquist Sampling

The presented signal-processing algorithms have all been formulated in discrete time, based on a finite sample rate f_s. The deterministic part of the signal, the reference signals, and the sin/cos signals were all based on continuous time signals sampled with the sample rate f_s. The stochastic properties of received-noise samples have been described by a discrete covariance matrix. For basically all of the derivations, there was no need to work with the Fourier transform of the signal. Even in Section 9.5, the presented FFT techniques are only an efficient implementation of a convolution, which can be expressed in the time domain. There has never been a need to access a specific spectral portion of the received signal, which would contain specific navigational information. Additionally, Appendix A.3 shows that a cross-correlation function between two wide-sense stationary processes is independent of the sampling rate. Overall, there was never any requirement imposed on the sample rate.

Strictly speaking, the navigational performance (in the sense of a low variance of the estimated values) is basically a function of the navigation signal autocorrelation function (and its derivatives) plus the C/N_0 value. The sample rate does not directly enter. Several options for choosing the sample rate are possible: oversampling, Nyquist sampling, and sub-Nyquist sampling. A dedicated discussion of sub-Nyquist sampling can be found in [9].

For this discussion, we assume a broadcast navigation signal of dual-sided bandwidth B. A real-valued signal is assumed. The signal is bandpass filtered (also with a bandwidth of B) and then (bandpass) sampled with a sample rate f_s. Noise is received with a uniform power spectral density of N_0. The scheme is depicted in Figure 6.6.

From the definition of a correlation function (it is basically a time average over many signal samples), the important fact is observed that the expected value of a correlator is independent of the sampling rate (see Appendix A.3). In other words, sampling does not affect the shape of any correlation function. Sampling also leaves the total noise power invariant. Before sampling, the total noise power is spread over a bandwidth B, whereas after ADC the total noise power is spread over $f_s/2$. In the case where $u = 2B/f_s$ is an integer, and in some other cases, the noise-power spectral density after ADC is flat [9].

Overall, the following sampling cases occur:

- $f_s > 2B$—Oversampling: The complete band is sampled and due to the oversampling the noise is colored after ADC. Optional spectral whitening produces white noise, but does not affect the signal, which has no frequency component within the regions affected by the whitening process. After spectral whitening we can assume $u = 1$.

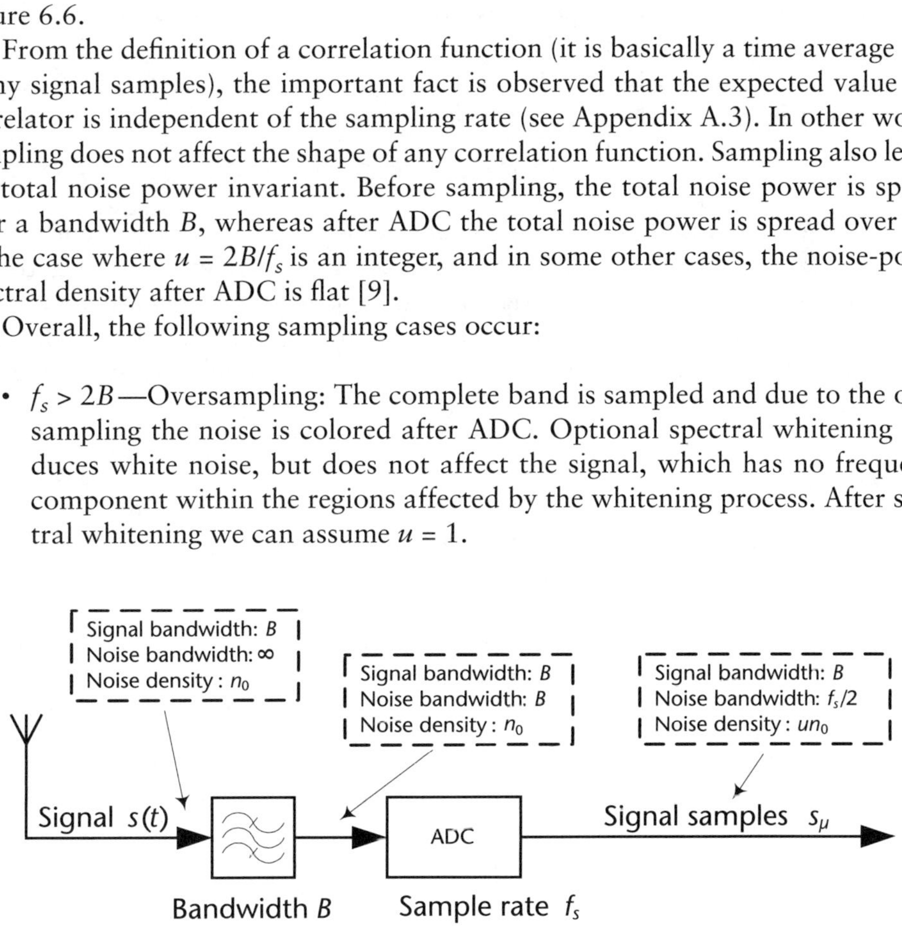

Figure 6.6 Sampling scheme for $u = 2B/f_s$ being an integer number larger than zero.

- $f_s = 2B$—Nyquist sampling: Signal and noise are sampled with the Nyquist rate. Ideal case, no losses, $u = 1$.
- $f_s < 2B$—Sub-Nyquist sampling: Aliasing of noise occurs, the signal autocorrelation function is not affected by the sub-Nyquist sampling, $u = 2B/f_s$.

When oversampling is used, the signal processing includes spectral regions, which are not covered by the navigation signal. These regions are ignored when correlating the received-signal samples with navigation signal replicas. However, oversampling can be quite useful because the high working sample rate allows small shifts of replica signals (e.g., early or late signals) with respect to a reference signal (e.g., the punctual signal). Typically it is much easier to shift a signal than to do a full regeneration of the signals (e.g., via resampling of Section 9.3). In fact, in hardware receivers, high oversampling rates are used to achieve small correlator spacings.

Nyquist sampling represents a kind of ideal sampling; the signal is optimally captured, the noise after ADC is white, and no noise-aliasing losses occur. A high-performance software receiver should work with Nyquist sample rates. Usually, finite analog-filter fall-off steepness effects can be ignored. The sample rate is (twice) the 3-dB bandwidth.

Sub-Nyquist sampling is the method of choice if the computational load is to be reduced, if the signal power is sufficiently high to cope with noise-aliasing losses, and if there is no need to shift reference signals. See the work by Pany and Eissfeller for further discussion [9]. Because of noise aliasing, the effective C/N_0 value is reduced by the undersampling factor u:

$$C/N_0 \to \frac{1}{u} C/N_0 \tag{6.56}$$

An important restriction for choosing the sample rate occurs if real-valued signals are used. In this case, the replica signal correlates with negative frequency components of the received signal. Those correlation values are required to average to zero during the correlation process, that is, (7.12) shall hold. This implies that after ADC the aliased doubled center frequency should be larger than 0 (see also Chapter 7).

References

[1] Amoroso, F., and Bricker, J. L., "Performance of the Adaptive A/D Converter in Combined CW and Gaussian Interference," *IEEE Trans. Commun.*, Vol. 34, No. 3, 1986, pp. 209–213.

[2] Proakis, J. G. and D. G. Manolakis, *Digital Signal Processing, Principles, Algorithms, and Applications,* 4th ed., Upper Saddle River, NJ: Prentice-Hall, 2007.

[3] van Dierendonck, A. J., "GPS Receivers," in *Global Positioning System: Theory and Applications, Vol. I,* pp. *329–407,* Parkinson, B. W., and J. J. Spilker, (eds.), Washington D.C.: American Institute of Aeronautics and Astronautics Inc., 1996.

[4] Chang, H., "Presampling Filtering, Sampling and Quantization Effects on Digital Matched Filter Performance," *Proc. Int. Telemetering Conference*, San Diego, CA, 1982, pp. 889–915.

[5] Spilker, J. J. Jr. and Natali, F. D., "Interference Effects and Mitigation Techniques," in *Global Positioning System: Theory and Applications, Vol. I,* pp. 717–771, Parkinson, B. W., and J. J. Spilker, (eds.), Washington, D.C.: American Institute of Aeronautics and Astronautics Inc., 1996.

[6] Betz, J. W., "Bandlimiting, Sampling, and Quantization for Modernized Spreading Modulations in White Noise," *Proc. Institute of Navigation National Technical Meeting (ION-NTM) 2008*, San Diego, CA, January 28–30, 2008.

[7] Betz, J. W. and Shnidman, N. R., "Receiver Processing Losses with Bandlimiting and One-Bit Sampling," *Proc. 20th Int. Technical Meeting of the Satellite Division of the Institute of Navigation (ION-GNSS) 2007*, Fort Worth, TX, September 25–28, 2007, pp. 1244–1256.

[8] Diniz, P. S. R., E. A. B. da Silva, and S. L. Netto, *Digital Signal Processing, System Analysis and Design*, Cambridge, U.K.: Cambridge University Press, 2002.

[9] Pany, T., and B. Eissfeller, "Code and Phase Tracking of Generic PRN Signals with Sub-Nyquist Sample Rates," *NAVIGATION, Journal of The Institute of Navigation*, Vol. 51, No. 2, 2004, pp. 143–159.

CHAPTER 7

Correlators

In Chapter 4, optimum parameter-estimation techniques were discussed and algorithms for MVUEs for the code phase, Doppler, signal power, and carrier phase were developed. The estimated parameters are derived from correlating the received-signal samples with different reference signals (waveforms), which are the received navigation signal at baseband $c(t)$ and its first derivative. This is only one possibility to choose the waveforms and ways to optimize the reference signals for multipath mitigation, for example, are shown in Section 8.2. Overall, a software receiver is very flexible with regards to signal correlation and the resulting tracking scheme shall be called waveform-based tracking.

This chapter introduces the waveform-based tracking scheme and compares it to the (hardware receiver oriented) correlator-based tracking scheme. Four different classes of waveforms are introduced that are used for carrier-phase, code-phase, and frequency estimation as well as for signal-quality monitoring. Difference correlators are analyzed that are the basis for the highly stable carrier-phase tracking scheme of Chapter 10. Also, codeless tracking techniques [for the encrypted GPS L2 P(Y) signal] are discussed, along with a description of how to adapt discriminator noise formulas for the case of colored noise.

7.1 Correlator and Waveform-Based Tracking

The correlator-based tracking scheme relies on the fact that a hardware receiver has only limited signal-generation capabilities but can work at a high sample rate and allows a high level of parallelization. Typically, a simple hardware GNSS receiver is only able to generate a carrier and binary PRN-code sequence. The product of both approximates the received signal $r(t)$. The PRN-code sequence is shifted several times by an integer number of samples and the shifted PRN-code replicas are correlated with the received signal at baseband (see Figure 7.1). The resulting correlators C_0, ..., C_b have different sample (or code-phase) shift values and are termed "prompt," "early," "late" (or "very early/late"). For basic tracking, three correlators are required (prompt, early, and late), but depending on the used multipath-mitigation scheme and on the signal-modulation scheme, significantly more correlators might be required.

For a software receiver that uses the multibit-correlation approach discussed in Section 9.4, it is reasonable to exploit this increased amplitude resolution to allow correlation with more complex reference signals instead of correlation only with the PRN-code sequence. The reference signals can then be optimized for certain criteria, as will be discussed in Section 8.2.

In the rest of this chapter, different correlators will be considered that correlate the received-signal samples with different classes (P, D, F, and W) of reference

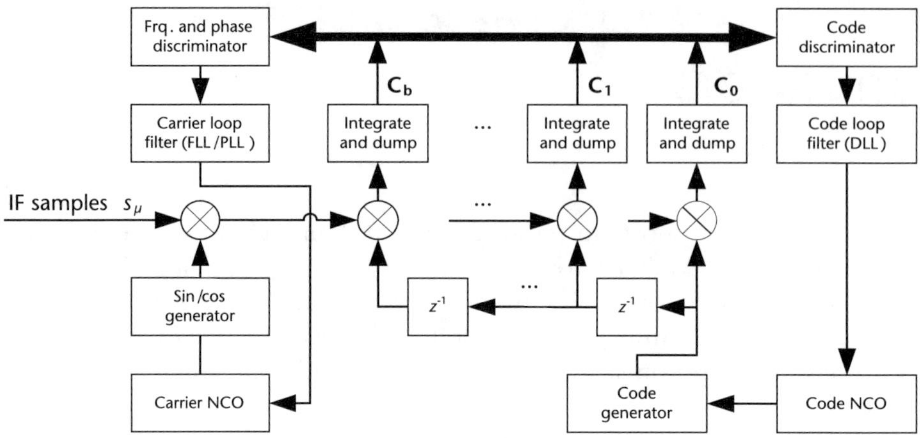

Figure 7.1 Correlator-based tracking.

signals. The P-correlator can be viewed as a matched filter because it correlates the received signal with the received-signal waveform and maximizes the postcorrelation SNR. It is mainly used for carrier tracking. The P-correlator corresponds to the prompt correlator in a hardware receiver. The D-correlator is applied for code tracking and corresponds to the difference of an early minus a late correlator in a hardware receiver. The F-correlator is used for frequency tracking and is realized as a time difference between two prompt correlators in a hardware receiver. Finally, the W-correlator is a gated correlator whose values are uncorrelated in code-phase dimension.

Overall, the proposed software-receiver tracking scheme is outlined in Figure 7.2 and makes use of only *two* complex-valued correlation operations. This saves computational power compared to the multicorrelator approach of Figure 7.1.

In the next sections, the noise performance of the different correlators will be derived based on formulas for a generic correlator.

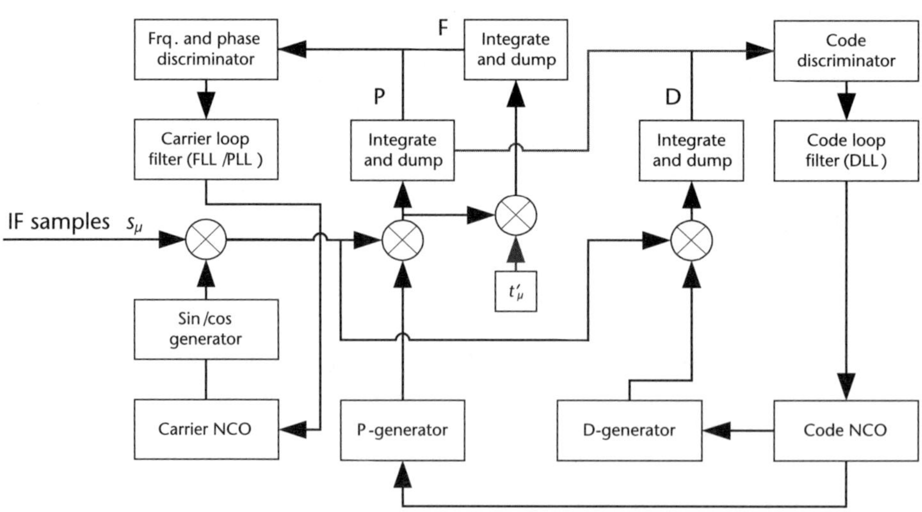

Figure 7.2 Waveform-based tracking.

7.2 Generic Correlator

The analysis of a generic correlator is based on the assumption of Section 1.8. The sequence of received-signal samples S_μ is the sum of a deterministic signal r_μ plus noise N_μ given by

$$S_\mu = r_\mu + N_\mu \quad \mu = 1,\ldots,L \tag{7.1}$$

The deterministic part is assumed to be of the form

$$r_\mu = ac(t_\mu - \tau)\exp\left\{i\left(\omega\left(t_\mu - \frac{L+1}{2f_s}\right) - \varphi\right)\right\} \tag{7.2}$$

where a is the signal amplitude, ω is the angular IF plus Doppler in radians per second, φ is the carrier-phase offset in radians, and τ is the code delay in seconds. The signal is sampled at times t_μ

$$t_\mu = \frac{\mu}{f_s} \Rightarrow \begin{cases} t_0 = 0 \\ t_L = T_{coh} \end{cases} \tag{7.3}$$

where f_s is the sample rate in samples per second. The function $c(t)$ represents the received and filtered navigation signal at baseband and is allowed to be complex-valued. The stochastic part N_μ shall be left unspecified now, but it typically includes thermal and quantization noise.

Furthermore,

$$t'_\mu = t_\mu - \frac{L+1}{2f_s} \tag{7.4}$$

is defined for a more compact notation.

The received signal is characterized by four fundamental signal parameters: a, τ, ω, and φ. Those parameters are assumed constant during the integration period T_{coh}. If the broadcast GNSS signal contains a navigation data message, the message bit/symbol is assumed constant during the integration period and can be included in the carrier phase φ.

The received samples are correlated with an internally generated sequence of samples given by

$$r_{rec,\mu} = c_{rec}(t_\mu - \tau_0)\exp\{i(\omega_0 t'_\mu - \varphi_0)\} \tag{7.5}$$

For the different classes of correlators (P, D, F, and W), the internal sequence is based on a different baseband signal c_{rec}. We will later substitute different signals, including the PRN code itself and early-late replicas, among others. Its fundamental parameters are denoted by a subscript "0." Those parameters are typically near the true signal parameters. They define the correlation point and should not be confused with the estimated signal parameters. Especially in the case of a multicorrelator, the estimated parameters and the correlation point differ significantly.

The output of the correlator C is defined as

$$C(S_\mu) = \sum_{\mu=1}^{L} r_{rec,\mu} S_\mu = \sum_{\mu=1}^{L} r_{rec,\mu} r_\mu + \sum_{\mu=1}^{L} r_{rec,\mu} N_\mu = C_{sig}(r_\mu) + C_{noise}(N_\mu) \tag{7.6}$$

and is split into a deterministic signal part C_{sig} and a stochastic noise part C_{noise}. The integration (summation) involves a number of L samples corresponding to a coherent integration time of $T_{coh} = L/f_s$.

Within this section, all expected values are calculated with respect to noise; that is,

$$\langle \ldots \rangle \equiv \langle \ldots \rangle_{\mathrm{N}} \tag{7.7}$$

If the input signal has only a real component, we write the correlator output as

$$\mathrm{C}(\mathrm{Re}\{S_\mu\}) = \frac{1}{2}\left[\mathrm{C}(S_\mu) + \mathrm{C}(\overline{S}_\mu)\right] \tag{7.8}$$

Where $\overline{S}_\mu$ are complex-conjugated-signal samples. Note that expressions involving complex-conjugated-signal samples are not implemented in the receiver. They serve only to obtain a more compact notation.

We start with the assumption of bandlimited white noise and that the Nyquist-sampling criterion is fulfilled for the sample rate f_s. It expresses itself for complex-valued signals as

$$f_s = B \tag{7.9}$$

and for real-valued signals as

$$2f_s = B \tag{7.10}$$

where B is the dual-sided white-noise bandwidth. In Section 7.6, the results will be extended for the case of colored noise. In both cases (real or complex input signal), the correlator output is a complex-valued random variable. Its expected value and its (co)variance are calculated in Section 7.2.1.

7.2.1 Expected Value

The expected value for the correlator output is the sum of the deterministic contribution and the noise contribution. The latter contribution vanishes, as will be demonstrated.

The deterministic part evaluates to (see Appendix A.4.4)

$$\begin{aligned} C_{sig}(r_\mu) &= \sum_{\mu=1}^{L} r_{rec,\mu} r_\mu = a \sum_{\mu=1}^{L} c(t_\mu - \tau) c_{rec}(t_\mu - \tau_0) \exp\{i(\omega t'_\mu - \varphi) + i(\omega_0 t'_\mu - \varphi_0)\} \\ &= aLR_{c,c_{rec}}(\tau - \tau_0) \exp\{-i(\varphi + \varphi_0)\} \kappa(\omega + \omega_0) \\ &= aT_{coh} f_s R_{c,c_{rec}}(\tau - \tau_0) \exp\{-i(\varphi + \varphi_0)\} \kappa(\omega + \omega_0) \end{aligned} \tag{7.11}$$

When a real-valued input signal is considered, it is reasonable to assume that its center frequency is far from 0 MHz. The replica signal frequency is similar to the received-signal center frequency and $|\mathrm{mod}(\omega + \omega_0, 2\pi f_s)| >> 0$ should be fulfilled, yielding

$$C_{sig}(r_\mu) \approx a f_s T_{coh} R_{c,c_{rec}}(\tau - \tau_0) \exp\{-i(\varphi + \varphi_0)\} \kappa(\omega + \omega_0) \approx 0 \tag{7.12}$$

For complex signals (whose center frequency might be around 0 MHz) the above equation does not hold.

Repeating the same derivation with the complex-conjugate received signal gives

$$
\begin{aligned}
&C_{sig}(\overline{r}_\mu) = \\
&= \sum_{\mu=1}^{L} r_{rec,\mu}\overline{r}_\mu = a\sum_{\mu=1}^{L} \overline{c}(t_\mu - \tau)c_{rec}(t_\mu - \tau_0)\exp\{-i(\omega t'_\mu - \varphi) + i(\omega_0 t'_\mu - \varphi_0)\} \qquad (7.13)\\
&\approx aT_{coh}f_s R_{\overline{c},c_{rec}}(\tau - \tau_0)\exp\{i(\varphi - \varphi_0)\}\kappa(\omega_0 - \omega)
\end{aligned}
$$

For a real-valued input signal, the correlation results for the normal and the complex-conjugate input signal must be summed. Because correlation is linear, we obtain under the assumption $|\mathrm{mod}(\omega + \omega_0, 2\pi f_s)| >> 0$

$$
\begin{aligned}
C_{sig}(\mathrm{Re}\{r_\mu\}) &= \frac{1}{2}\left[C_{sig}(r_\mu) + C_{sig}(\overline{r}_\mu)\right]\\
&= \frac{a}{2}T_{coh}f_s R_{\overline{c},c_{rec}}(\tau - \tau_0)\exp\{i(\varphi - \varphi_0)\}\kappa(\omega_0 - \omega)
\end{aligned} \qquad (7.14)
$$

We assume that the noise samples are unbiased. Consequently,

$$
\left\langle C_{noise}(N_\mu)\right\rangle = \left\langle C_{noise}(\overline{N}_\mu)\right\rangle = 0 \qquad (7.15)
$$

To summarize, the expected value for a complex input signal is

$$
\begin{aligned}
\left\langle C(S_\mu)\right\rangle &\approx aT_{coh}f_s R_{c,c_{rec}}(\tau - \tau_0)\exp\{-i(\varphi + \varphi_0)\}\kappa(\omega_0 + \omega)\\
\left\langle C(\overline{S}_\mu)\right\rangle &\approx aT_{coh}f_s R_{\overline{c},c_{rec}}(\tau - \tau_0)\exp\{i(\varphi - \varphi_0)\}\kappa(\omega_0 - \omega)
\end{aligned} \qquad (7.16)
$$

If a real-valued input signal is considered and $|\mathrm{mod}(\omega + \omega_0, 2\pi f_s)| >> 0$ then the expected value is

$$
\left\langle C(\mathrm{Re}\{S_\mu\})\right\rangle \approx \frac{aT_{coh}f_s R_{\overline{c},c_{rec}}(\tau - \tau_0)}{2}\exp\{i(\varphi - \varphi_0)\}\kappa(\omega_0 - \omega) \qquad (7.17)
$$

because $\kappa(\omega)$ typically vanishes for large frequencies (see Appendix A.4.4).

7.2.2 Covariance

To evaluate the covariance, assume two different correlators that are distinguished by an index "a" and the index "b." Both correlators are associated with different parameters and internal signals, specifically

$$C^a(S_\mu) \mapsto c^a_{rec}, \tau^a_0, \omega^a_0, \varphi^a_0$$
$$C^b(S_\mu) \mapsto c^b_{rec}, \tau^b_0, \omega^b_0, \varphi^b_0 \tag{7.18}$$

Both work with the same received signal samples.
The covariance is defined as

$$\begin{aligned}
&\operatorname{cov}\left\langle C^a(S_\mu), C^b(S_\mu)\right\rangle = \left\langle \overline{\left(C^a(S_\mu) - \left\langle C^a(S_\mu)\right\rangle\right)} \left(C^b(S_\mu) - \left\langle C^b(S_\mu)\right\rangle\right)\right\rangle \\
&= \left\langle \overline{\left(C^a(r_\mu + N_\mu) - \left\langle C^a(r_\mu + N_\mu)\right\rangle\right)} \left(C^b(r_\mu + N_\mu) - \left\langle C^b(r_\mu + N_\mu)\right\rangle\right)\right\rangle \\
&= \left\langle \overline{C^a(N_\mu)} C^b(N_\mu)\right\rangle
\end{aligned} \tag{7.19}$$

because the signal contribution cancels and the expected value of the noise contribution vanishes. Thus

$$\begin{aligned}
&\operatorname{cov}\left\langle C^a(S_\mu), C^b(S_\mu)\right\rangle = \\
&= \sum_{\mu,\upsilon=1}^{L} \left\langle \overline{N_\mu c^a_{rec}(t_\mu - \tau^a_0)\exp\{i(\omega^a_0 t'_\mu - \varphi^a_0)\}} N_\upsilon c^b_{rec}(t_\upsilon - \tau^b_0)\exp\left\{i(\omega^b_0 t'_\upsilon - \varphi^b_0)\right\}\right\rangle \\
&= 2\sum_{\mu=1}^{L} \overline{c^a_{rec}}(t_\mu - \tau^a_0) c^b_{rec}(t_\mu - \tau^b_0)\exp\{it'_\mu(\omega^b_0 - \omega^a_0) - i(\varphi^b_0 - \varphi^a_0)\} \\
&\approx 2 f_s T_{coh} R_{\overline{c^a_{rec}}, c^b_{rec}}(\tau^a_0 - \tau^b_0)\exp\{i(\varphi^a_0 - \varphi^b_0)\}\kappa(\omega^b_0 - \omega^a_0)
\end{aligned} \tag{7.20}$$

because, for the following, we assume that (1.16) and (1.28) hold. The covariance for the complex-conjugated input signal is given similarly as

$$\begin{aligned}
&\operatorname{cov}\left\langle C^a\left(\bar{S}_\mu\right), C^b\left(\bar{S}_\mu\right)\right\rangle = \left\langle \overline{\left(C^a\left(\bar{S}_\mu\right) - \left\langle C^a\left(\bar{S}_\mu\right)\right\rangle\right)} \left(C^b\left(\bar{S}_\mu\right) - \left\langle C^b\left(\bar{S}_\mu\right)\right\rangle\right)\right\rangle \\
&= \left\langle \overline{C^a\left(\bar{N}_\mu\right)} C^b\left(\bar{N}_\mu\right)\right\rangle \\
&= \sum_{\mu,\upsilon=1}^{L} \left\langle \overline{\bar{N}_\mu c^a_{rec}\left(t_\mu - \tau^a_0\right)\exp\left\{i\left(\omega^a_0 t'_\mu - \varphi^a_0\right)\right\}} \bar{N}_\upsilon c^b_{rec}\left(t'_\upsilon - \tau^b_0\right)\exp\left\{i\left(\omega^b_0 t'_\upsilon - \varphi^b_0\right)\right\}\right\rangle \\
&= 2\sum_{\mu=1}^{L} \overline{c^a_{rec}}\left(t_\mu - \tau^a_0\right) c^b_{rec}\left(t_\mu - \tau^b_0\right)\exp\left\{it'_\mu\left(\omega^b_0 - \omega^a_0\right) - i\left(\varphi^b_0 - \varphi^a_0\right)\right\} \\
&= \operatorname{cov}\left\langle C^a\left(S_\mu\right), C^b\left(S_\mu\right)\right\rangle
\end{aligned} \tag{7.21}$$

For mixed terms, we obtain

$$\begin{aligned}
&\operatorname{cov}\left\langle C^a(\bar{S}_\mu), C^b(S_\mu)\right\rangle = \left\langle \overline{\left(C^a(\bar{S}_\mu) - \left\langle C^a(\bar{S}_\mu)\right\rangle\right)}\left(C^b(S_\mu) - \left\langle C^b(S_\mu)\right\rangle\right)\right\rangle \\
&= \left\langle \overline{C^a(\bar{N}_\mu)} C^b(N_\mu)\right\rangle \\
&= \sum_{\mu,\,\upsilon=1}^{L} \left\langle \overline{\bar{N}_\mu c^a_{rec}\left(t_\mu - \tau^a_0\right)\exp\left\{i\left(\omega^a_0 t'_\mu - \varphi^a_0\right)\right\}} N_\upsilon c^b_{rec}\left(t_\upsilon - \tau^b_0\right)\exp\left\{i\left(\omega^b_0 t'_\upsilon - \varphi^b_0\right)\right\}\right\rangle \\
&= 2\sum_{\mu=1}^{L} \overline{c^a_{rec}}\left(t_\mu - \tau^a_0\right) c^b_{rec}\left(t_\mu - \tau^b_0\right)\exp\left\{it'_\mu\left(\omega^b_0 - \omega^a_0\right) - i\left(\varphi^b_0 - \varphi^a_0\right)\right\}\left\langle N_\mu N_\upsilon\right\rangle = 0
\end{aligned} \tag{7.22}$$

as the expected value of these products of the complex-valued Gaussian noise is zero:

$$\left\langle N_\mu N_\upsilon\right\rangle = \left\langle \bar{N}_\mu \bar{N}_\upsilon\right\rangle = 0 \tag{7.23}$$

For real-valued input signals, we obtain

$$\begin{aligned}
\operatorname{cov}\left\langle C^a(\mathrm{Re}\{S_\mu\}), C^b(\mathrm{Re}\{S_\mu\})\right\rangle &= \frac{1}{4}\operatorname{cov}\left\langle C^a(S_\mu + \bar{S}_\mu), C^b(S_\mu + \bar{S}_\mu)\right\rangle \\
&= \frac{1}{2}\operatorname{cov}\left\langle C^a(S_\mu), C^b(S_\mu)\right\rangle \\
&\approx f_s T_{coh} R_{\overline{c^a_{rec}}, c^b_{rec}}(\tau^a_0 - \tau^b_0)\exp\{i(\varphi^a_0 - \varphi^b_0)\}\kappa(\omega^b_0 - \omega^a_0)
\end{aligned} \tag{7.24}$$

7.2.3 Variance

The variance of the correlator output is a special case of the covariance assuming identical parameters for the *a*' and *b*' case.

For complex input signals, the equation

$$\operatorname{var}\left\langle C(S_\mu)\right\rangle = 2 f_s T_{coh} R_{\overline{c_{rec}}, c_{rec}}(0) \tag{7.25}$$

holds and for real-valued input signals

$$\operatorname{var}\left\langle C(\mathrm{Re}\{S_\mu\})\right\rangle = f_s T_{coh} R_{\overline{c_{rec}}, c_{rec}}(0) \tag{7.26}$$

7.3 Correlator Types with Illustration

Depending on the choice of the internally generated reference signal, many different types of correlators can be obtained. They are discussed in Section 7.3.1. For simplicity, the case of real-valued received signal samples is discussed.

7.3.1 P-Correlator

P-correlators are based on a transmitted-like reference signal and are used to determine carrier-phase tracking errors to extract navigation data symbols/bits; they are used in the acquisition process (perhaps using FFT techniques for correlation). A P-correlator reference signal $c_P(t)$ has to fulfill the requirement

$$R_{\bar{c},c_P}(0) \neq 0 \tag{7.27}$$

There exist different possibilities for the choice of the $c_P(t)$. If, for the P-correlator C_P, we use as internally generated signal the received signal at baseband itself,

$$c_P(t) = c(t) \tag{7.28}$$

then the CRLB (assuming no multipath is present) for the carrier-phase discriminator will be reached and optimal performance for navigation bit/symbol decoding and acquisition will be reached as well.

For this setting (7.28), the expected value and variance evaluate to

$$\begin{aligned}
&\left\langle C_P(\mathrm{Re}\{S_\mu\})\right\rangle \\
&= \sqrt{2C/N_0 B}T_{coh}R_{\bar{c},c}(\tau-\tau_0)\exp\{i(\varphi-\varphi_0)\}\kappa(\omega_0-\omega) \\
&\mathrm{var}\left\langle C_P(\mathrm{Re}\{S_\mu\})\right\rangle = 2BT_{coh}
\end{aligned} \tag{7.29}$$

Here, (A.76) has been used to express the amplitude as a function of C/N_0 and the dual-sided signal bandwidth B. The covariance of two P-correlators using different signal parameters can be easily calculated using (7.24). If both correlators work with the same signal, carrier phase, and angular frequency, this expression takes the form

$$\mathrm{cov}\left\langle C_P^a\left(\mathrm{Re}\{S_\mu\};\tau_0^a\right), C_P^b\left(\mathrm{Re}\{S_\mu\};\tau_0^b\right)\right\rangle = 2BT_{coh}R_{\bar{c}_P,c_P}\left(\tau_0^a-\tau_0^b\right) \tag{7.30}$$

The P-correlator values are, in general, highly correlated for similar code-phase offsets. If both correlators are based on different angular frequencies $\omega_0{}^a$ and $\omega_0{}^b$, they are uncorrelated if $\kappa(\omega_0{}^a - \omega_0{}^b) = 0$.

In practice, for different reasons, it might be convenient to choose a slightly different form of $c_P(t)$ from $c(t)$. For the case of GNSS signals, and if simplicity is required, the infinite-bandwidth representation of the PRN code is a good choice, because for most signals it can be represented by a 1-bit amplitude resolution. For carrier-phase multipath mitigation, a linear combination of the form "2P-E-L" of the infinite-bandwidth baseband representation of $c(t)$ may prove advantageous [1]. Assuming, however, an arbitrary waveform $c_P(t)$ for the P-correlator, the stochastic properties are summarized as

$$\begin{aligned}
&\left\langle C_P(\mathrm{Re}\{S_\mu\})\right\rangle = \sqrt{2C/N_0 B}T_{coh}R_{\bar{c},c_P}(\tau-\tau_0)\exp\{i(\varphi-\varphi_0)\}\kappa(\omega_0-\omega) \\
&\mathrm{var}\left\langle C_P(\mathrm{Re}\{S_\mu\})\right\rangle = 2BT_{coh}R_{\bar{c}_P,c_P}(0) \\
&\mathrm{cov}\left\langle C_P^a\left(\mathrm{Re}\{S_\mu\};\tau_0^a\right), C_P^b\left(\mathrm{Re}\{S_\mu\};\tau_0^b\right)\right\rangle \\
&= 2BT_{coh}R_{\bar{c}_P,c_P}\left(\tau_0^a-\tau_0^b\right)\exp\left\{i\left(\varphi_0^a-\varphi_0^b\right)\right\}\kappa\left(\omega_0^b-\omega_0^a\right)
\end{aligned} \tag{7.31}$$

7.3.2 F-Correlator

The F-correlator is used to determine Doppler-frequency tracking errors, a task usually performed in many GNSS receivers by differencing two carrier-phase estimates in time [2]. Using a dedicated correlator for this estimation can be advantageous because if Doppler frequency is estimated in a single step, the CRLB for the estimated frequency is achieved and the pull-in region is increased [3]. Furthermore, the computation of the F-correlator does *not* require significant extra computational load in a software receiver because the linear function multiplying the reference signal in (7.32) can be approximated by a piecewise constant function. The correlation loop of Section 9.4 can be split into several smaller loops, thereby allowing computation of the P- and F-correlator in one turn.

The F-correlator derives from the LSQ Doppler estimation strategy of Section 4.3 and the reference signal is given by a P-correlator reference signal multiplied with time:

$$c_F(t) = \left(t - \frac{T_{coh}}{2}\right) c_P(t) \tag{7.32}$$

The resulting F-correlator reference signal must still allow the separation of code and carrier correlation, as described in Section 1.8.5.

The expected value of the F-correlator is

$$\begin{aligned}\left\langle C_F(\mathrm{Re}\{S_\mu\})\right\rangle &= \left(-i\frac{\partial}{\partial\omega_0}\right)\left\langle C_P(\mathrm{Re}\{S_\mu\})\right\rangle \\ &= \sqrt{2C/N_0 B} T_{coh} R_{\bar{c},c_P}(\tau-\tau_0)\exp\{i(\varphi-\varphi_0)\}\left(-i\frac{\partial}{\partial\omega_0}\right)\kappa(\omega_0-\omega) \\ &= -\sqrt{2C/N_0 B} T_{coh} R_{\bar{c},c_P}(\tau-\tau_0)\exp\{i(\varphi-\varphi_0)\} i\kappa'(\omega_0-\omega)\end{aligned} \tag{7.33}$$

and the covariance

$$\begin{aligned}&\mathrm{cov}\left\langle C_F\left(\mathrm{Re}\{S_\mu\};\omega_0^a\right), C_F\left(\mathrm{Re}\{S_\mu\};\omega_0^b\right)\right\rangle = \\ &= \mathrm{cov}\left\langle\left(-i\frac{\partial}{\partial\omega_0^a}\right)C_P\left(\mathrm{Re}\{S_\mu\};\omega_0^a\right), \left(-i\frac{\partial}{\partial\omega_0^b}\right)C_P\left(\mathrm{Re}\{S_\mu\};\omega_0^b\right)\right\rangle \\ &= -\frac{\partial}{\partial\omega_0^a}\frac{\partial}{\partial\omega_0^b}\mathrm{cov}\left\langle C_P\left(\mathrm{Re}\{S_\mu\};\omega_0^a\right), C_P\left(\mathrm{Re}\{S_\mu\};\omega_0^b\right)\right\rangle \\ &= -2BT_{coh}R_{\bar{c}_P,c_P}\left(\tau_0^a-\tau_0^b\right)\exp\left\{i\left(\varphi_0^a-\varphi_0^b\right)\right\}\left(\frac{\partial}{\partial\omega_0^a}\frac{\partial}{\partial\omega_0^b}\kappa\left(\omega_0^b-\omega_0^a\right)\right)\end{aligned} \tag{7.34}$$

Because of the discussion of Section 1.8.5,

$$\begin{aligned}\mathrm{var}\left\langle C_F(\mathrm{Re}\{S_\mu\};\omega_0)\right\rangle &= 2BT_{coh}R_{\bar{c}_P,c_P}(0)\left(\left(i\frac{\partial}{\partial\omega}\right)\left(i\frac{\partial}{\partial\omega}\right)\kappa(\omega)\right)\Big|_{\omega_0=0} \\ &= 2BT_{coh}R_{\bar{c}_P,c_P}(0)(-\kappa''(0)) \\ &= 2BT_{coh}R_{\bar{c}_P,c_P}(0){T_{coh}}^2\frac{1}{12}\chi_{freq}\end{aligned} \tag{7.35}$$

7.3.3 D-Correlator

The D-correlator is used to determine the code-phase tracking error. In the most simple case, it is an early-minus-late replica of the infinite-bandwidth baseband representation of $c(t)$. For code multipath mitigation, a double-delta linear combination or a shaping correlator is known to be useful [1].

Generally, we require for the D-correlator reference signal $c_D(t)$

$$\begin{aligned} &R_{\bar{c}_D,c_P}(0) = 0 \\ &R_{\bar{c},c_D}(\tau) \approx R_{\bar{c},c_D}{}'(0)\,\tau \qquad \text{for } |\tau| < l \quad \text{and } R_{\bar{c},c_D}{}'(0) \neq 0 \end{aligned} \tag{7.36}$$

Here, $c_P(t)$ denotes the P-correlator reference signal and the $c(t)$ received signal. The symbol l is the required size of the code-phase linearity region.

If we choose for the D-correlator C_D as an internally generated signal the first derivative of the received signal at baseband:

$$c_{rec}(t) = c'(t) \tag{7.37}$$

then the CRLB is obtained if no multipath signal is present and

$$\begin{aligned} &\left\langle C_D(\mathrm{Re}\{S_\mu\})\right\rangle \\ &= \sqrt{2C/N_0}BT_{coh}R_{\bar{c},c}{}'(\tau-\tau_0)\exp\{i(\varphi-\varphi_0)\}\kappa(\omega_0-\omega) \\ &\mathrm{var}\left\langle C_D(\mathrm{Re}\{S_\mu\})\right\rangle = -2BT_{coh}R_{\bar{c},c}{}''(0) \\ &\mathrm{cov}\left\langle C_D^a\left(\mathrm{Re}\{S_\mu\};\tau_0^a\right), C_D^b\left(\mathrm{Re}\{S_\mu\};\tau_0^b\right)\right\rangle \\ &= -2BT_{coh}R_{\bar{c},c}{}''\left(\tau_0^a-\tau_0^b\right)\exp\left\{i\left(\varphi_0^a-\varphi_0^b\right)\right\}\kappa\left(\omega_0^b-\omega_0^a\right) \end{aligned} \tag{7.38}$$

For an arbitrary choice of the D-correlator reference signal, (7.17) and (7.24) are used to obtain stochastic correlator properties.

7.3.4 W-Correlator

The main purpose of the W-correlator is its use in CDMA signal-quality monitoring or multipath mitigation. If it is used within a multicorrelator, it gives uncorrelated measurements of the received-signal chip form. It stands in analogy to the so-called Vision correlator [4].

The reference signal for the W-correlator is defined as the PRN-code sequence of the received signal convoluted with an approximation of Dirac's delta function. Let us assume that the received-CDMA signal at baseband takes the form

$$c(t) = \sum_{k=1}^{N_{PRN}} h_k m(tf_c - k) \quad \text{for} \quad 0 \leq t < N_{PRN}/f_c \tag{7.39}$$

where N_{PRN} denotes the length of the PRN code and f_c represents the code rate in chip per second. The function m defines the used-modulation scheme [e.g., BPSK or

(M)BOC for GNSS signals], including filtering, and h_k is the PRN code spreading sequence. For example, if an infinite-bandwidth BPSK signal is considered, m takes the form

$$m(\tilde{\tau}) = \begin{cases} 1 & 0 \le \tilde{\tau} < 1 \\ 0 & \text{otherwise} \end{cases} \tag{7.40}$$

where $\tilde{\tau}$ is expressed in chips. The W-correlator reference signal is defined as

$$c_W(t) = \sum_{k=1}^{N_{PRN}} h_k \delta_\varepsilon (tf_c - k) \quad \text{for} \quad 0 \le t < N_{PRN}/f_c \tag{7.41}$$

where

$$\delta_\varepsilon(\tilde{\tau}) = \begin{cases} 1 & 0 \le \tilde{\tau} < \varepsilon \\ 0 & \text{otherwise} \end{cases} \tag{7.42}$$

approximates Dirac's delta function. Note that, for numerical reasons, the amplitude is set to 1 instead of choosing $1/\varepsilon$. Furthermore, it should be noted that if the reference signal and the PRN code is periodically extended for $t < 0$ or $t \ge N_{PRN} / f_c$, then

$$c(t) = \sum_{k=-\infty}^{\infty} h_k m(tf_c - k) \tag{7.43}$$

holds, because m has limited support and the series converges.

The correlation function evaluates to

$$\begin{aligned} R_{\bar{c},c_W}(\tau) &= \lim_{T\to\infty} \frac{1}{2T} \int_{-T}^{T} \bar{c}(t-\tau) c_{rec}(t) dt \\ &= \lim_{T\to\infty} \frac{1}{2T} \int_{-T}^{T} \sum_{k=-\infty}^{\infty} \overline{h_k m\left((t-\tau)f_c - k\right)} \sum_{l=-\infty}^{\infty} h_l \delta_\varepsilon (tf_c - l) dt \\ &= \lim_{T\to\infty} \frac{1}{2T} \sum_{k,l=-\infty}^{\infty} \bar{h}_k h_l \int_{-T}^{T} \overline{m\left((t-\tau)f_c - k\right)} \delta_\varepsilon (tf_c - l) dt \\ &= \lim_{N\to\infty} \frac{f_c}{2N} \sum_{k,l=-N}^{N} \bar{h}_k h_l \int_{-N/f_c}^{N/f_c} \overline{m\left((t-\tau)f_c - k\right)} \delta_\varepsilon (tf_c - l) dt \end{aligned} \tag{7.44}$$

where in the last step a common integration and summation interval have been introduced. The last step is valid in the limit $T \to \infty$ for a fixed value of τ. For simplicity, let us now assume that the approximate delta function behaves like Dirac's delta function, namely

$$\int_{-T}^{T} y(t) \delta_\varepsilon (at - b) dt \approx \frac{\varepsilon}{a} y\left(b/a\right) \quad \text{if} \quad -aT \le b \le aT \tag{7.45}$$

Then

$$\begin{aligned} R_{\bar{c},cw}(\tau) &\approx \lim_{N\to\infty} \frac{\varepsilon}{2N} \sum_{k,l=-N}^{N} \bar{h}_k h_l \overline{m\left((l/f_c-\tau)f_c-k\right)} \\ &= \lim_{N\to\infty} \frac{\varepsilon}{2N} \sum_{k=-N}^{N} \sum_{l=-k-N}^{-k+N} \bar{h}_k h_{k+l} \overline{m(l-\tau f_c)} \\ &= \lim_{N\to\infty} \frac{\varepsilon}{2N} \sum_{l=-\infty}^{\infty} \overline{m(l-\tau f_c)} \sum_{k=-N}^{N} \bar{h}_k h_{k+l} \\ &= \lim_{N\to\infty} \frac{\varepsilon}{2N} \sum_{l=-\infty}^{\infty} \overline{m(l-\tau f_c)}(2N+1)R_{prn}(l) \\ &= \varepsilon \sum_{l=-\infty}^{\infty} \overline{m\left(l-\tau f_c\right)}R_{prn}(l) \end{aligned} \tag{7.46}$$

where $R_{prn}(l)$ is the normalized autocorrelation function of the PRN code with $R_{prn}(0) = 1$. Assuming an ideal PRN code with all other values vanishing, we obtain

$$R_{\bar{c},cw}(\tau) \approx \varepsilon \overline{m(-\tau f_c)} \tag{7.47}$$

Thus, the W-correlator allows measuring the chip waveform (including filtering) of the used CDMA modulation scheme.

We obtain

$$R_{\overline{cw},cw}(\tau) \approx \varepsilon \delta_\varepsilon(-\tau f_c) \tag{7.48}$$

by simply substituting m with δ_ε in the above-listed derivation used to obtain (7.47).

The expected value, variance, and covariance of the W-correlator evaluate to

$$\begin{aligned} &\left\langle C_W(\mathrm{Re}\{S_\mu\})\right\rangle = \varepsilon\sqrt{2C/N_0 B}T_{coh} \exp\{i(\varphi-\varphi_0)\}\kappa(\omega_0-\omega)\overline{m(-\tau f_c)} \\ &\mathrm{var}\left\langle C_W(\mathrm{Re}\{S_\mu\})\right\rangle = \varepsilon 2BT_{coh} \\ &\mathrm{cov}\left\langle C_W^a\left(\mathrm{Re}\{S_\mu\};\tau_0^a\right), C_W^b\left(\mathrm{Re}\{S_\mu\};\tau_0^b\right)\right\rangle \\ &= 2BT_{coh}\varepsilon\delta_\varepsilon\left(\left(\tau_0^b-\tau_0^a\right)f_c\right)\exp\left\{i\left(\varphi_0^a-\varphi_0^b\right)\right\}\kappa\left(\omega_0^b-\omega_0^a\right) \end{aligned} \tag{7.49}$$

If the difference in the code phase expressed in chips exceeds ε, then the two W-correlators are uncorrelated. For many signal-analysis purposes, this is a major advantage compared to the P-correlator. The value of ε should be chosen carefully because the SNR (i.e., the ratio between the squared expected value divided by the variance) depends linearly on ε. If the value of ε is too small, it will make the W-correlator output noisy. For a multicorrelator, a good choice for ε is the multicorrelator spacing.

7.4 Difference Correlators

P-correlators, as described in the Section 7.3.3, can act as sufficient statistics and are a possible basis upon which to estimate the fundamental signal parameter's code phase, Doppler, carrier phase, and amplitude. On the contrary, difference correlators are a suitable basis when difference signal parameters (code phase and carrier phase) are of interest. Difference correlators are defined as the product of two correlators taking one as a complex-conjugate value.

Difference correlators are in two GNSS applications of importance: double-difference tracking and P-code-aided cross-correlation tracking of the encrypted GPS P(Y)-code. Double-difference tracking is described in Section 10.5.2. Cross-correlation techniques to track the encrypted P(Y)-code are summarized by Woo [5]. Using C/A-code-based aiding, the Y-code is tracked on L1 and L2 using very short (2-μs) coherent integration intervals aligned with an unknown W-code chip. Due to the C/A-code aiding, the Y-code correlation point is kept within the linear region (in the code phase and Doppler), but the correlator values contain the W-code chip and are too noisy to allow carrier-phase estimation and unwrapping. Instead, a W-code independent L1–L2 difference correlator is formed and the L1–L2 carrier phase difference is estimated, filtered, and unwrapped (note: the effective L1–L2 wavelength is 86 cm, being 3.5 times larger than the L2 wavelength). Adding the L1–L2 carrier-phase difference to the C/A-code-based L1 carrier phase gives an L2 carrier-phase estimate.

In Section 7.4.1, the thermal-noise performance of single-difference and in Section 7.4.2 double-difference prompt correlators will be investigated.

7.4.1 Single-Difference P-Correlators

For the analysis of single-difference P-correlators, we start from an undifferenced P-correlator and assume that Doppler errors can be ignored and that the replica-signal waveform $c_P(t)$ coincides sufficiently well with the received-signal waveform $c(t)$. According to Section 7.3.1, the P-correlator statistic is given as

$$\begin{aligned} C_P &= C_{P,sig} + C_{P,noise} \\ C_{P,sig} &\approx d\sqrt{2C/N_0 B T_{coh}} \exp\{i(\varphi - \varphi_0)\} \\ \mathrm{var}\langle C_{P,noise}\rangle &= 2BT_{coh} \end{aligned} \tag{7.50}$$

For convenience, the data bit d is explicitly introduced (instead of merging it with φ), assuming values ±1. The undifferenced-SNR value is defined as the ratio between the squared expected value and the variance:

$$SNR = \frac{\left|C_{P,sig}\right|^2}{\mathrm{var}\langle C_{P,noise}\rangle} = C/N_0 T_{coh} \tag{7.51}$$

To form the difference correlators, in the first step, the P-correlator is compensated for internal tracking effects to obtain a correlator value related only to the received-carrier phase [note: the expected P-correlator value of (7.50) is related to the difference of the received and internal carrier phase]; that is,

$$P^{k,n} = C_P^{k,n} \exp\left\{i\varphi_0^{k,n}\right\} \tag{7.52}$$

and the (receiver) single-difference correlator is formed according to

$$\Delta P^k = P^{k,m}\overline{P^{k,n}} \tag{7.53}$$

Here, we use the same index notation as in Section 10.5.2 (superscripts k and l for transmitters and subscripts m and n for receivers).

Using (7.50), we obtain

$$\begin{aligned}\Delta P^k = &\left(d^k \sqrt{2(C/N_0)^{k,m} B} T_{coh} \exp\left\{i\varphi^{k,m}\right\} + C_{P,noise}^{k,m}\right) \\ &\times\overline{\left(d^k \sqrt{2(C/N_0)^{k,n} B} T_{coh} \exp\left\{i\varphi^{k,n}\right\} + C_{P,noise}^{k,n}\right)}\end{aligned} \tag{7.54}$$

and the deterministic part of the single-difference correlator evaluates to

$$\begin{aligned}\Delta P_{sig}^k &= 2BT_{coh}{}^2 \sqrt{(C/N_0)^{k,m}(C/N_0)^{k,n}} \exp\left\{i\left(\varphi^{k,m} - \varphi^{k,n}\right)\right\} \\ &= 2BT_{coh}{}^2 \sqrt{(C/N_0)^{k,m}(C/N_0)^{k,n}} \exp\left\{i\Delta\varphi^k\right\}\end{aligned} \tag{7.55}$$

which can be used to obtain an estimate of the single-difference carrier-phase estimate $\Delta\varphi^{k}=\varphi^{k,m}-\varphi^{k,n}$. The data bit d has cancelled.

The stochastic part evaluates to

$$\begin{aligned}\Delta P_{noise}^k = &\, C_{P,noise}^{k,m} \overline{d^k \sqrt{2(C/N_0)^{k,n} B} T_{coh} \exp\left\{i\varphi^{k,n}\right\}} \\ &+ \overline{C_{P,noise}^{k,n}} d^k \sqrt{2(C/N_0)^{k,m} B} T_{coh} \exp\left\{i\varphi^{k,m}\right\} + C_{P,noise}^{k,m} \overline{C_{P,noise}^{k,n}}\end{aligned} \tag{7.56}$$

and because correlator values from different receivers are uncorrelated

$$\left\langle \Delta P_{noise}^k \right\rangle = 0 \tag{7.57}$$

The variance of the single-difference correlator evaluates to

$$\begin{aligned}\operatorname{var}\left\langle \Delta P_{noise}^k \right\rangle &= \left\langle \Delta P_{noise}^k \overline{\Delta P_{noise}^k} \right\rangle \\ &= \left\langle \begin{array}{l} 2(C/N_0)^{k,n} B \left|C_{P,noise}^{k,m}\right|^2 T_{coh}{}^2 \\ +2(C/N_0)^{k,m} B \left|C_{P,noise}^{k,n}\right|^2 T_{coh}{}^2 + \left|C_{P,noise}^{k,m}\right|^2 \left|C_{P,noise}^{k,n}\right|^2 \end{array} \right\rangle \\ &= 2(C/N_0)^{k,n} B2BT_{coh}{}^3 + 2(C/N_0)^{k,m} B2BT_{coh}{}^3 + (2BT_{coh})^2 \\ &= 4B^2 T_{coh}{}^2 \left((C/N_0)^{k,m} T_{coh} + (C/N_0)^{k,n} T_{coh} + 1\right)\end{aligned} \tag{7.58}$$

and is a function of the two C/N_0 values. Furthermore, the variance depends nonlinearly on the coherent integration time, which reflects the fact that a nonlinear operation has been used.

The SNR of the single-difference correlator evaluates to

$$SNR_{\Delta} = \frac{\left|\Delta P_{sig}^{k}\right|^2}{\mathrm{var}\left\langle \Delta P_{noise}^{k} \right\rangle_{\mathbf{N}}} = \frac{4B^2 T_{coh}{}^4 (C/N_0)^{k,m} (C/N_0)^{k,n}}{4B^2 T_{coh}{}^2 ((C/N_0)^{k,m} T_{coh} + (C/N_0)^{k,n} T_{coh} + 1)}$$
$$= \frac{T_{coh}{}^2 (C/N_0)^{k,m} (C/N_0)^{k,n}}{((C/N_0)^{k,m} T_{coh} + (C/N_0)^{k,n} T_{coh} + 1)} = \frac{SNR^{k,m} SNR^{k,n}}{(1 + SNR^{k,m} + SNR^{k,n})} \tag{7.59}$$

and in case both SNR values are equal and much larger than one:

$$SNR_{\Delta} \underset{SNR^{k,m}=SNR^{k,n}=SNR\gg 1}{\longrightarrow} = \frac{SNR}{2} \tag{7.60}$$

In the latter case, single-differencing reduces the SNR value by 3 dB. An equivalent 3-dB loss would also occur if single-differencing were done at the carrier-phase level.

7.4.2 Double-Difference P-Correlators

A double-difference P-correlator is used by forming the (transmitter) difference between two single-difference correlators

$$\nabla\Delta P = \Delta P^k \overline{\Delta P^l} \tag{7.61}$$

and evaluates according to (7.55) as

$$\nabla\Delta P = \left(2BT_{coh}{}^2 \sqrt{(C/N_0)^{k,m}(C/N_0)^{k,n}} \exp\left\{i\Delta\varphi^k\right\} + \Delta P_{noise}^k\right)$$
$$\times \overline{\left(2BT_{coh}{}^2 \sqrt{(C/N_0)^{l,m}(C/N_0)^{l,n}} \exp\left\{i\Delta\varphi^l\right\} + \Delta P_{noise}^l\right)} \tag{7.62}$$

The deterministic part of the double-difference correlator

$$\nabla\Delta P_{sig} = 4B^2 T_{coh}{}^4 \sqrt{(C/N_0)^{k,m}(C/N_0)^{k,n}(C/N_0)^{l,m}(C/N_0)^{l,n}} \exp\{i\nabla\Delta\varphi\} \tag{7.63}$$

is used to estimate the double-difference carrier phase $\nabla\Delta\varphi = \varphi^{k,m} - \varphi^{k,n} - \varphi^{l,m} + \varphi^{l,n}$.

The stochastic part is

$$\nabla\Delta P_{noise} = 2BT_{coh}{}^2 \sqrt{(C/N_0)^{k,m}(C/N_0)^{k,n}} \exp\left\{i\Delta\varphi^k\right\} \overline{\Delta P_{noise}^l}$$
$$+ \overline{2BT_{coh}{}^2 \sqrt{(C/N_0)^{l,m}(C/N_0)^{l,n}} \exp\left\{i\Delta\varphi^l\right\}} \Delta P_{noise}^k + \Delta P_{noise}^k \overline{\Delta P_{noise}^l} \tag{7.64}$$

and its expected value vanishes,

$$\left\langle \nabla\Delta P_{noise} \right\rangle = 0 \tag{7.65}$$

because the single-difference correlator to two different transmitters is assumed to be uncorrelated. The variance of the double-difference correlator is given as

$$\begin{aligned}\mathrm{var}\left\langle \nabla\Delta P_{noise}\right\rangle_{\mathrm{N}} &= 4B^2 T_{coh}{}^4 (C/N_0)^{k,m}(C/N_0)^{k,n}\,\mathrm{var}\left\langle \Delta P^{l}_{noise}\right\rangle \\ &+ 4B^2 T_{coh}{}^4 (C/N_0)^{l,m}(C/N_0)^{l,n}\,\mathrm{var}\left\langle \Delta P^{k}_{noise}\right\rangle \\ &+ \mathrm{var}\left\langle \Delta P^{l}_{noise}\right\rangle \mathrm{var}\left\langle \Delta P^{k}_{noise}\right\rangle\end{aligned} \tag{7.66}$$

and for identical SNR values being much larger than 1, the double-difference correlator SNR is 6 dB less than the undifferenced SNR; that is,

$$SNR_{\nabla\Delta} \underset{\substack{SNR^{k,m}=SNR^{k,n}= \\ SNR^{l,m}=SNR^{l,n}=SNR\gg 1}}{\longrightarrow} = \frac{SNR}{4} \tag{7.67}$$

7.5 Noisy Reference Signal for Codeless Tracking

This section describes the cross-correlation tracking scheme that can be used if the transmitted signal is not known and the same transmitted signal is received on two different paths (e.g., by two different receivers or on two different frequencies). Cross-correlation allows determination of differences of signal parameters between the two paths. For example, the encrypted GPS-P(Y) code signal on L2 can be tracked via cross-correlation of the received L2 signal with the L1 signal, as depicted in Figure 7.3. L1 signal parameters are obtained by tracking the GPS C/A-code [5].

Cross-correlation is an efficient technique to track the GPS-P(Y) code on L2 in a GPS-software radio because this technique avoids the time-consuming generation of a reference signal. The reference signal is simply the received signal on L1.

Cross-correlation can also be applied to track unknown signal structures by two receivers (see Figure 7.4). For example, VLBI uses cross-correlation to determine path-delay differences of radio-astronomy signals from quasars received by two radio telescopes. The estimated signal parameters relate to the geometric distance difference between the transmitter and the two receivers. The same principle has been applied to interferometric tracking of GNSS signals by Muth using a software-receiver

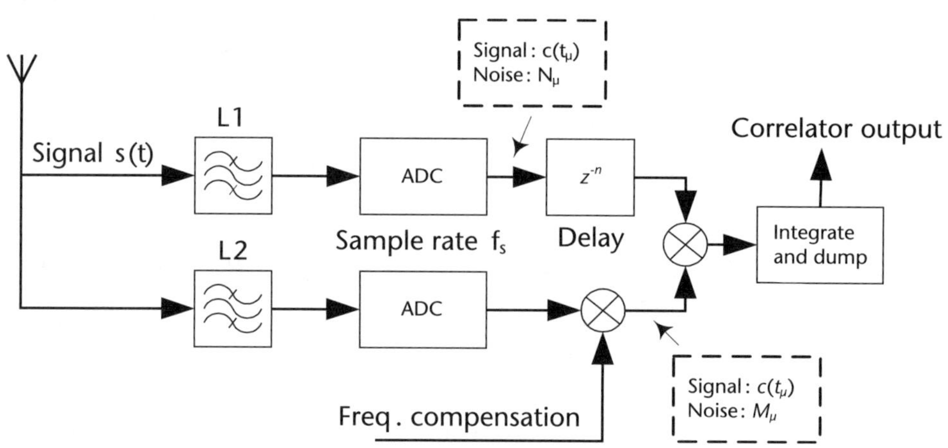

Figure 7.3 Cross-correlation tracking of GPS-P(Y) on L2.

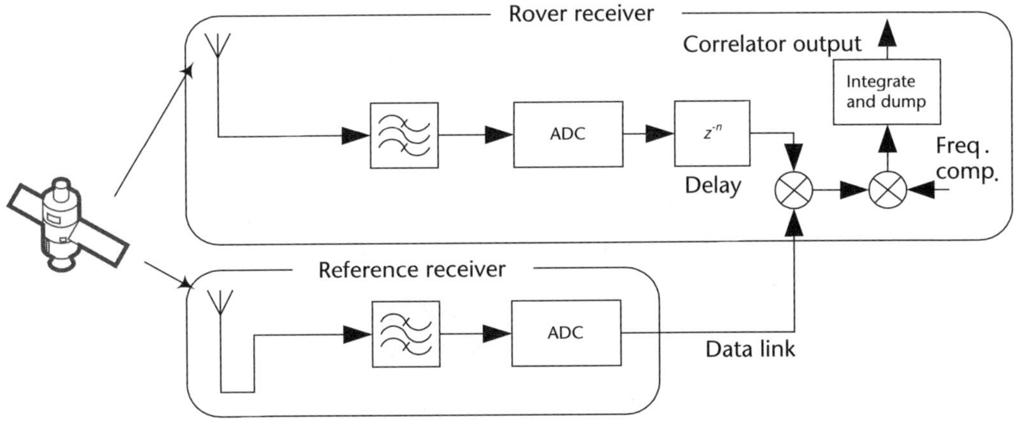

Figure 7.4 Cross-correlation used to track an unknown signal source.

approach [6]. This method has the advantage that the receiver does not need to know the broadcast PRN code or modulation scheme. If cross-correlation is used to track signals from two or more transmitters (e.g., from multiple GPS satellites), the distinction between the different signals can be difficult. In the case of L2-P(Y) cross-correlation tracking, different signals are distinguished only by their different Doppler frequencies. The correlator acts as a very narrow bandpass filter, centered at the Doppler frequency of the signal of interest (that is derived from C/A-code estimates). The Doppler frequency is selected by the "Freq. compensation" line in Figure 7.4. If two transmitted signals have the same Doppler frequency, the receiver cannot uniquely identify the two signals from one another and the "surge anomaly" occurs [7].

In the following example, we treat the cross-correlation correlator similar to a normal correlator, as discussed in Section 7.2 with the difference that the reference signal itself contains noise. We assume that signal samples of the received signal are modeled by (7.1) and (7.2). The analysis starts with complex-valued received input samples and uses (7.8) to obtain results for real-valued input samples.

The received samples are correlated with a reference signal given by

$$r_{rec,\mu} = c_{rec}(t_\mu - \tau_0)\exp\{i(\omega_0 t'_\mu - \varphi_0)\} + M_\mu \tag{7.68}$$

where the reference-signal and received-signal noise properties are

$$\begin{aligned} \langle N_\mu N_\upsilon \rangle_{\mathbf{N}} &= 0, \quad \langle N_\mu \bar{N}_\upsilon \rangle_{\mathbf{N}} = 2 \\ \langle M_\mu M_\upsilon \rangle_{\mathbf{M}} &= 0, \quad \langle M_\mu \bar{M}_\upsilon \rangle_{\mathbf{M}} = 2 \\ \langle M_\mu N_\upsilon \rangle_{\mathbf{N,M}} &= 0, \quad \langle M_\mu \bar{N}_\upsilon \rangle_{\mathbf{N,M}} = 0 \end{aligned} \tag{7.69}$$

The equation (7.68) for the reference signal models other possibly present signal components (e.g., image frequencies from sampling a real-valued signal or signals from other transmitters) by the noise term M_μ. In the following example, we assume that those signal components do not correlate with the received signal. This might not be true for specific situations of identical Doppler [7].

The output of the (cross-correlation) correlator X is defined as

$$\begin{aligned} X(\bar{S}_\mu) &= \sum_{\mu=1}^{L} r_{rec,\mu}\bar{S}_\mu = \sum_{\mu=1}^{L} r_{rec,\mu}\bar{r}_\mu + \sum_{\mu=1}^{L}\left(r_{rec,\mu}\bar{N}_\mu + \bar{r}_\mu M_\mu + \bar{N}_\mu M_\mu\right) \\ &= X_{sig}(r_\mu) + X_{noise}(r_\mu, N_\mu, M_\mu) \end{aligned} \tag{7.70}$$

and is split into a deterministic-signal part X_{sig} and a stochastic-noise part X_{noise}. The integration (summation) involves a number of n_{coh} samples corresponding to a coherent integration time of $T_{coh} = L/f_s$.

In the following example, all expected values are understood with respect to the reference-signal noise and the received noise:

$$\langle\ldots\rangle \equiv \langle\ldots\rangle_{\mathbf{N},\mathbf{M}} \tag{7.71}$$

7.5.1 Expected Value

The expected value for the correlator output is the sum of the deterministic contribution

$$\begin{aligned} X_{sig}(r_\mu) &= \sum_{\mu=1}^{L} r_{rec,\mu}\bar{r}_\mu = a\sum_{\mu=1}^{L} \bar{c}(t_\mu - \tau)c_{rec}(t_\mu - \tau_0)\exp\{-i(\omega t'_\mu - \varphi) + i(\omega_0 t'_\mu - \varphi_0)\} \\ &\approx aT_{coh}f_s R_{\bar{c},c_{rec}}(\tau - \tau_0)\exp\{i(\varphi - \varphi_0)\}\kappa(\omega_0 - \omega) \end{aligned} \tag{7.72}$$

and the noise contribution. The expected value of the latter contribution vanishes because both noise processes are uncorrelated and unbiased,

$$\left\langle X_{noise}\left(r_\mu, N_\mu, M_\mu\right)\right\rangle = \left\langle \sum_{\mu=1}^{L}\left(r_{rec,\mu}\bar{N}_\mu + \bar{r}_\mu M_\mu + \bar{N}_\mu M_\mu\right)\right\rangle = 0 \tag{7.73}$$

The deterministic contribution relates to the parameter difference of the code phases $\tau - \tau_0$, Doppler difference $\omega - \omega_0$, and carrier-phase difference $\varphi - \varphi_0$.

Thus, for complex-valued input samples, the expected value is

$$\begin{aligned} \left\langle X(\bar{S}_\mu)\right\rangle &\approx aT_{coh}f_s R_{\bar{c},c_{rec}}(\tau - \tau_0)\exp\{i(\varphi - \varphi_0)\}\kappa(\omega_0 - \omega) \\ \left\langle X(S_\mu)\right\rangle &\approx 0 \end{aligned} \tag{7.74}$$

and for real-valued input samples,

$$\left\langle X(\mathrm{Re}\{S_\mu\})\right\rangle \approx \frac{a}{2}T_{coh}f_s R_{\bar{c},c_{rec}}(\tau - \tau_0)\exp\{i(\varphi - \varphi_0)\}\kappa(\omega_0 - \omega) \tag{7.75}$$

7.5.2 Covariance

To evaluate the covariance, assume two different correlators, which are distinguished by an index "a" and an index "b." Both correlators are associated with different parameters, namely

$$
\begin{aligned}
X^a(S_\mu) &\mapsto c_{rec}, \tau_0^a, \omega_0^a, \varphi_0^a, M_\mu \\
X^b(S_\mu) &\mapsto c_{rec}, \tau_0^b, \omega_0^b, \varphi_0^b, M_\mu
\end{aligned}
\tag{7.76}
$$

Both work with the same received-signal samples. The underlying reference signal c_{rec} is identical, as is the noise process M_μ.

Furthermore, we only consider code-phase shifts as a multiple of the sampling interval:

$$
\tau_0^a = \frac{a}{f_s}, \quad \tau_0^b = \frac{b}{f_s} \tag{7.77}
$$

Similar to Section 7.2.2, the covariance is defined as

$$
\operatorname{cov}\left\langle X^a(S_\mu), X^b(S_\mu)\right\rangle = \left\langle \overline{X^a(r_\mu = 0, N_\mu, M_\mu)} X^b(r_\mu = 0, N_\mu, M_\mu)\right\rangle \tag{7.78}
$$

because the signal contribution cancels and the expected value of the received-noise contribution vanishes. Thus,

$$
\begin{aligned}
\operatorname{cov}\left\langle X^a(S_\mu), X^b(S_\mu)\right\rangle &= \operatorname{cov}\left\langle X^a(r_\mu = 0, N_\mu, M_\mu), X^b(r_\mu = 0, N_\mu, M_\mu)\right\rangle \\
&= \sum_{\mu,\upsilon=1}^{L} \left\langle \overline{N_\mu \left(c_{rec}\left(t_\mu - \tau_0^a\right) \exp\left\{i\left(\omega_0^a t'_\mu - \varphi_0^a\right)\right\} + M_{\mu-a}\right)} \right. \\
&\quad \left. \times N_\upsilon \left(c_{rec}\left(t_\upsilon - \tau_0^b\right) \exp\left\{i\left(\omega_0^b t'_\upsilon - \varphi_0^b\right)\right\} + M_{\upsilon-b}\right)\right\rangle \\
&= 2\sum_{\mu=1}^{L} \overline{c_{rec}}\left(t_\mu - \tau_0^a\right) c_{rec}\left(t_\mu - \tau_0^b\right) \exp\left\{it'_\mu\left(\omega_0^b - \omega_0^a\right) - i\left(\varphi_0^b - \varphi_0^a\right)\right\} \\
&\quad + \sum_{\mu,\upsilon=1}^{L} \left\langle \overline{N_\mu M_{\mu-a}} N_\upsilon M_{\upsilon-b}\right\rangle \\
&= 2\sum_{\mu=1}^{L} \overline{c_{rec}}\left(t_\mu - \tau_0^a\right) c_{rec}\left(t_\mu - \tau_0^b\right) \exp\left\{it'_\mu\left(\omega_0^b - \omega_0^a\right) - i\left(\varphi_0^b - \varphi_0^a\right)\right\} \\
&\quad + \sum_{\mu,\upsilon=1}^{L} \left\langle \overline{N_\mu} N_\upsilon\right\rangle \left\langle \overline{M_{\mu-a}} M_{\upsilon-b}\right\rangle \approx 2 f_s T_{coh} R_{\overline{c_{rec}}, c_{rec}}\left(\tau_0^a - \tau_0^b\right) \\
&\quad \times \exp\left\{i\left(\varphi_0^a - \varphi_0^b\right)\right\} \kappa\left(\omega_0^b - \omega_0^a\right) + 4\delta_{a,b} f_s T_{coh}
\end{aligned}
\tag{7.79}
$$

because (1.16) and (1.28) hold and we assume that the shifts a, b are much smaller than L. The covariance for the complex-conjugated input signal is given similarly as

$$
\operatorname{cov}\left\langle X^a\left(\overline{S}_\mu\right), X^b\left(\overline{S}_\mu\right)\right\rangle = \operatorname{cov}\left\langle X^a(S_\mu), X^b(S_\mu)\right\rangle \tag{7.80}
$$

For mixed terms, we obtain

$$\operatorname{cov}\left\langle X^a(\bar{S}_\mu), X^b(S_\mu)\right\rangle = \sum_{\mu,\,\upsilon=1}^{L}\left\langle \bar{N}_\mu\left(c_{rec}\left(t_\mu - \tau_0^a\right)\exp\left\{i\left(\omega_0^a t'_\mu - \varphi_0^a\right)\right\} + M_{\mu-a}\right)\right.$$
$$\left.\times N_\upsilon\left(c_{rec}\left(t_\upsilon - \tau_0^b\right)\exp\left\{i\left(\omega_0^b t'_\upsilon - \varphi_0^b\right)\right\} + M_{\upsilon-b}\right)\right\rangle$$
$$= 2\sum_{\mu=1}^{L}\overline{c_{rec}}\left(t_\mu - \tau_0^a\right)c_{rec}\left(t_\mu - \tau_0^b\right)\exp\left\{it'_\mu\left(\omega_0^b - \omega_0^a\right) - i\left(\varphi_0^b - \varphi_0^a\right)\right\} \tag{7.81}$$
$$\times\left\langle N_\mu N_\upsilon\right\rangle + \sum_{\mu,\upsilon=1}^{L}\left\langle N_\mu \bar{M}_{\mu-a} N_\upsilon M_{\upsilon-b}\right\rangle = 0$$

Based on the derivation of (7.24), for real-valued input signals we obtain

$$\operatorname{cov}\left\langle X^a\left(\operatorname{Re}\{S_\mu\}\right), X^b\left(\operatorname{Re}\{S_\mu\}\right)\right\rangle =$$
$$\approx f_s T_{coh} R_{\overline{c_{rec}},c_{rec}}\left(\tau_0^a - \tau_0^b\right)\exp\left\{i\left(\varphi_0^a - \varphi_0^b\right)\right\}\kappa\left(\omega_0^b - \omega_0^a\right) + 2\delta_{a,b} f_s T_{coh} \tag{7.82}$$

7.5.3 Variance

The variance of the correlator output is a special case of the covariance assuming identical parameters for the "a" and "b" cases.

For complex-valued input signals

$$\operatorname{var}\left\langle X(S_\mu)\right\rangle = 2 f_s T_{coh} R_{\overline{c}_{rec},c_{rec}}(0) + 4 f_s T_{coh} \tag{7.83}$$

holds, and for real-valued input signals

$$\operatorname{var}\left\langle X\left(\operatorname{Re}\{S_\mu\}\right)\right\rangle = f_s T_{coh} R_{\overline{c}_{rec},c_{rec}}(0) + 2 f_s T_{coh} \tag{7.84}$$

7.5.4 L2P(Y)-Code Carrier-Phase Discriminator Noise

In the following section, the set of formulas for the cross-correlation correlator are applied to determine the GPS-P(Y) L2–L1 cross-correlation carrier-phase discriminator noise as depicted in Figure 7.4. The discriminator estimates the L2–L1 carrier phase difference, which is added to the L1 carrier phase to obtain an L2 carrier-phase estimate. The L1 carrier-phase estimate is based on the C/A-code, whereas the L2–L1 carrier-phase estimate is based on the P(Y)-code. Consequently, both estimates are uncorrelated.

In Figure 7.4, the reference signal is obtained by multiplying the signal samples on L1 by a complex carrier. The frequency of the carrier needs to be selected in a way that the resulting carrier frequency ω_0 (including Doppler) matches approximately the L2 carrier frequency (including Doppler) for the satellite of interest. The necessary Doppler computations are based on GPS C/A-code measurements. Doppler frequency errors are tolerable if they are much smaller than the inverse of the coherent integration time. The reference signal is delayed by τ_0 to maximize the correlator output. The delay needs to be a multiple of the sampling period. The delay can be used to compute the L2–P(Y) code pseudorange, which is not discussed here. The parameter φ_0 of (7.68) is the carrier phase of the reference signal and is directly related to the received L1-carrier phase.

We assume that the deterministic part of the reference signal equals the transmitted signal [being the same P(Y)-code signal on L1 and L2] multiplied by an amplitude factor b:

$$c_{rec}(t) = bc(t) \tag{7.85}$$

The reference signal is complex-valued and b relates to the C/N_0 value on L1. According to Appendix A.2.1, the relationship is

$$C/N_{0,ref} = \frac{f_s b^2}{2} \tag{7.86}$$

The autocorrelation function of the reference signal becomes

$$R_{\overline{c}_{rec},c_{rec}}(\tau) = b^2 R_{\overline{c},c}(\tau) \tag{7.87}$$

The amplitude of the received signal relates to the C/N_0 value on L2

$$C/N_0 = \frac{f_s a^2}{4} \tag{7.88}$$

because a real-valued received signal is considered (see Appendix A.2.2).

The formula for the variance of a carrier-phase discriminator will be derived in Section 8.1.3 and is given by (8.31). For cross-correlation, it reads as

$$\begin{aligned}
\operatorname{var}\langle \Phi(\operatorname{Re}\{S_\mu\})\rangle &\approx \frac{\operatorname{var}\langle X_P(\operatorname{Re}\{S_\mu\})\rangle}{2\left|\langle X_P(\operatorname{Re}\{S_\mu\})\rangle\right|^2}\left(1+\frac{\operatorname{var}\langle X_P(\operatorname{Re}\{S_\mu\})\rangle}{2\left|\langle X_P(\operatorname{Re}\{S_\mu\})\rangle\right|^2}\right) \\
&= \frac{4f_sT_{coh}b^2R_{\overline{c},c}(0)+8f_sT_{coh}}{2(af_sT_{coh})^2\left|\kappa(\omega_0-\omega)bR_{\overline{c},c}(\tau-\tau_0)\right|^2}\left(1+\frac{4f_sT_{coh}b^2R_{\overline{c},c}(0)+8L}{2(af_sT_{coh})^2\left|\kappa(\omega_0-\omega)bR_{\overline{c},c}(\tau-\tau_0)\right|^2}\right) \\
&= \frac{(4T_{coh}2C/N_{0,ref}+8f_sT_{coh})f_s^2}{2(f_sT_{coh})^2\,4C/N_0 2C/N_{0,ref}\left|\kappa(\omega_0-\omega)R_{\overline{c},c}(\tau-\tau_0)\right|^2}\times \\
&\quad\times\left(1+\frac{(4T_{coh}2C/N_{0,ref}+8L)f_s^2}{2(f_sT_{coh})^2 4C/N_0 2C/N_{0,ref}\left|\kappa(\omega_0-\omega)R_{\overline{c},c}(\tau-\tau_0)\right|^2}\right) \\
&= \frac{2T_{coh}C/N_{0,ref}+2T_{coh}f_s}{4T_{coh}^2C/N_0C/N_{0,ref}\left|\kappa(\omega_0-\omega)R_{\overline{c},c}(\tau-\tau_0)\right|^2} \\
&\quad\times\left(1+\frac{2T_{coh}C/N_{0,ref}+2T_{coh}f_s}{4T_{coh}^2C/N_0C/N_{0,ref}\left|\kappa(\omega_0-\omega)R_{\overline{c},c}(\tau-\tau_0)\right|^2}\right) \\
&= \frac{C/N_{0,ref}+f_s}{2T_{coh}C/N_0C/N_{0,ref}\left|\kappa(\omega_0-\omega)R_{\overline{c},c}(\tau-\tau_0)\right|^2} \\
&\quad\times\left(1+\frac{C/N_{0,ref}+f_s}{2T_{coh}C/N_0C/N_{0,ref}\left|\kappa(\omega_0-\omega)R_{\overline{c},c}(\tau-\tau_0)\right|^2}\right)
\end{aligned} \tag{7.89}$$

where X_p denotes the prompt cross-correlation correlator.

If the C/N_0 value of the reference signal (e.g., on L1) is much larger than the sample rate, the phase-discriminator variance equals the case of a noiseless reference signal:

$$\operatorname{var}\left\langle \Phi(\operatorname{Re}\{S_\mu\})\right\rangle \underset{C/N_{0,ref} \gg f_s}{=} \frac{1}{2T_{coh}C/N_0\left|\kappa(\omega_0-\omega)R_{\bar{c},c}(\tau-\tau_0)\right|^2} \times\left(1+\frac{1}{2T_{coh}C/N_0\left|\kappa(\omega_0-\omega)R_{\bar{c},c}(\tau-\tau_0)\right|^2}\right) \tag{7.90}$$

In the general case of an arbitrary C/N_0 value of the reference signal, the phase-discriminator noise is identical to the noiseless reference-signal case if the received signal C/N_0 value is replaced by an effective value $C/N_{0,eff}$:

$$C/N_{0,eff} = C/N_0 \frac{C/N_{0,ref}}{C/N_{0,ref}+f_s} \tag{7.91}$$

Thus, the cross-correlation loss is given by

$$L_{xcorr} = \frac{C/N_{0,ref}}{C/N_{0,ref}+f_s} \tag{7.92}$$

and is a function of the reference-signal power and the sample rate f_s.

7.6 Incorporating Colored Noise

Within this section, we will extend the obtained mean and variance formulas for the case of colored noise. The approach is based on a signal transformation to reduce the colored-noise case to the white-noise case. Within this approach, artificial signals are introduced and the reader should keep in mind that those signals are just used for the theoretical analysis. They do not appear in the receiver.

7.6.1 White-Noise Transformation

The colored-noise analysis starts from a similar signal model as for the white-noise case. The signal samples S_μ are the sum of a deterministic part r_μ and a stochastic part N_μ,

$$S_\mu = r_\mu + N_\mu \tag{7.93}$$

The complex-valued noise samples are from a wide-sense stationary process and the covariance is

$$\left\langle N_\mu N_\upsilon\right\rangle_N = 0, \left\langle \bar{N}_\mu \bar{N}_\upsilon\right\rangle_N = 0, \left\langle N_\mu \bar{N}_\upsilon\right\rangle_N = 2Q(\mu-\upsilon) \tag{7.94}$$

The noise-power spectral density is related to the covariance function via the Fourier-series equation

$$\tilde{Q}(f) = \sum_{\mu=-\infty}^{\infty} Q(\mu)\exp\left\{-\frac{2\pi i f}{f_s}\mu\right\} \qquad {f_s}/{2} < f \le {f_s}/{2} \tag{7.95}$$

and its inverse

$$Q(\mu) = \frac{1}{f_s} \int_{f=-f_s/2}^{f_s/2} \tilde{Q}(f) \exp\left\{\frac{2\pi i f}{f_s}\mu\right\} df \tag{7.96}$$

Note that

$$1 = Q(0) = \frac{1}{f_s} \int_{f=-f_s/2}^{f_s/2} \tilde{Q}(f) df \tag{7.97}$$

must be fulfilled so that the C/N_0 definition can be maintained. For the C/N_0 definition in the case of colored noise, the noise-power density is interpreted as a *mean-noise power density*.

The complex-valued signal model from Section 1.8 is used:

$$r_\mu = ac\left(t_\mu - \tau\right)\exp\{i(\omega t'_\mu - \varphi)\} \tag{7.98}$$

In the following examples, we analyze in detail the expression of a correlator operating on the complex-conjugated samples. The discussion is analogous for a correlator operating on the nonconjugated samples. According to Section 7.2, the correlator is defined as

$$C\left(\bar{S}_\mu\right) = \sum_{\mu=1}^{L} r_{rec,\mu}\bar{S}_\mu = \mathbf{S}^* \cdot \mathbf{r}_{rec} \tag{7.99}$$

The expression is now rewritten, introducing artificial signals $\mathbf{S}'$, $\mathbf{r}'_{rec}$, and $\mathbf{N}'$

$$C\left(\bar{S}_\mu\right) = \mathbf{S}^* \cdot \mathbf{r}_{rec} = \mathbf{S}^* Q^{-1/2} \cdot Q^{1/2}\mathbf{r}_{rec} = \left(Q^{-1/2}\mathbf{S}\right)^* \cdot Q^{1/2}\mathbf{r}_{rec} = \mathbf{S}'^* \cdot \mathbf{r}'_{rec} \tag{7.100}$$

which are defined as

$$\begin{aligned} &\mathbf{S}' = \mathbf{r}' + \mathbf{N}' \\ &\mathbf{S}' = Q^{-1/2}\mathbf{S}, \quad \mathbf{r}' = Q^{-1/2}\mathbf{r}, \quad \mathbf{N}' = Q^{-1/2}\mathbf{N} \\ &\mathbf{r}'_{rec} = Q^{1/2}\mathbf{r}_{rec} \end{aligned} \tag{7.101}$$

by using the Hermitian colored-noise covariance matrix Q. Similar to the discussion of spectral whitening in Section 6.4.1, the artificially introduced noise is white; that is,

$$\left\langle N_\mu' N_\upsilon' \right\rangle_{\mathbf{N}} = 0, \left\langle \bar{N}_\mu' \bar{N}_\upsilon' \right\rangle_{\mathbf{N}} = 0, \left\langle N_\mu' \bar{N}_\upsilon' \right\rangle_{\mathbf{N}} = 2\delta_{\mu,\upsilon} \tag{7.102}$$

So far, the discussion has been carried out using finite-length sample vectors. It is, however, more convenient to confine the discussion to the level of continuous-time signals, thereby allowing simpler expressions when working in the frequency domain.

Let $W^x\{\ldots\}$ be a filter operating on the continuous time signals defined in frequency domain as

$$c(t) \to W^x\{c(t)\} \quad \Leftrightarrow \quad \tilde{c}(f) \to \left|\tilde{W}(f)^x\right| \tilde{\Phi}(f)^x \tilde{c}(f) \tag{7.103}$$

with x being a real number (acting as an exponent) and an arbitrary phase function defined in the frequency domain with

$$\left|\tilde{\Phi}(f)\right| = 1 \tag{7.104}$$

The filter's magnitude response is related to the colored-noise power spectral density via

$$\tilde{W}^{-2}(f) = \tilde{Q}(f) \tag{7.105}$$

The filter $W^x\{...\}$ is called a (continuous-time) *whitening filter*.

Applying the whitening filter to the received signal and the inverse whitening filter to the internally generated reference signal before sampling

$$\begin{aligned} S'(t) &= W\{S(t)\} \\ c_{ref}'(t) &= W^{-1}\{c_{ref}(t)\} \end{aligned} \tag{7.106}$$

leaves the correlation value invariant and allows the application of the white-noise theory. Only the correlation functions appearing in formulas on the stochastic correlator properties need to be adjusted.

The correlation function between an internal reference signal and the received signal remains unchanged because

$$R_{\overline{c},c_{rec}}(\tau) \rightarrow R_{\overline{W\{c\}},W^{-1}\{c_{rec}\}}(\tau) = R_{\overline{c},c_{rec}}(\tau) \tag{7.107}$$

and the correlation function between two reference signals changes to

$$R_{\overline{c_{rec}^a},c_{rec}^b}(\tau) \rightarrow R_{\overline{W^{-1}\{c_{rec}^a\}},W^{-1}\{c_{rec}^b\}} = R_{\overline{c_{rec}^a},W^{-2}\{c_{rec}^b\}}(\tau) \tag{7.108}$$

In Section 7.6.2, the technique will be illustrated for the example of the code-discriminator noise for an early–late tracker of a signal of finite bandwidth.

7.6.2 Early–Late Code Discriminator with Infinite Sample Rate

In the following example, the white-noise transformation will be used to determine the discriminator noise for an early–late tracker of a signal of finite bandwidth using an infinite sample rate (i.e., we work now with continuous time signals).

We assume that an infinite-bandwidth signal with the waveform at baseband $c_p(t)$ is transmitted and the same signal is also generated inside a receiver as a replica signal. The signal plus colored noise is received by the front end, which limits the bandwidth of both. First, the results for real-valued signal samples are analyzed and the latter results are also presented for a complex-valued signal. The spectra of the continuous-time signals are shown in Figure 7.5. The noise is, in both cases, bandlimited and eventually colored.

We now start with the analysis of the real-valued signal.

The signal is transmitted at a nominal frequency f_0 and a brick-wall filter is used within the front end. No signal or noise energy passes outside the passband that ranges from $-B$ to B. The spectrum of the real-valued signal and the noise is

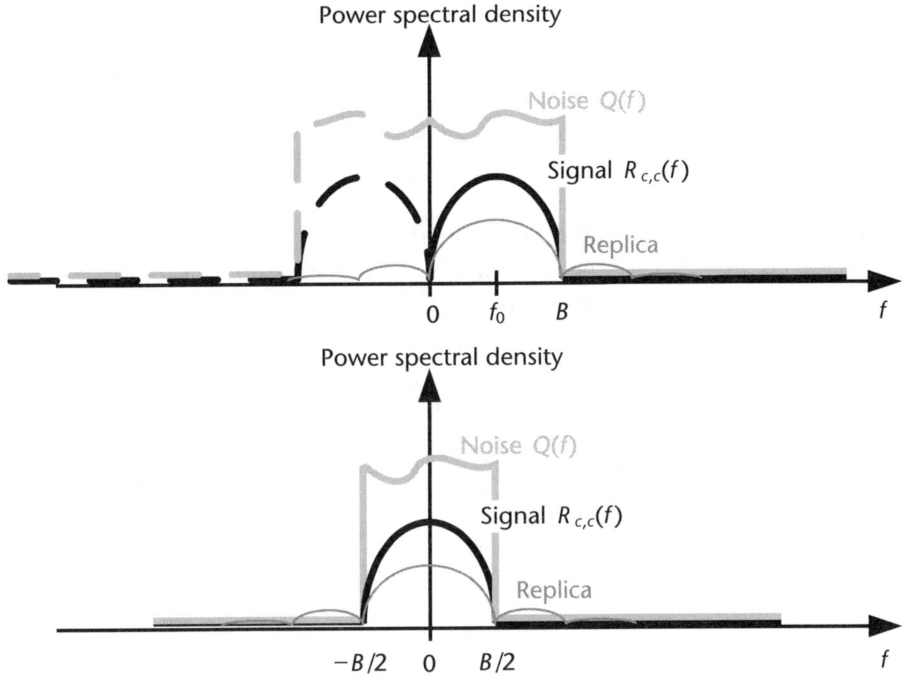

Figure 7.5 Continuous-time spectra (upper: real IF signal at f_0; lower: complex signal).

symmetric with respect to zero. Let f_0 be the nominal frequency of the received signal, with

$$f_0 = B/2. \tag{7.109}$$

The transmitted/replica signal at baseband $c_P(t)$ is of infinite bandwidth. Note that although a real-valued input signal is considered, the baseband waveform is generally complex-valued. The real-valued input signal covers twice the bandwidth as the complex signal at baseband. After the brick-wall filter, the positive frequency components of the received signal are represented at baseband as

$$\tilde{c}(f) = \begin{cases} \tilde{c}_P(f) & |f| < B/2 \\ 0 & \end{cases}$$

The received signal fits into the passband because it is shifted by $f_0 = B/2$. Within the passband, the noise-power spectral density is given as

$$\tilde{Q}(f) = \begin{cases} \tilde{Q}(f) & |f| < B \\ 0 & \end{cases} \tag{7.110}$$

To maintain the C/N_0 definition (in a mean-noise power-density sense), we require

$$\int_{f=-B}^{B} \tilde{Q}(f)df = 1 \tag{7.111}$$

In contrast to (7.97), assume here an infinite sample rate.

The cross-correlation function between the received signal and an internally generated signal is given by (see Section A.2.5)

$$R_{\bar{c},c_{rec}}(\tau) = \int_{f=-\infty}^{\infty} \tilde{R}_{\bar{c},c_{rec}}(f) e^{2\pi i f \tau} df = \int_{f=-B/2}^{B/2} \tilde{\bar{c}}(-f)\tilde{c}_{rec}(f) e^{2\pi i f \tau} df \tag{7.112}$$

because outside the passband, the received signal has only vanishing frequency components. The frequency-domain representation of the correlation function is the product of the frequency transform of both signals.

The correlation function between two internally generated signals changes according to (7.108) into

$$\begin{aligned} R_{\overline{c^a_{rec}}, W^{-2}\{c^b_{rec}\}}(\tau) &= \int_{f=-3B/2}^{B/2} \tilde{Q}\left(f + B/2\right) \tilde{R}_{\overline{c^a_{rec}}, c^b_{rec}}(f) e^{2\pi i f \tau} df \\ &= \int_{f=-3B/2}^{B/2} \tilde{Q}(f + f_0) \tilde{\bar{c}}^a_{rec}(-f) \tilde{c}^b_{rec}(f) e^{2\pi i f \tau} df \end{aligned} \tag{7.113}$$

The whitening filter is shifted by f_0, the center frequency, where the correlation takes place.

Ignoring for the moment the squaring loss (which is sufficient to demonstrate the white-noise transformation), the variance of the code discriminator, working on real-values samples with white noise, obtained from (8.14), is

$$\operatorname{var}\left\langle D\left(\operatorname{Re}\{S_\mu\}\right)\right\rangle = \left|\frac{1}{R_{\bar{c},c_D}'(0)}\right|^2 \frac{R_{\bar{c}_D,c_D}(0)}{2C/N_0 T_{coh}} \tag{7.114}$$

Applying the white-noise transformation, the code-discriminator variance changes to

$$\operatorname{var}\left\langle D(\operatorname{Re}\{S_\mu\})\right\rangle = \left|\frac{1}{R_{\bar{c},c_D}'(0)}\right|^2 \frac{R_{\bar{c}_D,Q\{c_D\}}(0)}{2C/N_0 T_{coh}} \tag{7.115}$$

Note that $Q\{\ldots\} = W^{-2}\{\ldots\}$.

The early–late reference signal is

$$c_D(t) = c_P\left(t - d/2\right) - c_P\left(t + d/2\right) \tag{7.116}$$

where d is the discriminator spacing in seconds.

The correlation function of the complex-conjugated received signal with the early–late reference signal is

$$\begin{aligned} R_{\bar{c},c_D}(\tau) &= R_{\bar{c}_P,c_D}(\tau) = R_{\bar{c}_P,c_P}\left(\tau - d/2\right) - R_{\bar{c}_P,c_P}\left(\tau + d/2\right) \\ &= \int_{f=-B/2}^{B/2} \tilde{R}_{\bar{c}_P,c_P}(f) \left(e^{2\pi i f\left(\tau - d/2\right)} - e^{2\pi i f\left(\tau + d/2\right)} \right) df \end{aligned} \tag{7.117}$$

The first derivative at the origin is

$$\begin{aligned} R_{\bar{c},c_D}{}'(0) &= 2\pi i \int_{f=-B/2}^{B/2} f\tilde{R}_{\bar{c}_P,c_P}(f)\left(e^{-d\pi i f} - e^{d\pi i f}\right)df \\ &= 4\pi \int_{f=-B/2}^{B/2} f\tilde{R}_{\bar{c}_P,c_P}(f)\sin(d\pi f)df \end{aligned} \tag{7.118}$$

The modified autocorrelation function of the early–late reference signal becomes

$$\begin{aligned} R_{\bar{c}_D,Q\{c_D\}}(0) &= 2R_{\bar{c}_P,Q\{c_P\}}(0) - R_{\bar{c}_P,Q\{c_P\}}(-d) - R_{\bar{c}_P,Q\{c_P\}}(d) \\ &= \int_{f=-3B/2}^{B/2} \tilde{Q}(f+f_0)\tilde{R}_{\bar{c}_P,c_P}(f)\left(2 - e^{-2\pi i f d} - e^{2\pi i f d}\right)df \\ &= 4\int_{f=-3B/2}^{B/2} \tilde{Q}(f+f_0)\tilde{R}_{\bar{c}_P,c_P}(f)\sin^2(\pi f d)df \end{aligned} \tag{7.119}$$

Overall, the code-discriminator variance working on real-valued samples with colored noise becomes

$$\operatorname{var}\left\langle D(\operatorname{Re}\{S_\mu\})\right\rangle = \frac{1}{2C/N_0T_{coh}} \frac{\int_{f=-3B/2}^{B/2} \tilde{Q}(f+f_0)\tilde{R}_{\bar{c}_P,c_P}(f)\sin^2(\pi f d)df}{\left|2\pi \int_{f=-B/2}^{B/2} f\tilde{R}_{\bar{c}_P,c_P}(f)\sin(\pi f d)df\right|^2} \tag{7.120}$$

Note that,

$$\tilde{R}_{\bar{c}_P,c_P}(f) = \tilde{\bar{c}}_P(-f)\tilde{c}_P(f) = \left|\tilde{c}_P(f)\right|^2 \tag{7.121}$$

If a *complex-valued input signal* is processed, the integration boundaries would be symmetric, yielding

$$\operatorname{var}\left\langle D(S_\mu)\right\rangle = \frac{1}{2C/N_0T_{coh}} \frac{\int_{f=-B/2}^{B/2} \tilde{Q}(f)\tilde{R}_{\bar{c}_P,c_P}(f)\sin^2(\pi f d)df}{\left|2\pi \int_{f=-B/2}^{B/2} f\tilde{R}_{\bar{c}_P,c_P}(f)\sin(\pi f d)df\right|^2} \tag{7.122}$$

Equations of type (7.122) have also been derived in the work by Betz and Kolodziejski using a different methodology [8].

Within the limits of vanishing discriminator spacing and white noise, the CLRB is reached:

$$\operatorname{var}\left\langle D(S_\mu)\right\rangle \underset{d\to 0}{=} \frac{1}{2C/N_0T_{coh}} \frac{\int_{f=-B/2}^{B/2} \tilde{R}_{\bar{c}_P,c_P}(f)\pi^2 f^2 d^2 df}{\left|2\pi \int_{f=-B/2}^{B/2} f\tilde{R}_{\bar{c}_P,c_P}(f)\pi f d df\right|^2}$$

$$= \frac{1}{2C/N_0T_{coh}} \frac{1}{(2\pi)^2 \int_{f=-B/2}^{B/2} f^2 \tilde{R}_{\bar{c}_P,c_P}(f)df} \tag{7.123}$$

7.7 Comparison of Finite and Infinite Sample Rates

The previous section showed how the white-noise transformation can be used to calculate the code-discriminator variance of a bandlimited signal using an infinite-sample rate (continuous time signals). There exists a different approach to compute the same code-discriminator variance, but this time using a finite-sample rate equal to the Nyquist rate. Both approaches give slightly different results, especially for small correlator spacing. This discrepancy was already pointed out by several authors [8, 9]. The theory of Betz and Kolodziejski states that the variance approaches the CRLB for $d \to 0$, whereas the work of Holmes states that the variance diverges for $d \to 0$. A numerical simulation supports Holmes [10].

In the following discussion, the code-discriminator variance of a bandlimited signal will be obtained under the assumption of Nyquist sampling and the result is contrasted to the infinite sample-rate case.

Complex-valued signal samples are assumed. The sample rate f_s equals the signal and noise bandwidth B. A flat noise-power spectral density is used. The signal spectra after sampling are depicted in Figure 7.6. Received signal and noise are

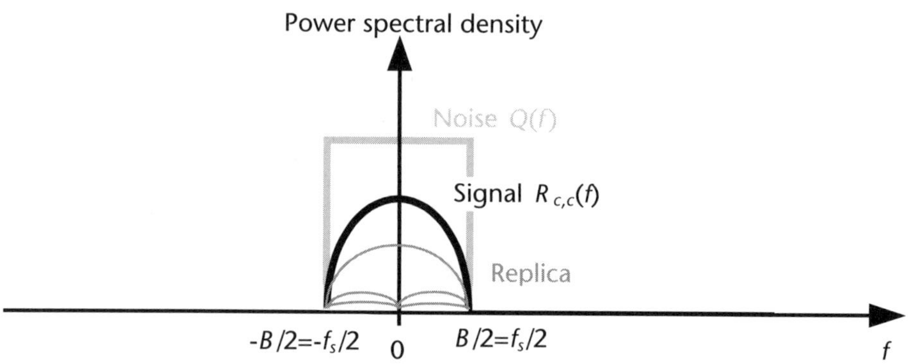

Figure 7.6 Signal spectra with Nyquist sampling (replica signal aliased).

sampled without aliasing, but the spectra of the internally generated infinite bandwidth replica shows aliasing effects.

After sampling, the noise-power spectral density is flat and the white-noise theory can be applied. Ignoring the squaring loss, the code-discriminator variance from (8.14) is expressed in the frequency domain as

$$\operatorname{var}\left\langle D(S_\mu)\right\rangle = \left|\frac{1}{R_{\bar{c},c_D}{}'(0)}\right|^2 \frac{R_{\bar{c}_D,c_D}(0)}{2C/N_0T_{coh}}$$

$$= \frac{1}{2C/N_0T_{coh}} \frac{\int\limits_{f=-\infty}^{\infty} \tilde{R}_{\bar{c}_P,c_P}(f)\sin^2(\pi f d)df}{\left|2\pi \int\limits_{f=-B/2}^{B/2} f\tilde{R}_{\bar{c}_P,c_P}(f)\sin(\pi f d)df\right|^2} \tag{7.124}$$

This expression compares to (7.122) and differs only in the integration boundaries of the nominator. In fact, in (7.124), the aliased early-minus-late reference-signal frequency components correlate with each other, whereas they do not correlate in the infinite sample-rate case (7.122).

To visualize the difference between the infinite and finite sample-rate case a BPSK power spectral density is assumed

$$\tilde{R}_{\bar{c}_P,c_P}(f) = T_c \frac{\sin^2(\pi f T_c)}{(\pi f T_c)^2} \tag{7.125}$$

where T_c is the chip period in seconds. Table 7.1 summarizes the parameter used for the evaluation of the formulas. It should be noted that (7.123) and (7.124) give the code variance in seconds-squared and the discriminator spacing is in seconds. For better visualization in Figure 7.7, the variance (actually the standard deviation) is plotted in meters and the discriminator spacing is in chips.

The integrals occurring in (7.123) and (7.124) can be solved analytically using a symbolic mathematics software package. The resulting expressions are not given here. From Figure 7.7, one clearly sees the divergence of the code-noise variance, if $d \rightarrow 0$ and if a finite sample rate is used.

Thus, an early–late tracker does *not* approach the CRLB if $d \rightarrow 0$ in a real-receiver implementation, which necessarily works with a finite sample rate. The

Table 7.1 Parameters for Nyquist/Infinite Sample-Rate Code-Variance Evaluation

Name	*Value*
Modulation	BPSK
Chip rate	1.023 Mchip/s
Bandwidth	20.46 MHz
C/N_0	45 dBHz
T_{coh}	20 ms

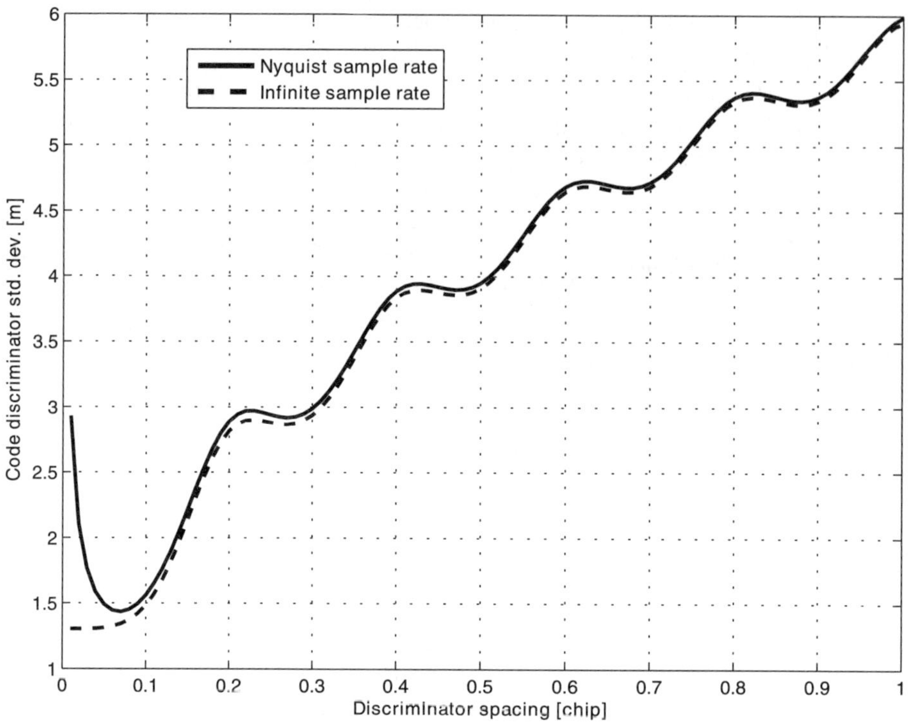

Figure 7.7 Code noise determined under the finite and infinite sample-rate assumption.

CRLB is reached if the code-tracking reference signal equals the first derivative of the received signal at baseband (see Section 4.3).

References

[1] Irsigler, M., *Multipath Propagation, Mitigation and Monitoring in the Light of Galileo and Modernized GPS*. University of Federal Armed Forces Munich, Werner-Heisenberg-Weg 39, D-85577 Neubiberg, http://www.unibw.de/unibib/digibib/ediss/bauv, 2008.

[2] van Dierendonck, A. J., "GPS Receivers," in *Global Positioning System: Theory and Applications, Vol. I*, pp. 329–407, Parkinson, B. W., and J. J. Spilker, (eds.), Washington, D.C.: American Institute of Aeronautics and Astronautics Inc., 1996.

[3] Won, J. H., T. Pany, and B. Eissfeller, "Implementation, Verification and Test Results of a MLE-based F-Correlator Method for Multi-Frequency GNSS Signal Tracking," *Proc. 20th Int. Technical Meeting of the Satellite Division of the Institute of Navigation (ION-GNSS) 2007*, Fort Worth, TX, September 25–28, 2007, pp. 2237–2249.

[4] Fenton, P. C., and J. Jones, "The Theory and Performance of NovAtel Inc.'s Vision Correlator," *Proc. ION-GNSS 2005*, Long Beach, September 13–16, 2005, pp. 2178–2186.

[5] Woo, K. T., "Optimum Semi-Codeless Carrier Phase Tracking of L2," *NAVIGATION, Journal of The Institute of Navigation*, Vol. 47, No. 2, 2000, pp. 82–99.

[6] Muth, B. J., "Software Based GNSS L1 Interferometric Positioning," *Proc. 21st Int. Technical Meeting of the Satellite Division of the Institute of Navigation (ION-GNSS) 2008*, Savannah, GA, September 16–19, 2008, pp. 3009–3019.

[7] Sleewagen, J.-M., "Surge Anomaly in Cross-Correlation GPS Measurements: Description and Analysis," *NAVIGATION, Journal of The Institute of Navigation*, Vol. 46, No. 2, 1999, pp. 119–125.

[8] Betz, J. W., and K. R. Kolodziejski, "Extended Theory of Early-Late Code Tracking for a Bandlimited GPS Receiver," *NAVIGATION, Journal of The Institute of Navigation*, Vol. 47, No. 3, 2000, pp. 211–226.

[9] Holmes, J. K., "Noncoherent Late Minus Early Power Code Tracking Loop Performance with Front End Filtering," *Proc. 20th Int. Technical Meeting of the Satellite Division of the Institute of Navigation (ION-GPS) 1997*, Kansas City, MO, September 16–19, 1997, pp. 583–591.

[10] Pany, T., et al., "Code and Carrier Phase Tracking Performance of a Future Galileo RTK Receiver," *Proc. European Navigation Conference (ENC-GNSS) 2002*, Graz, May 27–30, 2002.

CHAPTER 8

Discriminators

The different correlator types of Chapter 7 are the basis for estimating signal-parameter values. The correlator values are combined to form the code-phase, Doppler, and carrier-phase discriminator values. Nonlinear operations are required to remove the carrier-phase dependency of the correlators and to obtain noncoherent discriminators. The reference functions can be optimized to achieve a desired tracking performance in terms of multipath mitigation and thermal noise. This process is called S-curve shaping. Multiple correlator values can be used to simultaneously estimate direct and reflected signal parameters. Finally, this chapter outlines how to compute positioning accuracy from discriminator noise.

8.1 Noncoherent Discriminators

The D- (and F-correlator) values depend approximately linearly on the code-phase (and respectively on the Doppler) difference of the received and replica signal. However, the D- and F-correlator are also proportional to the complex-signal amplitude. As pointed out in Section 4.3, the complex-amplitude dependency can be removed by dividing the D- or F-correlator by the P-correlator. Care must be taken if the P-correlator value is near zero and eventually the discriminator values need to be clipped. The carrier-phase estimate is obtained via an **atan** operation without any clipping operation.

In the following, all expected values are with respect to noise; that is,

$$\langle \ldots \rangle = \langle \ldots \rangle_{\mathrm{N}} \tag{8.1}$$

8.1.1 Code Discriminator

The code-phase discriminator is obtained by dividing one D-correlator by one P-correlator. The reference signals associated with the different correlators are

$$\begin{aligned} C_D &\leftarrow c_D(t) \\ C_P &\leftarrow c_P(t) \end{aligned} \tag{8.2}$$

The reference signals fulfill the following requirements within the code phase linearity limit l in order to allow the construction of a linear code-phase discriminator,

$$
\begin{aligned}
&R_{\bar{c}_D,c_P}(0) = 0 \\
&R_{\bar{c},c_D}(\tau) \approx R_{\bar{c},c_D}{}'(0)\tau \qquad |\tau| \ll l \\
&R_{\bar{c},c_D}{}'(0) \neq 0 \\
&R_{\bar{c},c_P}(0) \neq 0
\end{aligned}
\tag{8.3}
$$

Because of the first condition, both correlators are *uncorrelated* if they are based on the same code-phase delay. Because the correlators use a large number of signal samples, the central-limit theorem applies and the correlator values are Gaussian random variables. Uncorrelated Gaussian random variables are independent of each other.

The code-phase discriminator is constructed for a real-valued incoming signal as

$$
D = \mathrm{Re}\left\{\alpha \frac{C_D\left(\mathrm{Re}\{S_\mu\}\right)}{C_P\left(\mathrm{Re}\{S_\mu\}\right)}\right\} = \mathrm{Re}\left\{\alpha \frac{C_D\left(\mathrm{Re}\{S_\mu\}\right)\overline{C_P\left(\mathrm{Re}\{S_\mu\}\right)}}{\left|C_P\left(\mathrm{Re}\{S_\mu\}\right)\right|^2}\right\}
\tag{8.4}
$$

where α is a normalization constant that will be determined later.

In a first-order approximation, the expected value of the code discriminator is

$$
\begin{aligned}
\langle D\rangle &= \left\langle \mathrm{Re}\left\{\alpha \frac{C_D\left(\mathrm{Re}\{S_\mu\}\right)}{C_P\left(\mathrm{Re}\{S_\mu\}\right)}\right\}\right\rangle = \mathrm{Re}\left\{\alpha \frac{\left\langle C_D\left(\mathrm{Re}\{S_\mu\}\right)\right\rangle}{\left\langle\left\langle C_P\left(\mathrm{Re}\{S_\mu\}\right)\right\rangle - \left\langle C_P\left(\mathrm{Re}\{S_\mu\}\right)\right\rangle + C_P\left(\mathrm{Re}\{S_\mu\}\right)\right\rangle}\right\} \\
&= \mathrm{Re}\left\{\alpha \frac{\left\langle C_D\left(\mathrm{Re}\{S_\mu\}\right)\right\rangle}{\left\langle C_P\left(\mathrm{Re}\{S_\mu\}\right)\right\rangle}\left\langle\left(1 - \frac{\left\langle C_P\left(\mathrm{Re}\{S_\mu\}\right)\right\rangle - C_P\left(\mathrm{Re}\{S_\mu\}\right)}{\left\langle C_P\left(\mathrm{Re}\{S_\mu\}\right)\right\rangle}\right)\right\rangle\right\}^{-1} \\
&\approx \mathrm{Re}\left\{\alpha \frac{\left\langle C_D\left(\mathrm{Re}\{S_\mu\}\right)\right\rangle}{\left\langle C_P \geq \left(\mathrm{Re}\{S_\mu\}\right)\right\rangle}\left\langle 1 + \frac{\left\langle C_P\left(\mathrm{Re}\{S_\mu\}\right)\right\rangle - C_P\left(\mathrm{Re}\{S_\mu\}\right)}{\left\langle C_P\left(\mathrm{Re}\{S_\mu\}\right)\right\rangle} - \frac{\left(\left\langle C_P\left(\mathrm{Re}\{S_\mu\}\right)\right\rangle - C_P\left(\mathrm{Re}\{S_\mu\}\right)\right)^2}{2\left\langle C_P\left(\mathrm{Re}\{S_\mu\}\right)\right\rangle^2}\right\rangle\right\} \\
&= \mathrm{Re}\left\{\alpha \frac{\left\langle C_D\left(\mathrm{Re}\{S_\mu\}\right)\right\rangle}{\left\langle C_P\left(\mathrm{Re}\{S_\mu\}\right)\right\rangle}\left(1 - \left\langle\frac{\left(\left\langle C_P\left(\mathrm{Re}\{S_\mu\}\right)\right\rangle - C_P\left(\mathrm{Re}\{S_\mu\}\right)\right)^2}{2\left\langle C_P\left(\mathrm{Re}\{S_\mu\}\right)\right\rangle^2}\right\rangle\right)\right\} \\
&= \mathrm{Re}\left\{\alpha \frac{R_{\bar{c},c_D}(\tau - \tau_0)}{R_{\bar{c},c_P}(\tau - \tau_0)}\left(1 - \left\langle\frac{\left(\left\langle C_P\left(\mathrm{Re}\{S_\mu\}\right)\right\rangle - C_P\left(\mathrm{Re}\{S_\mu\}\right)\right)^2}{2\left\langle C_P\left(\mathrm{Re}\{S_\mu\}\right)\right\rangle^2}\right\rangle\right)\right\}
\end{aligned}
\tag{8.5}
$$

because the D- and P-correlators are independent random variables. For high P-correlator SNR,

$$\langle D \rangle \approx \mathrm{Re}\left\{ \alpha \frac{R_{\bar{c},c_D}(\tau - \tau_0)}{R_{\bar{c},c_P}(\tau - \tau_0)} \right\} \tag{8.6}$$

The expected discriminator value should be equal to the code-tracking error, therefore

$$\begin{aligned} 1 \stackrel{!}{=} \frac{\partial}{\partial \tau} \langle D \rangle \big|_{\tau=\tau_0} &\approx \frac{\partial}{\partial \tau} \left(\mathrm{Re}\left\{ \alpha \frac{R_{\bar{c},c_D}{}'(0)(\tau - \tau_0)}{R_{\bar{c},c_P}(0) + R_{\bar{c},c_P}{}'(0)(\tau - \tau_0)} \right\} \right) \Bigg|_{\tau=\tau_0} \\ &= \mathrm{Re}\left\{ \alpha \frac{R_{\bar{c},c_D}{}'(0)}{R_{\bar{c},c_P}(0)} \right\} \end{aligned} \tag{8.7}$$

A normalization constant of

$$\alpha = \frac{R_{\bar{c},c_P}(0)}{R_{\bar{c},c_D}{}'(0)} + i\varepsilon \frac{R_{\bar{c},c_P}(0)}{R_{\bar{c},c_D}{}'(0)} \qquad \varepsilon \in \mathbb{R} \tag{8.8}$$

achieves, in the case of high signal power,

$$\langle D \rangle \overset{|\tau - \tau_0| \ll l}{=} \tau - \tau_0 \tag{8.9}$$

Typically, $\varepsilon = 0$.

For low P-correlator SNR values, the normalization constant needs to be adjusted to maintain a unity slope of the discriminator at the origin. This is ignored in many GNSS-receiver implementations.

For $\tau = \tau_0$, the variance of the complex-valued code-phase discriminator is obtained as

$$\mathrm{var}\left\langle \frac{C_D(\mathrm{Re}\{S_\mu\})}{C_P(\mathrm{Re}\{S_\mu\})} \right\rangle = \mathrm{var}\left\langle C_D(\mathrm{Re}\{S_\mu\}) \right\rangle \left\langle \left| C_P(\mathrm{Re}\{S_\mu\}) \right|^{-2} \right\rangle \tag{8.10}$$

because both correlators are independent and because the expected value of C_D vanishes. The inverse of the squared absolute value of the complex Gaussian random variable is approximated based on the assumption that its variations are much smaller than the mean value:

$$\begin{aligned} \left\langle |x|^{-2} \right\rangle &= \left\langle \left| x - \langle x \rangle + \langle x \rangle \right|^{-2} \right\rangle = \left| \langle x \rangle \right|^{-2} \left\langle \left| \frac{x - \langle x \rangle}{|\langle x \rangle|} + 1 \right|^{-2} \right\rangle \\ &= \left| \langle x \rangle \right|^{-2} \left\langle \left(\left| 1 + \frac{x - \langle x \rangle}{|\langle x \rangle|} \right|^{-1} \right)^2 \right\rangle = \left| \langle x \rangle \right|^{-2} \left\langle \left(\left| \left(1 + \frac{x - \langle x \rangle}{|\langle x \rangle|} \right)^{-1} \right| \right)^2 \right\rangle \end{aligned}$$

$$\approx |\langle x\rangle|^{-2}\left\langle\left|1-\frac{x-\langle x\rangle}{|\langle x\rangle|}\right|^2\right\rangle = |\langle x\rangle|^{-4}\left\langle\left||\langle x\rangle|-x+\langle x\rangle\right|^2\right\rangle$$

$$= |\langle x\rangle|^{-4}\left(|\langle x\rangle|^2 - 2|\langle x\rangle|\mathrm{Re}\langle x-\langle x\rangle\rangle + \left\langle|x-\langle x\rangle|^2\right\rangle\right)$$

$$= |\langle x\rangle|^{-2} + |\langle x\rangle|^{-4}\,\mathrm{var}\langle x\rangle \tag{8.11}$$

Thus, the code-discriminator variance is obtained by using (7.17) and (7.26) with $f_s = 2B$ and $a = (2C / N_0 / B)^{1/2}$ as

$$\mathrm{var}\left\langle\frac{C_D(\mathrm{Re}\{S_\mu\})}{C_P(\mathrm{Re}\{S_\mu\})}\right\rangle$$

$$\approx \mathrm{var}\langle C_D(\mathrm{Re}\{S_\mu\})\rangle\left(|\langle C_P(\mathrm{Re}\{S_\mu\})\rangle|^{-2} + |\langle C_P(\mathrm{Re}\{S_\mu\})\rangle|^{-4}\,\mathrm{var}\langle C_P(\mathrm{Re}\{S_\mu\})\rangle\right)$$

$$= \frac{2BT_{coh}R_{\overline{c_D},c_D}(0)}{\left|\kappa\sqrt{2C/N_0B}T_{coh}R_{\overline{c},c_P}(0)\right|^2} \tag{8.12}$$

$$\times\left(1+\left|\kappa\sqrt{2C/N_0B}T_{coh}R_{\overline{c},c_P}(0)\right|^{-2}2BT_{coh}R_{\overline{c_P},c_P}(0)\right)$$

$$= \frac{R_{\overline{c_D},c_D}(0)}{C/N_0T_{coh}\left|\kappa R_{\overline{c},c_P}(0)\right|^2}\left(1+\frac{R_{\overline{c_P},c_P}(0)}{C/N_0T_{coh}\left|\kappa R_{\overline{c},c_P}(0)\right|^2}\right)$$

with

$$\kappa = \kappa(\omega_0 - \omega) \tag{8.13}$$

accounting for Doppler correlation losses.

The variance of the real-valued code-phase discriminator is half the variance of the complex-valued discriminator and is for $\varepsilon = 0$

$$\mathrm{var}\langle D(\mathrm{Re}\{S_\mu\})\rangle = \frac{|\alpha|^2}{2}\mathrm{var}\left\langle\frac{C_D(\mathrm{Re}\{S_\mu\})}{C_P(\mathrm{Re}\{S_\mu\})}\right\rangle$$

$$= \left|\frac{R_{\overline{c},c_P}(0)}{R_{\overline{c},c_D}{}'(0)}\right|^2\frac{R_{\overline{c_D},c_D}(0)}{2C/N_0T_{coh}\left|\kappa R_{\overline{c},c_P}(0)\right|^2}\left(1+\frac{R_{\overline{c_P},c_P}(0)}{C/N_0T_{coh}\left|\kappa R_{\overline{c},c_P}(0)\right|^2}\right) \tag{8.14}$$

The variance of the code discriminator is mainly given by the ratio of the autocorrelation function at the origin of the code-tracking reference function $c_D(t)$

divided by the squared first derivative of the cross-correlation function of the complex-conjugated received signal $c(t)$ with the code-tracking reference function. The variance is minimal if $c_D(t) = c'(t)$ and $\varepsilon = 0$.

8.1.2 Doppler Discriminator

The Doppler discriminator is obtained by dividing the F-correlator by the P-correlator. The reference signals associated with the different correlators are

$$\begin{aligned} C_F &\leftarrow \left(t - \frac{T_{coh}}{2}\right) c_P(t) \\ C_P &\leftarrow c_P(t) \end{aligned} \tag{8.15}$$

The F-correlator reference function is directly derived from the P-correlator reference function. The P-correlator reference function fulfills

$$\begin{aligned} R_{\overline{c_F},c_P}(0) &= 0 \\ R_{\overline{c},c_P}(0) &\neq 0 \end{aligned} \tag{8.16}$$

Similar to the code-phase discriminator, the F- and P-correlator are independent Gaussian random variables.

The F-correlator is given by

$$F = \operatorname{Re}\left\{\beta \frac{C_F(\operatorname{Re}\{S_\mu\})}{C_P(\operatorname{Re}\{S_\mu\})}\right\} = \operatorname{Re}\left\{\beta \frac{C_F(\operatorname{Re}\{S_\mu\})\overline{C_P(\operatorname{Re}\{S_\mu\})}}{\left|C_P(\operatorname{Re}\{S_\mu\})\right|^2}\right\} \tag{8.17}$$

where β is a normalization constant to be determined later.

Similar to (8.5) the expected value of the F-correlator is

$$\begin{aligned} \langle F \rangle &= \left\langle \operatorname{Re}\left\{\beta \frac{C_F(\operatorname{Re}\{S_\mu\})}{C_P(\operatorname{Re}\{S_\mu\})}\right\}\right\rangle \\ &\approx \operatorname{Re}\left\{\beta \frac{\langle C_F(\operatorname{Re}\{S_\mu\})\rangle}{\langle C_P(\operatorname{Re}\{S_\mu\})\rangle}\left(1 - \frac{\left\langle\left(\langle C_P(\operatorname{Re}\{S_\mu\})\rangle - C_P(\operatorname{Re}\{S_\mu\})\right)^2\right\rangle}{2\langle C_P(\operatorname{Re}\{S_\mu\})\rangle^2}\right)\right\} \end{aligned} \tag{8.18}$$

Assuming a high P-correlator SNR and using (7.17), (7.33), and (A.76) yields

$$\begin{aligned} \langle F \rangle &\approx \operatorname{Re}\left\{\beta \frac{\langle C_F(\operatorname{Re}\{S_\mu\})\rangle}{\langle C_P(\operatorname{Re}\{S_\mu\})\rangle}\right\} = \operatorname{Re}\left\{-\beta \frac{\sqrt{2C/N_0B}T_{coh}R_{\overline{c},c_P}(\tau-\tau_0)(i\kappa'(\omega_0-\omega))}{aT_{coh}BR_{\overline{c},c_P}(\tau-\tau_0)\kappa(\omega_0-\omega)}\right\} \\ &= \operatorname{Re}\left\{-\beta \frac{\sqrt{2C/N_0B}(i\kappa'(\omega_0-\omega))}{\sqrt{2C/N_0B}\kappa(\omega_0-\omega)}\right\} = \operatorname{Re}\left\{-i\beta \frac{\kappa'(\omega_0-\omega)}{\kappa(\omega_0-\omega)}\right\} \end{aligned} \tag{8.19}$$

As a result of the discussion in Section 1.8.5, on separation of code and carrier correlation,

$$\begin{aligned}\langle F\rangle &\approx \mathrm{Re}\left\{-i\beta\frac{\kappa'(\omega_0-\omega)}{\kappa(\omega_0-\omega)}\right\}\overset{|\omega_0-\omega|\ll}{\approx}\mathrm{Re}\left\{-i\beta\frac{\kappa'(0)+\kappa''(0)(\omega_0-\omega)}{\kappa(0)+\kappa'(0)(\omega_0-\omega)}\right\}\\ &=\mathrm{Re}\left\{-i\beta\frac{\kappa''(0)(\omega_0-\omega)}{\kappa(0)}\right\}\approx\mathrm{Re}\left\{i\beta T_{coh}{}^2\frac{1}{12}\chi_{freq}(\omega_0-\omega)\right\}\end{aligned}\tag{8.20}$$

Consequently, the normalization constant is

$$\beta=\frac{12i}{T_{coh}^2\chi_{freq}}+\varepsilon\quad \varepsilon\in\mathbb{R}\tag{8.21}$$

Typically, $\varepsilon = 0$.

The normalization constant should be increased for low P-correlator SNR to compensate for the second-order expected value decrease in (8.18). For high P-correlator SNR, the expected value is

$$\langle F\rangle\overset{|\omega-\omega_0|\ll}{=}\omega-\omega_0\tag{8.22}$$

For $\omega = \omega_0$, the F-correlator variance is derived from

$$\mathrm{var}\left\langle\frac{C_F(\mathrm{Re}\{S_\mu\})}{C_P(\mathrm{Re}\{S_\mu\})}\right\rangle=\mathrm{var}\left\langle C_F(\mathrm{Re}\{S_\mu\})\right\rangle\left\langle\left|C_P(\mathrm{Re}\{S_\mu\})\right|^{-2}\right\rangle\tag{8.23}$$

which evaluates using (8.11) and (8.13) to

$$\begin{aligned}&\mathrm{var}\left\langle\frac{C_F(\mathrm{Re}\{S_\mu\})}{C_P(\mathrm{Re}\{S_\mu\})}\right\rangle\\ &=\mathrm{var}\left\langle C_F(\mathrm{Re}\{S_\mu\})\right\rangle\left(\left|\left\langle C_P(\mathrm{Re}\{S_\mu\})\right\rangle\right|^{-2}+\left|\left\langle C_P(\mathrm{Re}\{S_\mu\})\right\rangle\right|^{-4}\mathrm{var}\left\langle C_P(\mathrm{Re}\{S_\mu\})\right\rangle\right)\\ &=\frac{2BT_{coh}R_{\bar{c}_P,c_P}(0)T_{coh}^2\chi_{freq}}{12\left|\kappa\sqrt{2C/N_0B}T_{coh}R_{\bar{c},c_P}(\tau-\tau_0)\right|^2}\\ &\quad\times\left(1+\left|\kappa\sqrt{2C/N_0B}T_{coh}R_{\bar{c},c_P}(\tau-\tau_0)\right|^{-2}2BT_{coh}R_{\bar{c}_P,c_P}(0)\right)\\ &=\frac{R_{\bar{c}_P,c_P}(0)T_{coh}^2\chi_{freq}}{12C/N_0T_{coh}\left|\kappa R_{\bar{c},c_P}(\tau-\tau_0)\right|^2}\left(1+\frac{R_{\bar{c}_P,c_P}(0)}{C/N_0T_{coh}\left|\kappa R_{\bar{c},c_P}(\tau-\tau_0)\right|^2}\right)\end{aligned}\tag{8.24}$$

The Doppler-discriminator variance is finally given as

$$\begin{aligned}
\operatorname{var}\left\langle F(\mathrm{Re}\{S_\mu\})\right\rangle &= \frac{|\beta|^2}{2}\operatorname{var}\left\langle \frac{C_F(\mathrm{Re}\{S_\mu\})}{C_P(\mathrm{Re}\{S_\mu\})}\right\rangle \\
&= \frac{12^2}{2T_{coh}^4 \chi_{freq}^{\;2}} \frac{R_{\bar{c}_P,c_P}(0)T_{coh}^2\chi_{freq}}{12C/N_0T_{coh}\left|\kappa R_{\bar{c},c_P}(\tau-\tau_0)\right|^2}\left(1+\frac{R_{\bar{c}_P,c_P}(0)}{C/N_0T_{coh}\left|\kappa R_{\bar{c},c_P}(\tau-\tau_0)\right|^2}\right) \\
&= \frac{6R_{\bar{c}_P,c_P}(0)}{C/N_0\chi_{freq}T_{coh}^3\left|\kappa R_{\bar{c},c_P}(\tau-\tau_0)\right|^2}\left(1+\frac{R_{\bar{c}_P,c_P}(0)}{C/N_0T_{coh}\left|\kappa R_{\bar{c},c_P}(\tau-\tau_0)\right|^2}\right)
\end{aligned} \tag{8.25}$$

This formula compares to the Doppler CRLB (4.84). The variance is larger than the CRLB if the replica waveform does not match the received waveform or if Doppler or code-phase mismatch occurs.

8.1.3 Phase Discriminator

The carrier-phase discriminator is directly derived from the P-correlator. The P-correlator uses as reference function

$$C_P \sim c_P(t) \tag{8.26}$$

which is required to fulfill

$$R_{\bar{c},c_P}(0) \neq 0 \tag{8.27}$$

The carrier phase discriminator shall either be based on the argument of the complex-valued P-correlator (four-quadrant arctan) or on the two-quadrant arctan function

$$\begin{aligned}
\Phi &= \mathrm{angle}\{C_P\} \\
\Phi &= \mathrm{atan}\frac{\mathrm{Im}\{C_P\}}{\mathrm{Re}\{C_P\}}
\end{aligned} \tag{8.28}$$

The first equation is used if the signal is not modulated with data-bit information and the second one is used if the signal is modulated with a binary data-bit stream using a BPSK modulation. Both discriminators are treated identically in the following example.

Following the same derivation as in Section 4.3.2.8, the expected value of the carrier-phase discriminator is

$$\left\langle \Phi \right\rangle \overset{|\varphi-\varphi_0|\ll}{=} \varphi-\varphi_0 \tag{8.29}$$

for high P-correlator SNR. The magnitude of the expected value reduces at lower SNR values. A Monte Carlo simulation of the expected value of the carrier phase for a different SNR value can be found in the work by Dierendonck [1]. It is an evaluation of the formula

$$\langle \Phi \rangle = \left\langle \arg\left\{10^{SNR/20} 2\exp\{i(\varphi-\varphi_0)\} + N\right\}\right\rangle \quad \mathrm{Re}\{N\},\mathrm{Im}\{N\} \sim N(0,1) \tag{8.30}$$

for fixed values of the carrier-phase error $\varphi - \varphi_0$ and the SNR value. N is a complex-valued random variable whose real and imaginary part are of unity variance. A Monte Carlo simulation of (8.30) is shown in Figure 8.1. The slope of the discriminator is plotted in Figure 8.2 and has been evaluated at a carrier-phase error of $\varphi - \varphi_0 = 0.1$ rad. Below an SNR value of 0 dB, the slope gradually decreases.

Because of (4.88) the carrier phase discriminator variance in a first-order approximation is

$$\begin{aligned} \mathrm{var}\left\langle \Phi(\mathrm{Re}\{S_\mu\})\right\rangle &\approx \frac{\mathrm{var}\left\langle C_P(\mathrm{Re}\{S_\mu\})\right\rangle}{2\left|\left\langle C_P(\mathrm{Re}\{S_\mu\})\right\rangle\right|^2}\left(1+\frac{\mathrm{var}\left\langle C_P(\mathrm{Re}\{S_\mu\})\right\rangle}{2\left|\left\langle C_P(\mathrm{Re}\{S_\mu\})\right\rangle\right|^2}\right) \\ &= \frac{R_{\bar{c}_P,c_P}(0)}{2C/N_0T_{coh}\left|\kappa R_{\bar{c},c_P}(\tau-\tau_0)\right|^2}\left(1+\frac{R_{\bar{c}_P,c_P}(0)}{2C/N_0T_{coh}\left|\kappa R_{\bar{c},c_P}(\tau-\tau_0)\right|^2}\right) \end{aligned} \tag{8.31}$$

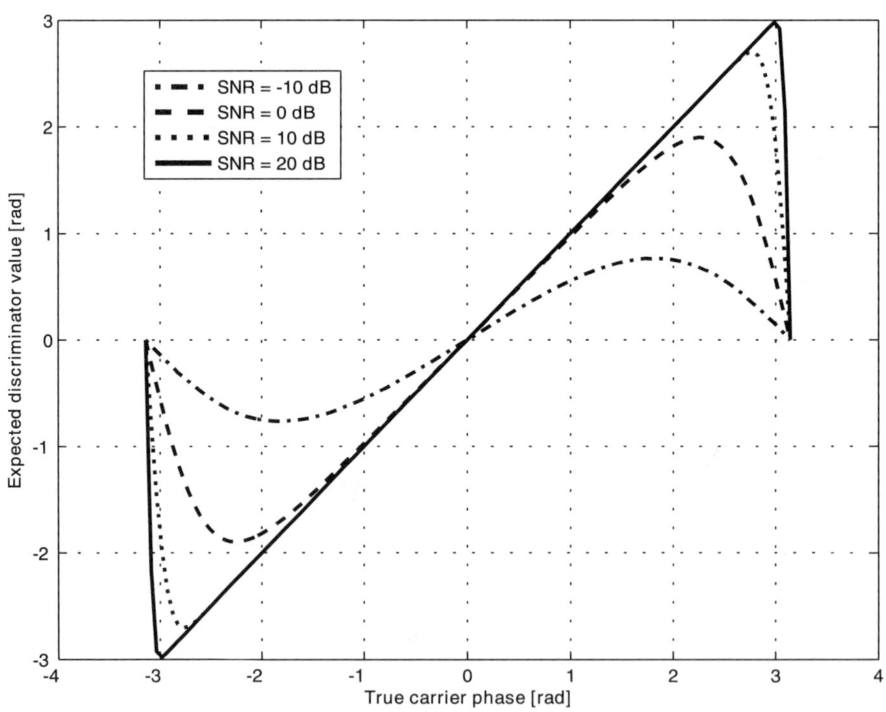

Figure 8.1 Expected value of the four-quadrant carrier phase discriminator for different SNR values.

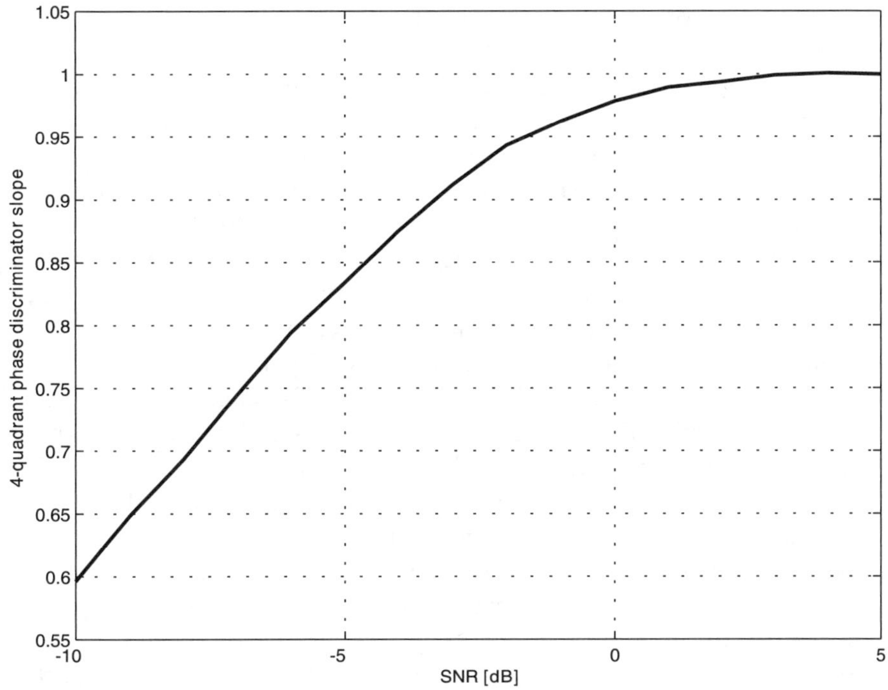

Figure 8.2 Slope of the four-quadrant carrier-phase discriminator for different SNR values.

8.1.4 Clipping

The formulas for the proposed code-phase and Doppler discriminators involve division by the P-correlator value. Division is a mathematically unsafe operation because a division by zero may occur. On the contrary, the correlation point should lie within the linear region of both discriminators.

The discriminator values can therefore be bounded by the extension of the linear code-phase or Doppler region (or more generally, by the maximum possible code-phase and Doppler difference). The discriminator values are *clipped* to stay within the mentioned thresholds. Clipping is not required for the carrier-phase discriminator.

Clipping is a nontrivial assumption on the tracking-loop performance, which is required to keep the correlation point in the linear region. If this requirement is fulfilled, clipping is a useful constraint on the admissible code-phase and Doppler range.

8.2 S-Curve Shaping

The formulas of the previous Section 8.1 allow computation of the code-phase, carrier-phase, and frequency tracking performance for a given set of reference signals. The reference signals need to fulfill the conditions (8.3), (8.16), and (8.27), but can otherwise be chosen arbitrarily. Most important, the waveform of the signals can be chosen flexibly and, by optimizing the waveform, the tracking performance can be improved.

The method of using a flexible reference signal is tailored for the software-receiver approach. First, the software-receiver algorithms of Chapter 9 work with a 16-bit sample resolution, allowing for the represention of sophisticated waveforms. By contrast, a simple hardware receiver may work with 1- or 2-bit resolution to represent the reference signals. Second, the generation of one P-correlator reference signal and of one D-correlator reference signal is sufficient for tracking in most applications. This corresponds to two complex-valued correlators per channel. By contrast, some other receiver structures need three (early, prompt, late) or five (very early, early, prompt, late, very late) complex-valued correlators per channel.

In the context of code tracking, the presented tracking scheme is also called a *code-continuous reference waveform* (CCRW) tracking scheme and was first introduced in an article by Weill [2]. The CCRW scheme falls into the class of nonparametric estimators because no multipath signal parameters are estimated, just line-of-sight signal parameters [3]. The scheme allows realization of most linear code discriminators, including the early–late and double-delta discriminator, being well known for optimum code performance with the GPS C/A-signal. Those correlator types can also be applied for modernized GNSS signals using the BOC, CBOC, or AltBOC modulation scheme as has been pointed out in Irsigler and Eissfeller's work [4]. In Section 7.6.2, the early–late discriminator was investigated for arbitrary signal waveforms and colored noise. As another example, the double-delta code-tracking reference waveform is given by

$$c_D(t) = c_p\left(t - d/2\right) - c_p\left(t + d/2\right) - \frac{1}{2}\left(c_p(t-d) - c_p(t+d)\right) \tag{8.32}$$

assuming that $c_p(t)$ is the (infinite-bandwidth) signal waveform at baseband. Two prominent discriminators cannot be realized in the CCRW scheme: the early-power-minus-late-power discriminator [1] or the BOC(n,n) side-peak cancellation technique [5]. Those two techniques require, however, more correlators than the CCRW.

In the following section, the code discriminator will be discussed and several performance criteria will be presented. A method will be outlined to choose the reference function for a given code-discriminator function (S-curve). The same discussion can also be carried out for phase and frequency discriminators, which will, however, not be discussed in this text.

8.2.1 Code-Discriminator Performance Characteristics

A code discriminator can be optimized for one or more of the following performance criteria:

- Minimum variance caused by thermal noise;
- Low multipath errors;
- High stability;
- Linearity in the tracking region.

Furthermore, it is typically required that the code discriminator shows only one stable tracking point (i.e., a zero crossing with positive slope) within the range of admissible code-tracking errors.

Thermal noise becomes minimal if the code-tracking reference function equals the first derivative of the received signal at baseband [6]. The resulting tracking scheme is quite similar to an early–late tracking scheme, with good but not optimal multipath performance. Furthermore, the linear region of a first-derivative correlator is approximately equal to the inverse of the signal bandwidth. The thermal-noise performance of the first-derivate correlator is optimal and the CRLB is reached.

The other performance measures are a direct function of the code-discriminator's S-curve and will be discussed in Section 8.2.2.

8.2.2 Optimum S-Curve

The graph obtained by plotting the expected value of the code discriminator (8.6) as a function of code-tracking error $\tau - \tau_0$ is commonly called an *S-curve* due to its S-like shape. A typical S-curve is plotted in Figure 8.3. The S-curve uniquely determines the multipath envelope, the size of the linear tracking region, and the pull-in range. The multipath envelope bounds the absolute value of the code-discriminator errors due to multipath [3]. The linear region determines the allowable range, where the theory of linearized tracking loops can be applied. The pull-in region covers the code-phase error values, where the discriminator output has the same sign as the code-phase error. If the code-phase error is within the pull-in range, the tracking loop will drive the code-phase error to zero.

In the following example, the maximum multipath error will be estimated under the assumption that the direct line-of-sight signal is present and one multipath signal exists. Furthermore, the code-phase delay of the direct line-of-sight signal $\Delta\tau$ and of the multipath signal $\Delta\tau + \Delta\tau_m$, with respect to the internally generated signal, is assumed to be within the linear region of the S-curve. For arbitrary delays, the multipath error $\Delta\tau$ is given by solving

$$\left\langle D(\Delta\tau)\right\rangle_N + \alpha\left\langle D(\Delta\tau + \Delta\tau_m)\right\rangle_N = 0 \tag{8.33}$$

where α is the ratio between the multipath signal amplitude and the direct line-of-sight amplitude. In case both delays are within the linear region, the code multipath error $\Delta\tau$ is

$$\Delta\tau + \alpha(\Delta\tau + \Delta\tau_m) = 0 \quad \Leftrightarrow \quad \Delta\tau = \frac{-\alpha}{(1+\alpha)}\Delta\tau_m \tag{8.34}$$

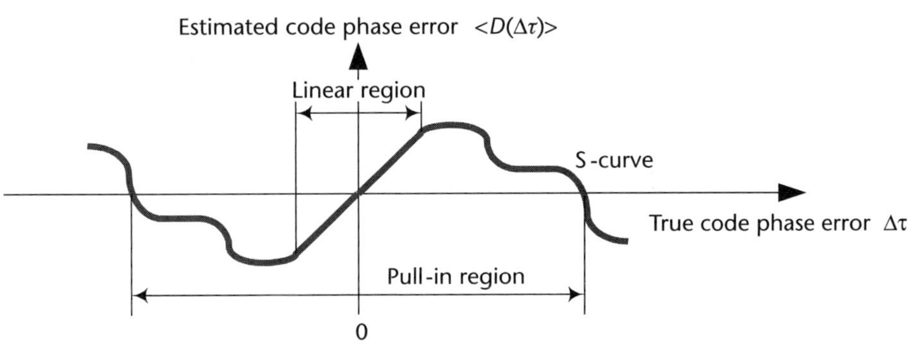

Figure 8.3 Code-discriminator S-curve.

Within the linear region, multipath errors are independent on the used correlation scheme and no multipath mitigation is possible. The size of the linear region needs therefore to be limited to achieve lower multipath errors. On the other hand, a large linear region results in a high tracking stability. A trade-off between the two performance figures has to be done depending on the specific target application.

8.2.3 Frequency-Domain S-Curve Shaping

The definition of the code discriminator (8.4) relies on the correlation function of the incoming signal $c(t)$ with a code-tracking reference signal $c_D(t)$. The correlation in frequency domain is written as

$$\tilde{R}_{\bar{c},c_D}(f) = \bar{c}(-f)c_D(f) \tag{8.35}$$

In the following discussion, a method will be presented that shows how the reference signal can be computed to achieve a desired correlation function $R_{\bar{c},c_D}$. The desired correlation function must be known beforehand. The method will be illustrated for the D-correlator, but can identically be applied also for the P-correlator, which is used to remove the carrier-phase dependency from the code discriminator.

The presented method relies on inverting (8.35) by dividing the cross-correlation Fourier transform by the Fourier transform of the incoming signal. The division can only be performed for nonvanishing signal frequency components. Frequency components with a vanishing (or very small) signal magnitude must not be present in the S-curve. In other words, the S-curve and received signal need to be *spectrally compatible* in the sense that the spectral content of the signal must be sufficient to represent the S-curve.

The D-correlator reference signal to obtain the S-curve $R_{\bar{c},c_D}$ is given by

$$c_D(f) = \begin{cases} \dfrac{\tilde{R}_{\bar{c},c_D}(f)}{\bar{c}(-f)} & |\bar{c}(-f)| > \gamma \\ 0 & |\bar{c}(-f)| \leq \gamma \end{cases} \tag{8.36}$$

where γ is a suitable constant.

The computation of the reference signal is straightforward and is illustrated here with two examples that work with discrete- and finite-length Fourier transforms. The computation's settings are summarized in Table 8.1. The method is ap-

Table 8.1 Frequency-Domain S-Curve Shaping Settings

Parameter	*Value*
Sample rate	16.384 MSamples/s
Fourier size	16,384
Signal time	1 ms
Spectral threshold γ	10% of max. spectral magnitude
Number of bits to represent reference signal	4

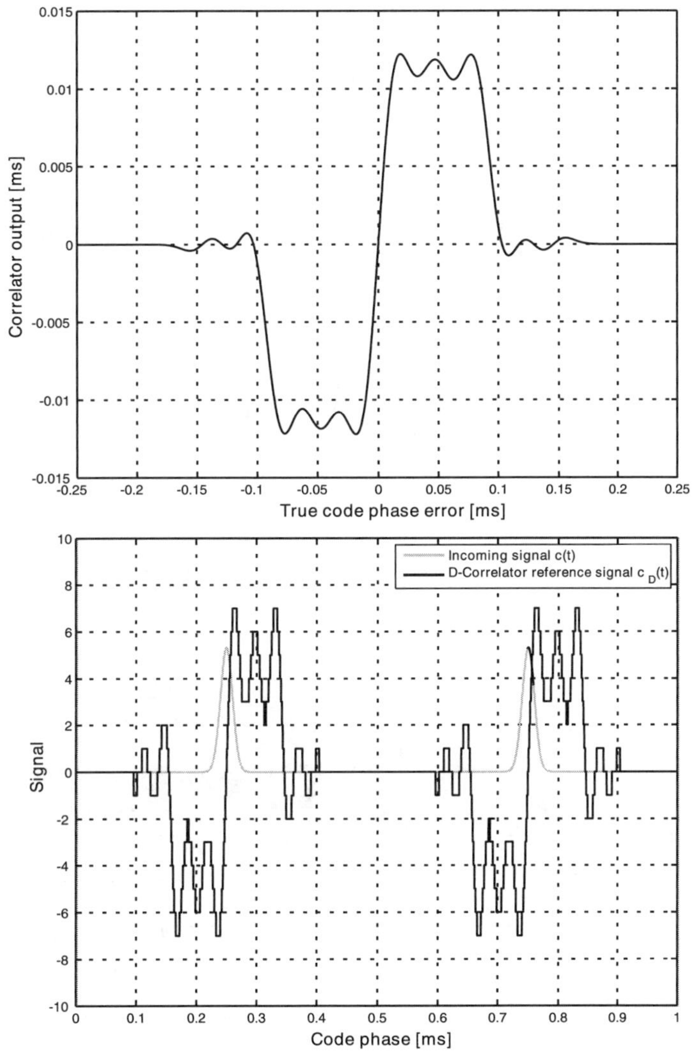

Figure 8.4 S-curve shaping applied to obtain a wide pull-in range Gaussian double-pulse discriminator.

plied for an infinite-bandwidth GPS C/A-code signal (PRN1) and for the Gaussian pulse signal described by Table 1.3.

As target S-curve, a wide-range pull-in discriminator is chosen. This choice only serves to give an example and to illustrate the method. The target S-curve is given by

$$\tilde{R}_{\bar{c},c_D}\left(\frac{\Delta n}{f_c}\right)f_c = \begin{cases} \Delta n & 0 \leq |\Delta n| < 10 \text{ chips} \\ 10\,\text{sgn}\{\Delta n\} & 10 \leq |\Delta n| < 90 \text{ chips} \\ 100\,\text{sgn}\{\Delta n\} - \Delta n & 90 \leq |\Delta n| < 100 \text{ chips} \\ 0 & \text{otherwise} \end{cases} \tag{8.37}$$

Here, the unit [chip] refers to the C/A-code and 1 chip = 0.978 μs. The resulting reference signal and the obtained cross-correlation function is plotted in Figure 8.4

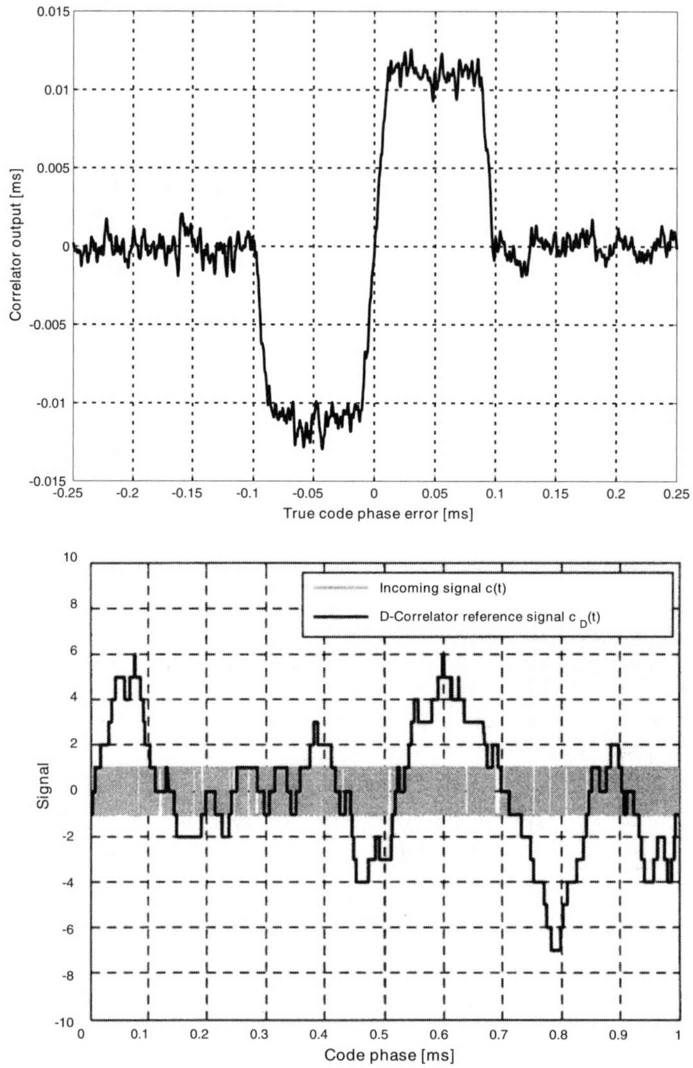

Figure 8.5 S-curve shaping applied to obtain a wide pull-in range C/A-(PRN1) code discriminator.

for the Gaussian double-pulse signal and in Figure 8.5 for the GPS C/A-code. In both cases, the reference signal is converted to a 4-bit signal (by simple rounding), allowing for a convenient representation in the software receiver.

The method of spectral S-curve shaping reproduces the target S-curve reasonably well. In both cases, deviations from the target S-curve occur because the signal spectrum does not cover the whole frequency range of the S-curve. In the case of the Gaussian pulse signal, the reference function has a clearly visible relationship to the received signal waveform. For the GPS C/A-code case, the reference function is structurally very different compared to the PRN spreading code sequence (see the incoming signal in Figure 8.5). If this reference function is used for code tracking, high code noise is expected. For the GPS C/A-code case, we propose a different S-curve shaping scheme (fitting) described in the following discussion.

8.2.4 Discussion

Spectral S-curve shaping is a convenient method to calculate a reference signal to achieve a given correlation function. It gives a stable tracking scheme (e.g., no BOC ambiguity) with defined multipath characteristics. One disadvantage of spectral S-curve shaping is that the S-curve needs to be specified for all code-phase values (not only for the region of interest), which eventually increases the thermal-noise variance.

In another work, a different S-curve shaping method is described that illustrates how to compute the code-tracking reference function to obtain a given S-curve [7]. The method relies on fitting a set of shifted autocorrelation functions to the target S-curve. The fitting is applied only in the region of interest. This fitting problem can be well-solved for high-bandwidth signals. For low-bandwidth signals, the solution becomes unstable and the desired S-curve cannot be reproduced. For high-bandwidth signals, near-optimum multipath mitigation can be achieved. For BOC signals, the pull-in range can be increased to avoid unstable tracking points. For BPSK signals, this method reproduces the double-delta correlator, which is known to have optimum multipath performance. This method has also been applied in a work by Paonni for the MBOC-modulation family [8].

Neither spectral shaping methods (spectral and fitting) take into account the thermal-noise performance. For example, the reference function plotted in Figure 8.5 shows almost no relationship to the PRN-code sequence and worse thermal-noise performance is expected. Generally, the parameters of the target S-curve (e.g., the size of the linear region) need to be chosen empirically to achieve good thermal-noise performance. Development of a modified S-curve shaping scheme that also includes thermal-noise performance is currently ongoing. The resulting code discriminator depends on the weights put on the thermal-noise errors and the multipath errors to assess the total performance.

8.3 Multipath Estimating Techniques

This section introduces a class of *parametric* discriminators simultaneously estimating line-of-sight and multipath signal parameters [3]. The method is based on the complex least-square adjustment described in Appendix A.1 together with the correlator models of Section 7.2. The presented discriminator tries first to detect the presence of multipath signals in addition to the line-of-sight signal and then eventually adapts the number of parameters to be estimated.

The multipath-estimating discriminator is advantageous over the non parametric approach because of the following facts:

- Multipath mitigation is applied only when needed, thereby avoiding an unnecessary signal power loss.
- Strong multipath signals (causing potentially large errors) are better detected and better mitigated than weak signals. A multipath power-independent performance can be achieved.
- Under certain assumptions, the method is theoretically optimal and naturally exploits the benefits of signals with a large Gabor bandwidth (e.g., modernized GNSS signals). This has been demonstrated in [9].

- The method intrinsically provides multipath mitigation for the code phase, Doppler, and carrier phase simultaneously.
- The LSQ procedure provides accuracy estimates for the discriminator values, which are based on the actual signal amplitude.

The major disadvantage of the presented scheme is the increased complexity. First, additional correlators are needed. Second, the LSQ scheme requires matrix operations. Within a software radio, however, both disadvantages are tolerable. The required additional correlators are obtained by using already-generated reference signals shifted in Doppler and code-phase direction. Thereby, the time-consuming signal-generation process is circumvented and correlation itself can be performed efficiently, as will be shown in Section 9.8. The required matrix operations are performed fast because the considered processors support floating-point operations by hardware. The most time-consuming operation is the inversion of the normal matrix, which is, for one multipath signal and one line-of-sight signal, a complex 6×6 matrix. The algorithm involves a fixed number of iterations (maximum 1–2).

A multipath-estimating delay-lock loop was introduced in the article by van Nee as an unweighted correlation-function fitting procedure [10]. Here, we generalize those results by using arbitrary reference waveforms, by using a full complex weighted least-squares scheme, and also by allowing for Doppler estimates. Furthermore, practical implementation aspects are given. The presented scheme tries to minimize the involved computational resources (just one or two propagation paths are considered) compared to a more rigorous technique that estimates for many multipath parameters [11]. The presented scheme minimizes the influence of one strong multipath signal caused by a specular reflection. This multipath type may introduce high and constant biases in the pseudorange measurements, which are hard to reduce and have a strong impact on the navigation solution if the specular reflection is constant in time. By contrast, diffuse reflections or diffractions typically have less power and have generally a more random character. Filtering (e.g., a carrier- or Doppler-aided DLL) reduces the influence of diffuse or timely variable reflections and no multipath-estimating discriminator is needed.

8.3.1 The LSQ Equations

To set up the LSQ equations, a parameter vector **q** is considered. It is used to set up a model of the received signal and **q** contains the high-rate pseudorange parameters of the line-of-sight signal and of the multipath signals. The parameter vector **q** is given as

$$\mathbf{q} = \begin{pmatrix} a_1 & \tau_1 & \omega_1 & \cdots & a_M & \tau_M & \omega_M \end{pmatrix}^T \tag{8.38}$$

where m is the signal index (ranging from 1 ... M), a_m is the complex-signal amplitude

$$a_m = \left|a_m\right| \exp\{-i\varphi_m\} \tag{8.39}$$

τ_m is the code phase in seconds, ω_m is the angular Doppler frequency in radians per second, and φ_m is the carrier phase in radians. The code phase is defined at the beginning of the interval, the Doppler is constant throughout the interval, and the carrier-phase value φ_m is defined in the middle of the integration interval, corresponding to a signal model

$$r_\mu = \sum_{m=1}^{M} a_m c(t_\mu - \tau_m)\exp\{i\omega_m t'_\mu\} = \sum_{m=1}^{M} a_m c(t_\mu - \tau_m)\exp\left\{i\omega_m\left(t_\mu - \frac{T_{coh}}{2}\right)\right\} \quad (8.40)$$

The number of multipath signals is $M - 1$. If no multipath signal is present, $M = 1$.

The received signal is correlated with different reference signals shifted by a code-phase and Doppler offset. Overall, B correlators are involved and the complex-correlation values are summarized in a vector **C**:

$$\mathbf{C} = (C^1 \quad \cdots \quad C^B)^T \quad (8.41)$$

The correlator b uses as reference signal $c_{rec}{}^b$, and the code phase at the beginning of the correlation interval is $\tau_0{}^b$. The angular Doppler frequency $\omega_0{}^b$ is constant during the integration interval and the carrier phase $\varphi_0{}^b$ vanishes at the midpoint of the integration interval (note that the carrier-phase estimate is derived from a_m):

$$C^b(S_\mu) \mapsto c_{rec}^b, \tau_0^b, \omega_0^b, \left(\varphi_0^b = 0\right) \quad (8.42)$$

Using the results of Section 7.2, a functional and stochastic model of the correlator values is defined that serves as the basis for the following discussion.

The LSQ discriminator is given by a weighted fit of this model to the correlator values, thereby minimizing

$$\hat{\mathbf{q}} = \arg\min_{\mathbf{q}} \left(\mathbf{C} - \mathbf{C_q}\right)^* Q_\mathbf{v}{}^{-1}\left(\mathbf{C} - \mathbf{C_q}\right) \quad (8.43)$$

The correlator model $\mathbf{C_q}$ is calculated by (7.17) and the covariance between two correlator values $Q_\mathbf{v}$ is calculated by (7.24).

The LSQ equation (8.43) is linearized around the linearization point $\mathbf{q}_0$,

$$\mathbf{q}_0 = \begin{pmatrix} a_{1,0} & \tau_{1,0} & \omega_{1,0} & \cdots & a_{M,0} & \tau_{M,0} & \omega_{M,0} \end{pmatrix}^T \quad (8.44)$$

and the correlator values are modeled for the linearization point as

$$\mathbf{C}_{\mathbf{q}_0} = \left(\left\langle C^1(\mathrm{Re}\{S_\mu\};\mathbf{q}_0)\right\rangle \quad \cdots \quad \left\langle C^B(\mathrm{Re}\{S_\mu\};\mathbf{q}_0)\right\rangle\right)^T \quad (8.45)$$

The differences of the true signal parameters with respect to the linearization point are

$$\Delta\mathbf{q} = \mathbf{q} - \mathbf{q}_0 \quad (8.46)$$

and the correlator differences are

$$\Delta\mathbf{C} = \mathbf{C} - \mathbf{C}_{\mathbf{q}_0} \quad (8.47)$$

For clarification, it should be noted that the LSQ discriminator involves two parameter sets called "points": the *correlation* point and the *linearization* point. The correlation point is given by (8.42) and represents the NCO values used to compute the correlation values. The correlation point is kept constant throughout the LSQ procedure and the time-consuming correlation process needs not to be repeated. This is in contrast to the discussion of Section 4.3 or the MLE approach of Won [12]. On the other hand, the linearization point is used to linearize the correlator model (7.17). The linearization point may vary during the iterations of the LSQ procedure. The length of the vectors representing the correlation point (equal to $2B$) and the linearization point (equal to $3M$) are different. The linearization point should be near the true values to ensure that the linearization is a reasonable approximation. Furthermore, at least one correlation point ($\tau_0{}^b$, $\omega_0{}^b$) should lie near the true parameter values.

The linearized LSQ equation is given as

$$\chi/_2 = (\Delta\mathrm{C} - A\Delta\mathbf{q})^* Q_{\mathbf{v}}{}^{-1}(\Delta\mathrm{C} - A\Delta\mathbf{q}) \to \min \tag{8.48}$$

and the value of χ is used to verify that the assumed model (i.e., the number of multipath parameters) is correct.

Using the amplitude definition (8.39) the bth correlator value is modeled from (7.17) as

$$\begin{aligned}
&\left\langle C^b(\mathrm{Re}\{S_\mu\};\mathbf{q}_0)\right\rangle = \\
&= \sum_{m=1}^{M} \frac{\left|a_{m,0}\right| \exp\{i\varphi_{m,0}\} T_{coh} f_s R_{\bar{c},c^b_{rec}}(\tau_{m,0} - \tau_0^b)}{2} \kappa(\omega_0^b - \omega_{m,0}) \\
&= \sum_{m=1}^{M} \frac{\bar{a}_{m,0} T_{coh} f_s R_{\bar{c},c^b_{rec}}(\tau_{m,0} - \tau_0^b)}{2} \kappa(\omega_0^b - \omega_{m,0})
\end{aligned} \tag{8.49}$$

The design matrix takes the form

$$A = (A_1 \quad \dots \quad A_M) \tag{8.50}$$

with A_m as submatrices given by

$$A_m = \begin{pmatrix}
\dfrac{\partial\left\langle C^1(\mathrm{Re}\{S_\mu\};\mathbf{q}_0)\right\rangle}{\partial(\Delta\bar{a}_m)} & \dfrac{\partial\left\langle C^1(\mathrm{Re}\{S_\mu\};\mathbf{q}_0)\right\rangle}{\partial(\Delta\tau_m)} & \dfrac{\partial\left\langle C^1(\mathrm{Re}\{S_\mu\};\mathbf{q}_0)\right\rangle}{\partial(\Delta\omega_m)} \\
\vdots & \vdots & \vdots \\
\dfrac{\partial\left\langle C^B(\mathrm{Re}\{S_\mu\};\mathbf{q}_0)\right\rangle}{\partial(\Delta\bar{a}_m)} & \dfrac{\partial\left\langle C^B(\mathrm{Re}\{S_\mu\};\mathbf{q}_0)\right\rangle}{\partial(\Delta\tau_m)} & \dfrac{\partial\left\langle C^1(\mathrm{Re}\{S_\mu\};\mathbf{q}_0)\right\rangle}{\partial(\Delta\omega_m)}
\end{pmatrix} \tag{8.51}$$

The derivatives are computed with respect to the parameter improvements $\Delta\mathbf{q}$ and are evaluated at the linearization point.

The derivatives are explicitly given as

$$\begin{aligned}
\frac{\partial}{\partial(\Delta\bar{a}_m)}\left\langle C^b(\mathrm{Re}\{S_\mu\})\right\rangle &= \frac{T_{coh}f_s R_{\bar{c},c^b_{rec}}(\tau_{m,0}-\tau^b_0)}{2}\kappa(\omega^b_0-\omega_{m,0}) \\
\frac{\partial}{\partial(\Delta\tau_m)}\left\langle C^b(\mathrm{Re}\{S_\mu\})\right\rangle &= \frac{\bar{a}_m T_{coh}f_s R_{\bar{c},c^b_{rec}}{}'(\tau_{m,0}-\tau^b_0)}{2}\kappa(\omega^b_0-\omega_{m,0}) \\
\frac{\partial}{\partial(\Delta\omega_m)}\left\langle C^b(\mathrm{Re}\{S_\mu\})\right\rangle &= -\frac{\bar{a}_m T_{coh}f_s R_{\bar{c},c^b_{rec}}(\tau_{m,0}-\tau^b_0)}{2}\kappa'(\omega^b_0-\omega_{m,0})
\end{aligned} \tag{8.52}$$

The covariance between two correlator values is derived from (7.24) as

$$Q_{\mathrm{v};b,c} = \mathrm{cov}\left\langle C^b(\mathrm{Re}\{S_\mu\}), C^c(\mathrm{Re}\{S_\mu\})\right\rangle = f_s T_{coh} R_{\overline{c^b_{rec}},c^c_{rec}}(\tau^b_0-\tau^c_0)\kappa(\omega^c_0-\omega^b_0) \tag{8.53}$$

The covariance matrix is independent of the linearization point and remains constant throughout the LSQ iteration process.

8.3.2 Calibration

A LSQ discriminator relies on an accurate model of the correlation values. In a real receiver implementation, this model has to be derived from real measurements in order to correctly account for the front-end filter characteristics. For example in van Nee's work, correlation values of a GPS receiver are averaged to get a mean correlation function [10]. Errors in the correlation model will degrade the LSQ-discriminator performance in terms of introduced biases and a less accurate fit. Furthermore, the correlator residuals might be mistaken as multipath signals. Calibrated correlator models are also necessary for signal-quality monitoring and correlator calibration can be found in Irsigler's thesis [13].

8.3.3 General Procedure

The LSQ discriminator estimates the code phase, the Doppler, the carrier phase, and the amplitude from a number of correlator values. The basic equations of the complex LSQ scheme are described in Appendix A.1. Unfortunately, the LSQ adjustment has several caveats. The most important problem arises from the fact that the normal matrix becomes singular if line-of-sight and multipath delays are identical, as has been discussed in Section 4.3.6.

The LSQ procedure starts from given correlator values. At first, the algorithm assumes the presence of no multipath signal ($M = 1$). Using appropriate initial values, the LSQ adjustment is iterated for a predefined maximum number of iterations or until the parameter vector converges. Then a χ^2-test is used to check if the no-multipath assumption is valid or not. If it is not, the same procedure is repeated, assuming this time the line-of-sight signal and one multipath signal ($M = 2$).

Especially for the case $M = 2$, the LSQ algorithm may diverge. This is the case if no multipath signal is present and the χ^2-test incorrectly decides for the $M = 2$ hypothesis. Then, as a fallback, the $M = 1$ parameters are used as an estimation result. The correlation point must be chosen such that, for $M = 1$, the LSQ algorithm always converges. A fallback to the $M = 1$ case also occurs if the estimated parameter variances for the $M = 2$ case exceed fixed-threshold values. This is done to avoid cases of a nearly singular normal matrix that occurs for (nearly) identical line-of-sight and multipath code-phase values.

It is proposed to use a complex LSQ algorithm to reduce the dimensions of the involved matrices by a factor of two. The complex LSQ algorithm implies that complex-valued code-phase and Doppler improvements are also estimated. The imaginary parts of these improvements are discarded when the linearization point is updated.

The proposed algorithm stops at $M = 2$ because, for higher M values, the algorithm tends to be unstable and the occurrence of more than one specular reflection is unlikely. For the $M = 2$ case, the algorithm identifies the line-of-sight signal parameters as the set of parameters having the smaller code-phase delay. Formally, the scheme can be extended straightforwardly for $M > 2$.

8.3.4 Correlator Placement

The initial code-phase and Doppler values of at least one reference signal used to compute the correlation values need to be near the true values for each signal. The tracking loop ensures that one set of correlators (the prompt correlator set) follows the line-of-sight signal. Other correlators that are necessary to detect and mitigate the multipath signal are placed around the prompt set with fixed code-phase and Doppler offsets. The precise placing depends on the expected multipath environment. Typically, the other correlators are delayed with respect to the line-of-sight correlators.

Ideally, all correlators together form a sufficient statistic for all possible code-phase and Doppler values, as discussed in Section 4.4. There, the multicorrelator approach and the first-derivative approach are described. The first-derivative approach works with triples of colocated P-, D-, and F-correlators, and each triple allows proper linearization of the correlator-value model within the linearity region around the nominal code-phase and Doppler values of the correlator triple. The union of all linearity regions should cover the true code phase and Doppler values of the line-of-sight and multipath values.

8.3.5 Initial Values

Choosing a proper LSQ initialization is important to avoid divergence problems in the first LSQ step. The initialization values are the first guess of the difference between true values and the correlation-point values.

For the line-of-sight signal parameters, the initial code-phase and Doppler errors are assumed to vanish because the tracking loop tries to follow the line-of-sight signal. The code-phase and Doppler values of the multipath signal are chosen according to the position of the correlator with the maximum power (after compensating for the line-of-sight signal effect). The position is chosen during the $M = 1$ step after the observed-minus-computed correlator values are computed.

Initial values for the complex-signal amplitude are obtained by performing one LSQ step. After this step, only the complex amplitude is updated and the code-phase and Doppler values are fixed to their initial values. After the first step, the code-phase and Doppler values also are updated. This initialization step ensures that the division by the complex amplitude to remove the carrier-phase dependency is done with a good signal-amplitude estimate.

8.3.6 Number of Required Iterations

From a theoretical point of view, the LSQ scheme is only optimal for high signal-power values. This fact has some consequences on the number of LSQ iterations when working with nonhigh signal power. We argue that a single LSQ iteration provides generally better results than iteration of the LSQ equations until convergence.

In Section 4.7, it was shown that iterating the LSQ procedure until convergence does not necessarily provide improved results compared to a single iteration. The reason is that, for weak signals, the noise produces many minima in the cost function. The subsequent iteration tends to converge to one of those minima, but the further the linearization point drifts away from the correlation point, the more inaccurate the correlation model will be. Furthermore, the normal matrix diverges if the code phase of the line-of-sight and of the multipath signal converge with each other.

On the other hand, a single iteration provides a numerically stable estimation procedure. A single iteration is also optimal if one correlation point is near the true values. A single iteration is also a good engineering solution. The only exception occurs for the case of high signal power, where a second iteration is required to make the estimates unbiased.

8.3.7 Multipath Detection

After performing the no-multipath step ($M = 1$), a χ^2-test (described in Appendix A.1.5) is done to verify that the no-multipath hypothesis is correct. The reader may ask why a test is necessary to verify the presence of a multipath because, theoretically, the $M = 2$ LSQ procedure should result in the no-multipath case in a vanishing multipath amplitude. Again, the reason lies in the fact that, for the no-multipath-case, the $M = 2$ LSQ procedure tends to yield identical line-of-sight and multipath code-phase values, making the normal matrix diverge.

Choosing a high significance level for the $M = 1$ χ^2-test ensures that the LSQ procedure also converges well for $M = 2$. On the other hand, a lower significance level can be used (to avoid cases of undetected multipath) and the nonconvergence of the $M = 2$ LSQ procedure can be used as an indicator that no multipath is present. Then a fallback to the $M = 1$ paramcters is used.

It should be pointed out that the number of multipath signals (and eventually coarse estimates of the code phase and Doppler of the multipath signals) can also be obtained by other methods, like RAIM, C/N_0 monitoring, and dedicated metrics. An excellent overview of those techniques is given in the work by Irsigler [13].

8.3.8 Discussion

The presented multipath-estimating discriminator is an adaptive method to mitigate multipath errors and has an optimal thermal-noise performance if no multipath signal is present or if the multipath signal is easily detectable. If the multipath signal remains undetected, the performance is still comparable to a narrow early–late discriminator. The method seems not to have any disadvantage in terms of performance compared to a nonparametric discriminator. The major drawbacks are the increased computational demands and the calibration efforts.

The method demonstrates also that the nonrandom parameter approach discussed in Chapter 4 does *not* provide an optimal estimator for arbitrary signal power levels if the linearity conditions are not fulfilled. The MLE or LSQ scheme is optimal only in the limit of infinite signal power. A typical LSQ problem is that the low-power multipath remains undetected. The nonoptimality also manifests itself as unstable LSQ iterations, especially if multipath-signal parameters are estimated. Many local minima of the cost function (8.43) exist. Although there exist methods to minimize this multidimensional function that reliably converge to one of these local minima [11], the estimation principle itself remains suboptimal.

As discussed in Section 4.8, Bayesian techniques are, by definition, also optimal for low-signal-power values, but rely on correct stochastic-parameter (especially multipath) models. In the author's opinion, this information is difficult to provide in practical situations, especially for multipath-signal parameters. The receiver must be aware of its surroundings (e.g., rural, indoor, urban). A universal multipath model would be highly desirable. The Bayesian approach also needs more computational resources when, for example, a particle filter is used. However, results presented by others demonstrate that the Bayesian approach may outperform an epoch-per-epoch LSQ estimation because it exploits the relationship between multipath-signal-parameter probability density functions at different epochs [14].

The LSQ discriminator and Baysian techniques rely on an accurate correlator model that might be difficult to obtain in a real implementation because it needs the receiver to be calibrated for given filters, amplifiers, and antennas.

8.4 From Discriminator Noise to Position Accuracy

This chapter provides formulas to compute variances of high-rate pseudorange estimates. For example, (8.14) is the code-discriminator variance and (8.25) is the F-correlator Doppler variance. As already mentioned, the high-rate pseudorange estimates are smoothed to obtain low-rate pseudorange estimates, which are then processed to obtain position estimates. Both steps are complex and many different tracking-loop algorithms exist, along with many positioning methods. They are not covered in this work, but there exists a canonical way to obtain the variance of the position solution from the high-rate pseudorange variance. This method shall be outlined to provide the reader with a method to evaluate the influence of various signal-processing techniques of the positioning solution. Furthermore, the method allows a simplified performance evaluation of vector-tracking stability described in Section 4.3.3.3.

The canonical method accounts only for thermal noise and ignores other relevant error sources (e.g., unmodeled multipath, transmitter-position errors, atmospheric delays, and so forth). Loop filters are characterized by their bandwidth B_L (see Section 4.3.3.1).

The canonical tracking-loop concept was introduced in the work by Betz and Kolodziejski [15]. The method was introduced to convert unsmoothed TOA estimates (i.e., high-rate pseudorange estimates) into smoothed TOA estimates (i.e., low-rate pseudorange estimates). The conversion from low-rate pseudorange estimates to positions depends on the transmitter geometry and the geometric placement of the transmitters is modeled by a dilution-of-precision factor (DOP). All estimates are considered to be unbiased.

Following Section 4.3.3.1, let q denote a high-rate pseudorange parameter that is available with a rate in seconds given by T_{coh}. The variance of the corresponding low-rate pseudorange parameter p is given by

$$\mathrm{var}\langle p\rangle|_{\mathrm{N}} = 2B_L T_{coh}\, \mathrm{var}\langle q\rangle|_{\mathrm{N}} \tag{8.54}$$

where B_L denotes the noise-equivalent loop bandwidth in hertz. This relation holds if $0 < B_L T_{coh} < 1$; refer to the work by Kazemi for a recent discussion on tracking-loop stability [16]. Furthermore, the tracking loop is required to model the dynamics sufficiently well so that transient errors can be ignored.

We assume now that, for one-epoch multiple identical-variance, independent and unbiased low-rate pseudorange estimates of the same type for different transmitters are available. An LSQ estimator is used to calculate a (generalized) position-parameter estimate (e.g., Doppler estimates might be used to obtain one velocity-component estimate or code-phase estimates to obtain a coordinate estimate). The variance of the (generalized) position estimate is given by

$$\mathrm{var}\langle x\rangle|_{\mathrm{N}} = \mathrm{DOP}\,\mathrm{var}\langle p\rangle|_{\mathrm{N}} \tag{8.55}$$

where the DOP factor accounts for the geometric placement of the transmitters. DOP factors are described in many text books of satellite navigation [17]. The DOP factor is the diagonal element of the parameters' covariance matrix obtained under the assumption of unity low-rate pseudorange variance.

References

[1] van Dierendonck, A. J., "GPS Receivers," in *Global Positioning System: Theory and Applications, Vol. I*, pp. 329–407, Parkinson, B. W., and J. J. Spilker, (eds.), Washington, D.C.: American Institute of Aeronautics and Astronautics Inc., 1996.

[2] Weill, L. R., "GPS Multipath Mitigation by Means of Correlator Reference Waveform Design," *Proc. Institute of Navigation National Technical Meeting (ION-NTM) 1997*, Santa Monica, CA, September 14–16, 1997, pp. 197–206.

[3] Kaplan, E. D., and C. J. Hegarty, (eds.), *Understanding GPS: Principles and Applications*, 2nd ed., Norwood, MA: Artech House, 2006.

[4] Irsigler, M., and B. Eissfeller, "Comparison of Multipath Mitigation Techniques with Consideration of Future Signal Structures," *Proc. 16th Int. Technical Meeting of the Satellite Division of the Institute of Navigation (ION-GPS/GNSS) 2003*, Portland, OR, September 9–12, 2003, pp. 2585–2592.

[5] Julien, O., et al., "A New Unambiguous BOC(n,n) Signal Tracking Technique," *Proc. European Navigation Conference (ENC-GNSS) 2004*, Rotterdam, May 16–19, 2004.

[6] Pany, T., and B. Eissfeller, "Code and Phase Tracking of Generic PRN Signals with Sub-Nyquist Sample Rates," *NAVIGATION, Journal of The Institute of Navigation*, Vol. 51, No. 2, 2004, pp. 143–159.

[7] Pany, T., M. Irsigler, and B. Eissfeller, "S-Curve Shaping: A New Method for Optimum Discriminator Based Code Multipath Mitigation," *Proc. 18th Int. Technical Meeting of the Satellite Division of the Institute of Navigation (ION-GNSS) 2005,* Long Beach, CA, September 13–16, 2005, pp. 2139–2154.

[8] Paonni, M., et al., "Looking for an Optimum S-Curve Shaping of the Different MBOC Implementations," *NAVIGATION, Journal of The Institute of Navigation*, Vol. 55, No. 4, 2008, pp. 255–266.

[9] Ávila Rodríguez, J. Á., T. Pany, and G. Hein, "Bounds on Signal Performance Regarding Multipath-Estimating Discriminators," *Proc. 19th Int. Technical Meeting of the Satellite Division of the Institute of Navigation (ION-GNSS) 2006*, Fort Worth, TX, September 26–29, 2006, pp. 1710–1722.

[10] van Nee, R. D. J., et al., "The Multipath Estimating Delay Lock Loop: Approaching Theoretical Accuracy Limits," *Proc. IEEE Position Location and Navigation Symposium*, Las Vegas, NV, April 11–15, 1994, pp. 246–251.

[11] Nunes, F. D., F. M. G. Sousa, and J. M. N. Leitao, "BOC/MBOC Multicorrelator Receiver with Least-Squares Multipath Mitigation Technique," *Proc. 21st Int. Technical Meeting of the Satellite Division of the Institute of Navigation (ION-GNSS) 2008*, Savannah, GA, September 16–19, 2008, pp. 652–662.

[12] Won, J. H., T. Pany, and B. Eissfeller, "Implementation, Verification and Test Results of a MLE-Based F-Correlator Method for Multi-Frequency GNSS Signal Tracking," *Proc. 20th Int. Technical Meeting of the Satellite Division of the Institute of Navigation (ION-GNSS) 2007,* Fort Worth, TX, September 25–28, 2007, pp. 2237–2249.

[13] Irsigler, M., *Multipath Propagation, Mitigation and Monitoring in the Light of Galileo and Modernized GPS*. University of Federal Armed Forces Munich, Werner-Heisenberg-Weg 39, D-85577 Neubiberg, http://www.unibw.de/unibib/digibib/ediss/bauv, 2008.

[14] Lentmaier, M., et al., "Dynamic Multipath Estimation by Sequential Monte Carlo Methods," *Proc. 20th Int. Technical Meeting of the Satellite Division of the Institute of Navigation (ION-GNSS) 2007*, Fort Worth, TX, September 25–28, 2007, pp. 1712–1721.

[15] Betz, J. W., and K. R. Kolodziejski, "Extended Theory of Early-Late Code Tracking for a Bandlimited GPS Receiver," *NAVIGATION, Journal of The Institute of Navigation*, Vol. 47, No. 3, 2000, pp. 211-226.

[16] Kazemi, P. L., "Optimum Digital Filters for GNSS Tracking Loops," *Proc. 21st Int. Technical Meeting of the Satellite Division of the Institute of Navigation (ION-GNSS) 2008*, Savannah, GA, September 16–19, 2008, pp. 2304–2313.

[17] Hofmann-Wellenhof, B., H. Lichtenegger, and E. Wasle, *GNSS: Global Navigation Satellite Systems: GPS, GLONASS, Galileo & More*, Vienna: Springer, 2008.

CHAPTER 9

Receiver Core Operations

Within a GNSS SDR, there are a few operations that consume most of the processing power. They shall be described and analyzed in the following chapter. A specific characteristic of the core operations is that their implementation needs fixed-point arithmetic to run efficiently. By contrast, calculation of the navigation solution or tracking-loop update (just to name two examples) can be done with the full floating-point power available to the processor. Furthermore, the core operations run within a one-dimensional loop and are, from the algorithmic point of view, rather simple. In some sense they represent a heritage from GNSS hardware receivers. An important exception in this context is the use of Fourier techniques in a GNSS SDR. Fourier techniques are more complex from the algorithmic point of view and may also be implemented with floating-point arithmetic.

It should also be mentioned that the choice of the algorithms is, in some sense, specific to the SDR implementation developed at the University of Federal Armed Forces (Munich/Germany), but approaches used by other research institutes are also discussed.

9.1 Test-System Configuration

Implementation of core algorithms is always, to a certain extent, platform-specific. The platform determines the optimal implementation strategy and determines how much performance can be expected. The performance measurements shown in the next sections are based on a system shown in Table 9.1.

The system is a laptop that was purchased at the beginning of 2006. It is a single-core CPU system based on Intel's Netburst architecture. This architecture has already been replaced by the Core 2 Duo architecture at the time of this writing.

The benchmark results and power consumption of the test system have been partly obtained by a tool from SiSoftware [1]. At the time of this writing, single-CPU systems with multiple cores (e.g., quad-core processors) were already available whose main memory bandwidth outperformed the presented system by a factor of four and whose integer-processing performance was 24 times faster but whose power consumption was also five times higher.

A detailed performance analysis of the core algorithms was obtained by using the run-time analysis tool by Intel that provides the number of clock ticks a single CPU instruction needs during execution [2]. It should be noted that those numbers are estimated based on CPU internal timer readings and are reproducible only on the test-system configuration. Even in this case, a slight variation of the estimated number of clock ticks is common.

Table 9.1 Test-System Configuration

Parameter	*Test System (Laptop)*	*Highest Values from [1]*
Computer model	Dell Inspirion 9300	—
CPU	Intel Pentium M (780)	—
Number of cores	1	4
CPU architecture	IA-32	—
CPU clock speed	2.26 GHz	—
Main memory	1 GB	—
L1/L2 cache bandwidth	23 GB/s resp. 11 GB/s	330 GB/s resp. 85 GB/s
L1/L2 (data) cache size	32 kB, resp. 2 MB	128 kB, resp. 4 MB
Main memory bandwidth	2.5 GB/s	10.2 GB/s
CPU power consumption	5–27W	140W
CPU chip technology	90 nm	—
Battery capacity	80 Wh	—
Battery weight	461g	—
External electrical power consumption in idle mode	34W	—

9.2 Signal-Sample Bit Conversion

Spread-spectrum navigation signals within a SDR are usually represented by integer samples of a given bit size. The term "signals" refers to both the received navigation signal(s) and the internally generated signals. Usually a low number of bits is sufficient to represent those signals. For example, in Section 6.1 it is stated that received GNSS signals can be represented with a loss of 0.2 dB by using 3 bits in the absence of interference. Internally generated sine/cosine carriers can be represented by a 1-bit amplitude (i.e., essentially as a square wave) causing a loss of 0.9 dB.

There are, however, two hardware-related factors that determine more rigorously how many bits shall be used to represent the signals. They are:

- Bits provided by the ADC or maximum bandwidth of the ADC to CPU data link;
- Integer formats supported by the CPU's vector instructions.

The first factor usually limits the number of bits to less than four (see Table 2.1). Only in special cases (e.g., if a commercial ADC board is used) may the SDR be forced to work with a larger number of input bits (e.g., 8, 14, or 16 bits are common COTS ADC board formats that would allow it to cope with high-power interference). The bit format provided by the ADC card may not use a two's-complement format, which is the CPU internal format. For example, sign/magnitude formats are commonly used ADC output formats.

The second factor is specific for the CPU and the algorithms used for correlation. For example, CPUs supporting the SSE-SSE4 instructions allow efficient multiplication of two 16-bit numbers 8-wise in parallel. Another method for signal multiplication works most efficiently if both signals are represented as 1-bit values. Then a simple XOR operation can be used to multiply two 1-bit numbers 128-wise in parallel with SSE/MMX instructions.

Consequently, there is generally a need to convert the format of the input ADC bits to the internal working format and the corresponding algorithm shall be described in Section 9.2.1.

9.2.1 Algorithm

The presented algorithm performs a bit-format conversion. The presented algorithm is based on a lookup table, in contrast to an approach that would use arithmetic or dedicated bit-conversion instructions.

The lookup-table approach uses the input values (of a given bit size and format) as indices of a precomputed table, and the table entries represent the output values (of a different bit size and format). The scheme is depicted in Figure 9.1 for a conversion of 2-bit ADC values into 16-bit values using the 2's complement format (see also Section 9.2.2 for a more detailed description). In this example, the ADC measures the voltage symmetrically around zero. For example, one can assume that an ADC value of "00" represents a voltage measurement between 0V and 0.1V, the value "01" is a voltage above 0.1V, the value "11" is a voltage between 0V and –0.1V, and "10" is a voltage below –0.1V. The ADC values shall be converted in a format represented by 16-bit 2's-complement integers: –3, –1, 1, and 3. Those integer values are symmetrically distributed around zero and are spaced equidistantly.

By using a lookup table, one avoids arithmetic calculations that otherwise have to be performed for the conversion. Those arithmetic calculations are normally not directly supported by the CPU and would require a very time-consuming indirect implementation. One disadvantage of the lookup-table approach is that it requires frequent memory accesses and that it cannot be vectorized because the CPU's core has only one single memory bus. The impossibility to vectorize this algorithm can represent a bottleneck in the SDR.

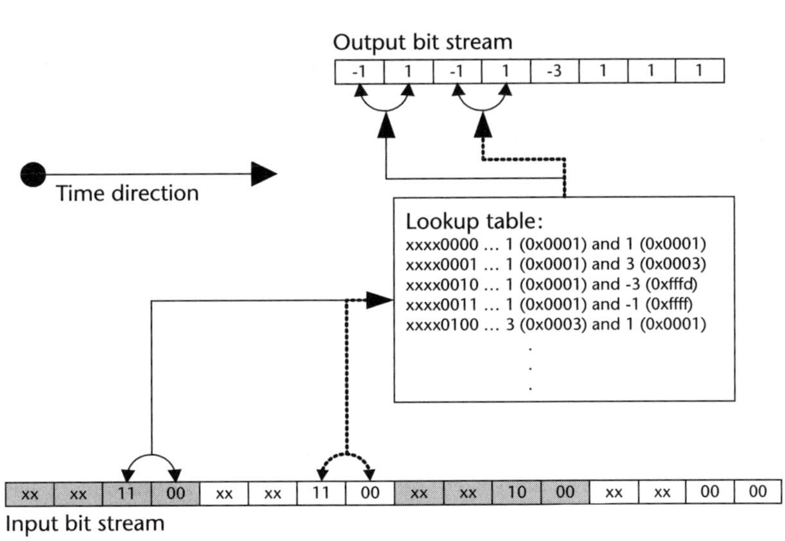

Figure 9.1 Block diagram for lookup-table-based bit conversion.

9.2.2 Numerical Performance

The following reference-assembler implementation on the chosen test system is shown in Table 9.2 and is based on Figure 9.1. It demonstrates the conversion of 2-bit values into 16-bit values. Within the input stream, each byte contains, on the four LSB positions, two 2-bit values. The upper four bits are ignored as they belong to a different input stream (e.g. to a different GNSS-frequency band). The two 2-bit values are converted into two 16-bit values in one step. The 16-bit values are represented as one 32-bit value. The algorithm runs in a one-dimensional loop and in each step one input byte is read and the corresponding 32-bit output value is stored in the output signal.

The assembly code of Table 9.2 shows only the core part and must be initialized. The CPU register *eax* is initialized in a way that *eax* points to the beginning of the input stream. The register *ebx* is set to "*-length,*" where *length* is the number of bytes of the input stream. The register *edx* points to the beginning of the output stream, and *ecx* points to the lookup table. The lookup table consists of 256 values and each value is represented by 32 bits (i.e., by the two 16-bit output values). The indices of this table are the 8-bit input values and it should be noted that only the four lower bits count. Consequently, the 256 32-bit output values take only 16 possible values.

Table 9.2 shows that one loop cycle takes 10.96 clock ticks on the test system. Within one cycle, two values are converted; assuming a CPU clock frequency as shown in Table 9.1, 412 Msamples can be converted per second. This value is comparably low, especially if a triple-frequency GNSS SDR with 40.96 MHz sample rate is considered. In this case, 122.88 Msamples per second have to be converted and the conversion itself (which is a more or less trivial operation) consumes 28% of the maximum CPU processing load of the test system. Table 9.1 also demonstrates that the most time-consuming (or clock-tick consuming) instruction is the lookup operation, which needs 7.81 clock ticks on average. Overall, the CPU is not able to pipeline the loop sequence well, nor is the compiler able to generate vectorized code. The lookup table itself can be kept in the CPU's L1 cache, but the input and output streams have to be read/written from/to the main memory.

Table 9.2 Assembly Language Code Snippet for Bit Size Conversion

Instruction	*Description*	*Clock Ticks*
`loop_begin:`		
`movzx esi, BYTE PTR [ebx+eax+length]`	Load next input 8-bit value into *esi*	0.98
`mov esi, DWORD PTR [ecx+esi*4]`	Perform lookup table operation. The register *esi* is the index of the table and it also receives the lookup table values	7.81
`mov DWORD PTR [edx], esi`	Store resulting 32-bit value contained in *esi* in the output stream	1.03
`add edx, 4`	Advance output stream by 32 bit	1.14
`add ebx, 1`	Advance input stream by 8 bit	—
`jne loop_begin`	go to beginning of loop, if *ebx* != 0	—
Total of clock ticks		10.96
Number of instructions		6

9.2.3 Discussion and Other Algorithms

The bit conversion is trivial from the mathematical point of view and therefore no further analysis is needed. Also, no precision is lost when the number of bits is increased.

The case of bit reduction occurs when, for example, 8-bit values from a commercial ADC card are packed and stored as 1-bit values. In this case, the sign-bit from the 8-bit values is taken and eight sign-bits are combined into one byte. The lookup-table approach is not applicable in this case because of the fact that it would need a huge (i.e., 64-bit-long) table index and eventually explicit programming would need to be used. The more general case of m-bit to n-bit bit reduction is also trivially realised by taking the upper n-bits of the m-bit value if both values use the sign/magnitude representation and the sign bit is the MSB.

9.3 Resampling

An efficient way of generating reference signals in a navigation SDR is to store precomputed signals in memory and to resample them to the sample rate actually needed. Typically, sine/cosine signals need to be generated, along with the PRN-code sequence and more general reference signals (see Section 8.2).

Because of time constraints, typically nearest-neighbor resampling is used because the amplitude accuracy of the reference signals is not very high. Remember that sine/cosine signals can be represented as 1-bit signals, which are square waves.

9.3.1 Algorithm

The algorithm investigated in the following section mimics a hardware-receiver-like NCO structure. The NCO is represented as a 64-bit-long variable whose upper 32-bits act as an index to the precomputed signal table and whose lower 32-bit values define the NCO phase and NCO-rate resolution. The algorithm (which is depicted in Figure 9.2) is initialized with the start-NCO value and the NCO increment. The algorithm runs in a loop for the required number of samples and, for each loop iteration, the lookup-table entry corresponding to the index given by the upper 32 bits of the NCO value is stored in the output stream. Afterwards, the NCO value is incremented by the NCO increment value and the next loop iteration starts.

The algorithm generates a signal with a constant rate (e.g., constant Doppler). If time-variable rate signals are needed, it is recommended to approximate those signals by piecewise constant rate signals simply for numerical performance reasons.

9.3.2 Numerical Performance

In Table 9.3 an optimized-assembler implementation of the algorithm depicted in Figure 9.2 is given for the test system of Table 9.1. The routine is initialized with

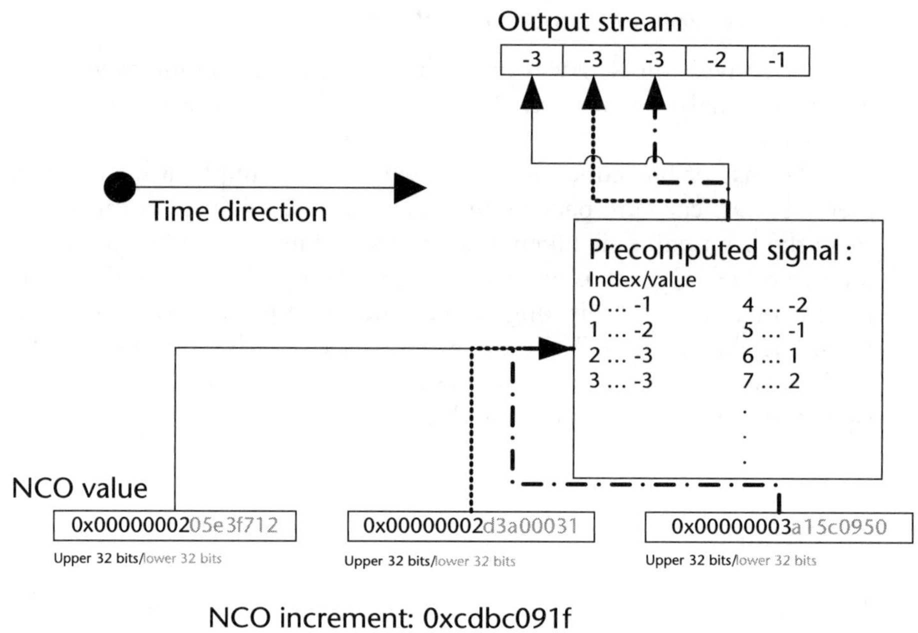

Figure 9.2 Block diagram of NCO-based resampling.

the following values. The register *edx* points to the beginning of the lookup table. The lower 32 bits of the NCO increment is stored in *eax*, the upper 32 bits of the increment is stored in *edi*. The NCO value itself is stored in *ebx* (upper 32 bits) and *esi* (lower 32 bits). The number of samples to be generated is stored in the memory address *no_samples*; *ecx* initially points to the beginning of the output stream.

Table 9.3 shows that one loop iteration takes 5.64 CPU cycles on the test system. Assuming a CPU clock frequency as shown in Table 9.1, 400.7 Msamples can be generated per second. This value is quite low and represents the major bottleneck in a software correlator GNSS receiver. For example, if correlation in one channel is considered, than the sine/cosine carrier signals have to be generated, as well as the P- and the D-correlator signals (see Section 7.3). In this way, the test system of Table 9.1 would be fully loaded by generating reference signals for 10 channels if a sample rate of 10 Msamples/s is used; correlation or any other task would be impossible.

It should be noted again that this algorithm cannot be directly vectorized because it heavily utilizes the memory bus. As a consequence, only one sample is generated per loop iteration. Further performance improvements can only be achieved if reference signals are already stored in the main memory with the required sample rate and need not be resampled.

9.3.3 NCO Resolution

The presented algorithm works with a fixed-point NCO realization and, thus, NCO phase and rate have a limited resolution, which is discussed in this section. Let f_s

Table 9.3 Assembly Language Code Snippet for Nearest-Neighbor Resampling

Instruction	*Description*	*Clock Ticks*
`loop_begin:`		
`mov ebp, DWORD PTR [edx+ebx*4]`	Retrieve value from lookup table. The register [*edx*] points to beginning of lookup table, *ebx* is the index	0.98
`add esi, eax`	Increment the lower 32 bits of NCO value stored in *esi* (*eax* contains the lower 32 bits of the increment)	1.33
`mov WORD PTR [ecx], bp`	Store 16-bit value in output stream	0.54
`mov ebp, DWORD PTR [no_samples]`	Load frame length (multiplied with 2) into *ebp*	1.04
`adc ebx, edi`	Increment upper 32 bits of NCO value stored in *ebx* (*edi* contains the upper 32 bits of the increment)	0.64
`add ecx, 0x2h`	Increment output sample index	0.83
`cmp ecx, ebp`	Check if end of frame is reached	0.19
`jb loop_begin`	If not, go to beginning of loop	0.09
Total of clock ticks		5.64
Number of instructions		8

denote the sample rate of the signal to be generated in 1/sec. The values of the precomputed lookup table shall be denoted as "value" and this term may represent a PRN-code sample or a sine/cosine value. It is common practice to store the reference signals "over sampled" [i.e., a single chip of a PRN code or a complete sine period is usually represented by a fixed number of (multiple) values]. This shall be denoted by the equation

$$\#\,\text{values} = o\#\,\text{samples} \tag{9.1}$$

where o denotes the oversampling factor. Here, "samples" stands either for "chip" or for "cycles."

The NCO-phase resolution is determined by o and by the number of bits in the lower part of the NCO (in the above example, 32) but which will now be denoted as m for the sake of generality. The NCO resolution in [samples] is given by

$$\Delta\phi = 1/(2^m o) \tag{9.2}$$

Because the NCO increment is an integer multiple of the NCO resolution, the NCO rate resolution in samples per second is given by

$$\Delta\phi = \frac{f_s}{o2^m} \tag{9.3}$$

For example, if a precomputed sine/cosine table is stored as a table of $o = 256$ values (containing exactly one period), then the presented algorithm allows a phase resolution of 9.1×10^{-13} cycles, which is, for most applications, sufficiently high. For an exemplary sample rate of 40.96 MHz, the NCO-rate resolution evaluates to 37 μHz. If a 16-bit NCO fine resolution had been used, then the NCO-rate resolution would evaluate in this case to 2.44 Hz, which could cause significant distortions in GNSS frequency tracking.

For PRN-code generation, a typical example might be to store the PRN-code sequence (or derived reference signals) with an oversampling factor of $o = 20$. This is, for example, required to achieve a $d = 0.1$ early–late narrow correlator. The corresponding 32-bit NCO code-phase resolution is then 1.2×10^{-11} chip. If a sample rate of 40.96 MHz is again considered, the NCO-rate resolution evaluates to 0.00048 chip/s. Assuming a C/A-code chip length of 293m, this corresponds to a range-rate resolution of 14 cm/s.

9.3.4 Discussion and Other Algorithms

Resampling of signals is a potential bottleneck in a GNSS radio, as has been demonstrated above. Again, this results from the inability of the CPU to support efficient lookup-table operations. Other methods to generate the reference signals (e.g., either by direct trigonometric floating-point functions for sine/cosine signals, or realization of linear-shift registers for PRN-code generation in assembler) are, however, much more inefficient.

In the presented algorithm, the output values are 16-bit values and storing them in memory is quite efficient. By contrast, if reference signals are represented by 1-bit values packed into bytes, then the single bits (looked up from the lookup table) need to be shifted via bit operations into a CPU register. If the register is filled with the required number of bits (e.g., 8), it is then written into the main memory.

The most efficient way to solve the resampling problem is to store the signals already with the required sample rate. Then the required parts of the resampled signal can be simply copied. In the case of the packed 1-bit signal representation, there exists, however, another problem. Because direct bit access is not possible, the minimal access unit is one byte. Consequently, the signal to be generated needs to be aligned to eight values with the precomputed signal. The precomputed signal is stored eight-wise, each copy shifted by one 1-bit (i.e., one value) with respect to its predecessor.

9.4 Correlators

Correlators are fundamental for a GNSS SDR because they define a sufficient statistic and all parameter estimates derive from them (see Section 4.4). Computationally, they represent a dot-product operation written as

$$C = \sum_{i=0}^{N-1} s_i r_i \tag{9.4}$$

The two vectors s_i and r_i may, for example, represent the incoming-IF signal and the internally generated reference signal.

9.4.1 SDR Implementation

Fortunately, the dot-product operation is essential to multimedia applications and modern CPUs do support specific commands to vectorize this operation. For example, in the SSE2 command set of the Pentium M, the command `pmaddwd` can be found, which combines the multiplication and addition into one command. It multiplies eight 16-bit values element-wise with another eight 16-bit values. Two of the resulting 32-bit products are added and, finally, four 32-bit products are stored in a CPU register (which is a special XMM vector register being 128 bits wide).

The implementation shown in Table 9.4 of the correlation is a direct application of the command `pmaddwd`. Before running the routine, the register *edi* points to the beginning of the first signal and *esi* points to the beginning of the second signal. The register *eax* is initialized to zero and *ebx* contains the length of the correlation (i.e., N). The vector register *xmm0* (which contains four 32-bit values) is set to zero at the beginning and contains the four partial sums of the correlation result C after the routine ends. Those four partial sums have then to be added to obtain C.

Table 9.4 shows that one loop iteration takes 15.07 CPU cycles on the test system. Eight values of the first signal are correlated with eight values of the second signal within one iteration. Assuming a CPU clock frequency as shown in Table 9.1, 2 × 1,200 Msamples can be correlated per second. Here, we assume that the first signal is kept within the L1 CPU cache, whereas the second signal is read from the main memory. If both signals can be read from the L1 CPU cache, one loop iteration takes 4.69 clock ticks, resulting in a correlation capacity of 3,855 Msamples/s. For this case, the clock ticks of the single instructions are written within parentheses in the last column of Table 9.4.

The reciprocal throughput (i.e., the minimum number of clock cycles per instruction) for the Intel Netburst architecture is reported to be 1 for the `paddd`, 2 for the `pmaddwd`, and 1 for the `movdqa` command [3]. For the Intel Core 2 Duo architecture, these values reduce to 0.5, 1, and 1, respectively.

Table 9.4 Assembly-Language Code Snippet for Correlation

Instruction	*Description*	*Clock Ticks*
`loop_begin:`		
`movdqa xmm1, XMMWORD PTR [edi+eax*2]`	Load next eight 16-bit values of signal 1 into register *xmm1*	0.96 (1.06)
`pmaddwd xmm1, XMMWORD PTR [esi+eax*2]`	Multiply (and partly add) *xmm1* with next eight 16-bit values of signal 2. The result are four part sums, each 32-bit, stored in *xmm1*	0.01 (0.08)
`paddd xmm0, xmm1`	Add part sums of *xmm1* to *xmm0*	12.19 (2.19)
`add eax, 0x8h`	Increment signal index by eight values	1.19 (0.47)
`cmp eax, ebx`	Check if end of signal is reached	0.59 (0.80)
`jb loop_begin`	If not, jump to loop begin	0.13 (0.09)
Total of clock ticks		15.07 (4.69)
Number of instructions		6

In cases where the second signal is read from main memory, the bottleneck is again the memory-bus bandwidth. Table 9.1 reports a value of 2.5 GB/s, which explains the fact that the correlation is limited by 1,200 Msamples/s. For a correlation speed of 1,200 Msamples/s, a number of 2.4 GB has to be transferred from the main memory to the CPU as each sample of the second signal is 2 bytes long.

9.4.2 Discussion and Other Algorithms

From Table 9.4, it is evident that the performance of the presented correlation algorithm heavily depends on whether the signals are read from main memory or if they are already within the CPU caches. A factor of nearly three is gained. Therefore, one of the key issues in a GNSS SDR is to organize the data flow such that the CPU caches are exploited efficiently.

Another algorithm published for PC-based GNSS SDR in an article by Ledvina proposes the already-mentioned approach to work with a 1-bit sample representation [4]. This bit effectively corresponds to the sign of the signal sample (e.g., a 0 corresponds to a negative sign, a 1 to a positive sign). The advantage of doing so is that the multiplication of two 1-bit values boils down to a `xor` operation, as can be seen from Table 9.5.

The Intel Netburst Archicture supports the command `pxor`, which allows performing an xor operation with two 128-bit registers with a throughput of two cycles, the Intel Core 2 Duo should give a throughput of 0.33 [3]. For the case of the Intel Core 2 Duo, on average for each CPU clock tick 2×384 1-bit values can be multiplied with each other. Those values have to be loaded from memory, but for the packed 1-bit representation the load on the memory bus is considerable lower than for the 16-bit representation. A little trickier is the addition of the 1-bit product values. Here, two efficient approaches are possible. The first approach uses explicit CPU commands that count the number of "1" bits in a register. Such commands have become available recently with the introduction of the SSE4 command set in Intel and AMD CPUs. The second approach subdivides the 128-bit register into eight 16-bit values. Each 16-bit value acts as an index of a lookup table, whose entries correspond to the number of "1"-bits of the 16-bit index values. By doing so, the sum of sixteen 1-bit values can be performed with one lookup-table operation.

Comparing the 16-bit approach with the 1-bit approach a few important differences are noted. The 1-bit approach shows its biggest advantage in the high memory bus efficiency and fully exploits the efficient correlation by a `xor` command. The biggest disadvantage of the 1-bit approach is the cumbersome data handling. The 16-bit approach allows a convenient realization of high-end signal-processing algorithms with minimum implementation losses; the correlation can be performed well, but the memory bandwidth demands are high.

Table 9.5 Definition of the xor Operation

A	*B*	*xor(A,B)*
0	0	1
0	1	0
1	0	0
1	1	1

9.5 Fast Fourier Transform

Fast Fourier transform techniques are mostly used in a GNSS SDR for signal acquisition and exploit the convolution theorem. In fact, they are essential in a SDR for acquisition, allowing the realization of a large number of effective correlators needed for fast and sensitive acquisition. However, signal-preprocessing or postcorrelation FFT tracking algorithms also benefit from an efficient FFT implementation. In contrast to the other core algorithms above, FFT implementations can be complex, but because they are used for myriads of other applications, very efficient libraries exist [2, 5]. Therefore, in this chapter no assembler language snippet is presented and performance benchmarks are based on the library [2]. On the other hand, the use of Fourier techniques for correlation can be quite tricky and optimization at the algorithmic level will be shown in the following section.

9.5.1 Algorithm

The FFT is a widely known algorithm for efficiently evaluating a DFT. For the following example, we assume that the DFT is defined as

$$\tilde{s}_k = \sum_{n=0}^{N-1} s_n \exp\left\{-2\pi i \frac{nk}{N}\right\} \Leftrightarrow \tilde{s}_k = \mathrm{FFT}\{s_n\} \tag{9.5}$$

where s_n is the time domain representation of a signal of length N and $\tilde{s}_k$ is the frequency-domain representation. The idea of the FFT became popular with the publication by Cooley and Tukey, but had already been discovered by C.F. Gauß in 1805 [6]. It is based on the following identity

$$\begin{aligned}\tilde{s}_k &= \sum_{m=0}^{N/2-1} s_{2m} \exp\left\{-2\pi i \frac{2mk}{N}\right\} + \sum_{m=0}^{N/2-1} s_{2m+1} \exp\left\{-2\pi i \frac{(2m+1)k}{N}\right\} \\ &= \sum_{m=0}^{N/2-1} s_{2m} \exp\left\{-2\pi i \frac{2mk}{N}\right\} + \exp\left\{-2\pi i \frac{k}{N}\right\} \sum_{m=0}^{N/2-1} s_{2m+1} \exp\left\{-2\pi i \frac{2mk}{N}\right\}\end{aligned} \tag{9.6}$$

The DFT of (9.5) is written as two DFTs whose individual length is half of the original length. The first DFT is called *even DFT* (as it operates on elements with even indices), the second one is called *odd DFT*. Both vectors are scaled and added together to get the initially required DFT result (this is usually called a *butterfly operation*). The process of splitting the large DFT into two smaller DFTs can be recursively repeated, ending in an efficient FFT algorithm. The use of FFT techniques is widespread and explained in many textbooks (e.g., [7]). A more detailed explanation shall not be given here. It should be noted that although mixed-radix DFT algorithms exist, the use of a radix-2 algorithm as presented above is still the most common implementation. Consequently, the vector length N needs to be an integer power of two.

It is a common practice to assume that the calculation of an FFT of size N needs

$$\#Ops_{FFT} = 5N \log_2 N \tag{9.7}$$

operations, which could either be multiplications or additions [5]. Modern desktop-PC CPUs can perform multiplications with the same speed as additions.

9.5.2 Convolution Theorem

Although very basic, the convolution theorem for a finite size DFT shall be repeated here as it is of utmost importance for the following sections. Referring to (9.5), the inverse DFT is defined as

$$s_n = \frac{1}{N}\sum_{k=0}^{N-1} \tilde{s}_k \exp\left\{2\pi i \frac{nk}{N}\right\} \Leftrightarrow s_n = \frac{1}{N}\text{IFFT}\{\tilde{s}_k\} \tag{9.8}$$

It is well known that the following identify holds

$$\sum_{n=0}^{N-1} \exp\left\{2\pi i \frac{nk}{N}\right\} = N\delta_{k,0} \tag{9.9}$$

where $\delta_{k,l}$ is the Kronecker delta, which evaluates to one if $k = l$ or otherwise to zero. If we assume periodic signals s and r; that is,

$$s_{n+aN} = s_n \quad a \in \mathbb{Z} \tag{9.10}$$

and similarly for r, then we can write a circular correlation function as

$$\begin{aligned} h_m &= \sum_{n=0}^{N-1} s_n r_{n+m} = \sum_{n=0}^{N-1} \frac{1}{N^2} \sum_{k,l=0}^{N-1} \tilde{s}_k \tilde{r}_l \exp\left\{\frac{2\pi i}{N}(nk + l(n+m))\right\} \\ &= \frac{1}{N^2} \sum_{k,l=0}^{N-1} \tilde{s}_k \tilde{r}_l \exp\left\{\frac{2\pi i}{N} ml\right\} \sum_{n=0}^{N-1} \exp\left\{\frac{2\pi i}{N} n(k+l)\right\} \\ &= \frac{1}{N^2} \sum_{k,l=0}^{N-1} \tilde{s}_k \tilde{r}_l \exp\left\{\frac{2\pi i}{N} ml\right\} N\delta_{k,-l} \\ &= \frac{1}{N} \sum_{k=0}^{N-1} \tilde{s}_k \tilde{r}_{-k} \exp\left\{-\frac{2\pi i}{N} mk\right\} \\ &= \frac{1}{N} \sum_{k=0}^{N-1} \tilde{s}_{-k} \tilde{r}_k \exp\left\{\frac{2\pi i}{N} mk\right\} \end{aligned} \tag{9.11}$$

The Fourier transform of the correlation values h_m is thus given by

$$\tilde{h}_k = \tilde{s}_{-k} \tilde{r}_k \tag{9.12}$$

The DFTs of the signals s and r are also periodic in the sense of

$$\tilde{s}_{k+aN} = \tilde{s}_k \quad a \in \mathbb{Z} \tag{9.13}$$

With the definition

$$\text{FLIP}\{\tilde{s}_k\} = \tilde{s}_{-k} \tag{9.14}$$

the convolution theorem can be summarized as

$$\sum_{n=0}^{N-1} s_n r_{n+m} = \frac{1}{N}\text{IFFT}\{\text{FLIP}\{\text{FFT}\{s_n\}\}\text{FFT}\{r_n\}\} \tag{9.15}$$

For real-valued signals, the following identity holds

$$\tilde{s}_{-k} = \overline{\tilde{s}_k} \Leftrightarrow \text{FLIP}\{\text{FFT}\{s_n\}\} = \overline{\text{FFT}\{s_n\}} \tag{9.16}$$

and the convolution theorem can be simplified for a real-valued signal s as

$$\sum_{n=0}^{N-1} s_n r_{n+m} = \frac{1}{N}\text{IFFT}\{\overline{\text{FFT}\{s_n\}}\text{FFT}\{r_n\}\} \tag{9.17}$$

9.5.3 Time-Domain Correlation and Data Preparation

A time-domain correlator computes the cross-correlation values between two signals s and r and is defined as

$$\Sigma_\lambda = \sum_{m=0}^{M-1} s_m r_{m+\lambda} \tag{9.18}$$

The first signal s is of length M. Equation (9.18) evaluates the cross-correlation function for a number of $N - M + 1$ points, where N is the length of the second signal r. The cross-correlation values shall be indexed by λ ranging from zero to $N - M$. The time-correlator scheme is depicted in Figure 9.3. It is important to note that time domain cross-correlation does not make any assumptions on the signal periodicity; it simply slides the first signal along the second one and evaluates

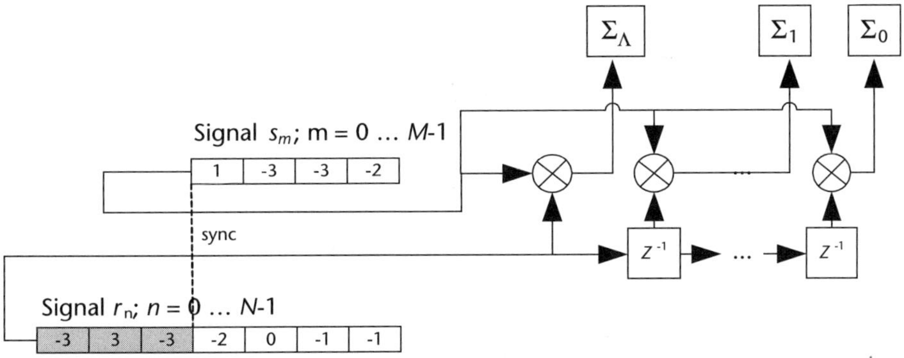

Figure 9.3 Time-domain correlator (the symbol Σ denotes an accumulator).

the correlation value. Furthermore, it should be noted that, in a GNSS-hardware receiver, the first signal is usually the received GNSS signal and the second one is internally generated. The second signal is usually generated with the same sample rate as the first signal and it is shifted sample by sample to obtain the cross-correlation values, although in principle it would be possible to generate the internal signal with arbitrary offsets. This is done to allow an easy realization in an ASIC and demands a high sampling rate to obtain a fine resolution of the cross-correlation function.

A frequency-domain correlator produces the exact same result as a time-domain correlator, although it exploits the convolution theorem. Because the convolution theorem implies periodic signals and a time-domain correlator does not, data preprocessing is required.

9.5.3.1 Zero Padding

A frequency-domain correlator using zero padding artificially elongates the signals to be correlated to avoid a signal overlap and the elongated signals are considered to be periodic. Zero values are appended to the signals, as shown in Figure 9.4. The number of zeros that have to be added derives from the requirement that (9.18) will be evaluated for a number of values of λ ranging from zero to $N - M$. Furthermore, it must be considered that both zero-padded signals must be of identical lengths and that this length allows for running a given FFT algorithm (i.e., for the common case of a radix-2 FFT, the length needs to be an integer power of two).

More specifically, the first signal s has to be zero-padded to achieve the same length as r and both signals have to be further zero-padded to achieve the desired FFT length Λ.

Applying the convolution theorem to the zero-padded signals, as shown in Figure 9.4, gives Λ correlation values. However, only the values with an index

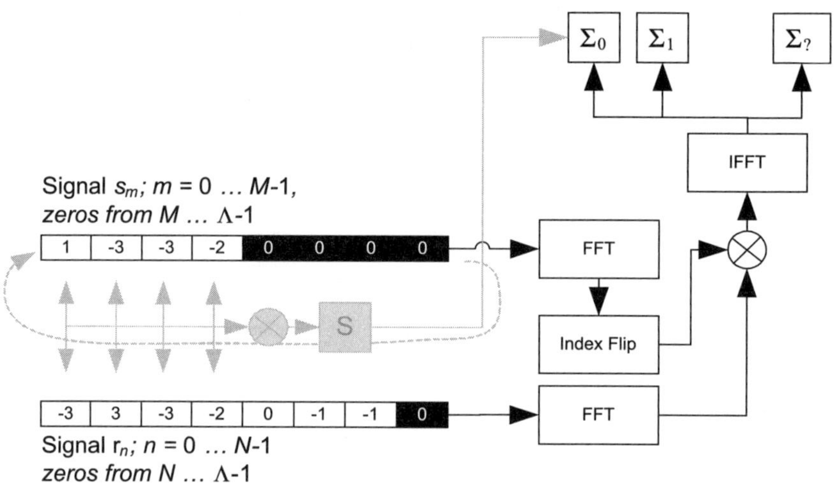

Figure 9.4 Frequency-domain correlator with zero padding (light-gray elements symbolize the time domain correlation for comparisons; the symbol Σ denotes an accumulator).

λ ranging from zero to $N - M$ are useful. The remaining values contain correlation values where the unpadded original signals only partly overlap and are thus useless.

9.5.3.2 Circular Correlation with Resampling

Zero padding can be avoided if the signal r of Figure 9.4 is periodic with a length of $N = M$ (i.e., it has the same length as the signal s). In this case, it suffices to generate only one period of the signal r. The signal s does not need to be periodic, but the correlation function of r with s is again periodic as the correlation function of any signal with a periodic signal is periodic. Of particular importance is the case when $N = M$ already matches the required FFT size (i.e., it is already an integer power of two in the case of a radix-2 FFT). Then, zero padding can be completely avoided and the maximum computational efficiency of a frequency-domain correlator is obtained. A block diagram for circular frequency-domain correlation is shown in Figure 9.5.

A critical factor is to achieve the required FFT length for a given correlation problem. For example, in signal acquisition, the length of the signal s is determined by the sample rate and the coherent-integration time. The coherent-integration time is mostly an integer multiple of the PRN-code period. In general, the length of the signal s does *not* fit the FFT size requirement. For signal acquisition (and also for other applications) it is possible to resample the signal (see Sections 9.3 and 5.3) to fulfill the FFT length requirement. This causes, however, resampling losses that are typically <1 dB and whether they can be tolerated must be decided on a case-by-case basis.

All output correlation values of the circular-correlation method are useful, in contrast to the zero-padding method. In fact, if the circular-correlation method can

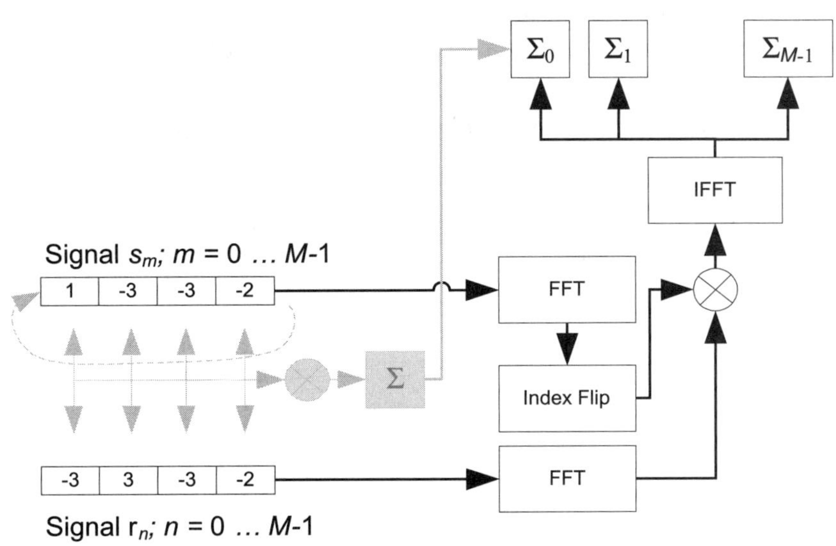

Figure 9.5 Frequency-domain correlator with circular correlation (light-gray elements symbolize the time domain correlation for comparisons; the symbol Σ denotes an accumulator).

be applied it allows reduction of the FFT size by a factor of two compared to zero padding.

9.5.4 Spectral Shifting

If one of the frequency-domain signal representations (here, e.g., $\tilde{r}_k$) is circularly shifted prior to the multiplication, the correlation function, including a defined frequency offset, is obtained.

Referring to (9.12), the shifted spectra $\tilde{r}'_k$ is defined as

$$\tilde{r}_k'(p) = \mathrm{SHIFT}_p\{\tilde{r}_k\} = \tilde{r}_{k+p} \tag{9.19}$$

and the inverse FFT evaluates to

$$\begin{aligned} r_n'(p) &= \frac{1}{\Lambda}\sum_{k=0}^{\Lambda-1} \tilde{r}_k'(p)\exp\left\{2\pi i\frac{nk}{\Lambda}\right\} = \frac{1}{\Lambda}\sum_{k=0}^{\Lambda-1} \tilde{r}_{k+p}\exp\left\{2\pi i\frac{nk}{N}\right\} \\ &= \frac{1}{\Lambda}\sum_{k=0}^{\Lambda-1} \tilde{r}_k\exp\left\{2\pi i\frac{n(k-p)}{\Lambda}\right\} = \exp\left\{-2\pi i\frac{np}{\Lambda}\right\}\frac{1}{\Lambda}\sum_{k=0}^{\Lambda-1} \tilde{r}_k\exp\left\{2\pi i\frac{nk}{\Lambda}\right\} \\ &= \exp\left\{-2\pi i\frac{np}{\Lambda}\right\} r_n \end{aligned} \tag{9.20}$$

where Λ is the used IFFT size. The value of Λ depends on whether zero padding or circular correlation is used. Thus, the spectrally shifted signal (9.19) in the frequency domain corresponds to the signal multiplied with a complex sine/cosine carrier in the time domain.

The method of spectral shifting can be applied to a zero-padding or a circular-correlation correlator. Spectral shifting modifies both techniques, as shown by the block diagram in Figure 9.6.

The obtained correlation function finally evaluates to

$$\begin{aligned} h_m(p) &= \frac{1}{\Lambda}\mathrm{IFFT}\{\mathrm{FLIP}\{\mathrm{FFT}\{s_n\}\}\mathrm{SHIFT}_p\{\mathrm{FFT}\{r_n\}\}\} \\ &= \sum_{n=0}^{N-1} s_n r_{n+m}'(p) = \sum_{n=0}^{N-1} s_n r_{n+m}\exp\left\{-2\pi i\frac{(n+m)p}{\Lambda}\right\} \end{aligned} \tag{9.21}$$

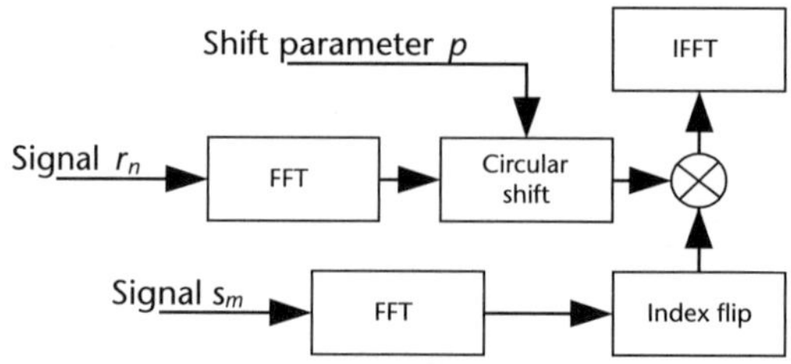

Figure 9.6 Frequency-domain correlator with spectral shifting.

It equals the normal correlation function, but a frequency term proportional to p / Λ is included. This term is usually used in signal acquisition to search the Doppler frequency. It should be emphasized that the frequency resolution, which can be obtained by spectral shifting, is inversely proportional to the FFT length. This is of special importance for the zero-padding case, in which a finer frequency resolution can be obtained if more zeroes are padded to the signals.

9.5.5 Limited-Size Inverse FFT

Sagiraju recently pointed out that the inverse-Fourier transform of (9.17) can be considerably simplified if the peak of the desired correlation function is known a priori to be confined within the index values λ ranging from 0 to $\Lambda_1 - 1$ (or equivalently within $-\Lambda_1/2 + 1 \ldots \Lambda_1/2$), where $|\Lambda_1| < N - M + 1$ [8]. Before the IFFT, the spectrum is block-averaged and the IFFT is applied on the block-averaged values, thereby reducing the computational burden.

More specifically, if it is possible to write

$$\Lambda = \Lambda_1 \Lambda_2 \qquad \Lambda_1, \Lambda_2 \in \mathbb{N} \tag{9.22}$$

then the procedure depicted in Figure 9.7 can be applied prior to the inverse-Fourier transform.

The product of both signal spectra $\tilde{h}_k$ defined in (9.12) is subdivided into blocks of size Λ_2 and the average value of each block is calculated. In this way, Λ_1 values are obtained, which are subsequently the inputs for the IFFT.

Referring to (9.12), this procedure is mathematically expressed as follows. First, the decimated product spectrum is defined as

$$\tilde{d}_\kappa = \frac{1}{\Lambda_2} \sum_{k=\kappa\Lambda_2}^{(\kappa+1)\Lambda_2 - 1} \tilde{h}_k \qquad \kappa = 0 \ldots \Lambda_1 - 1 \tag{9.23}$$

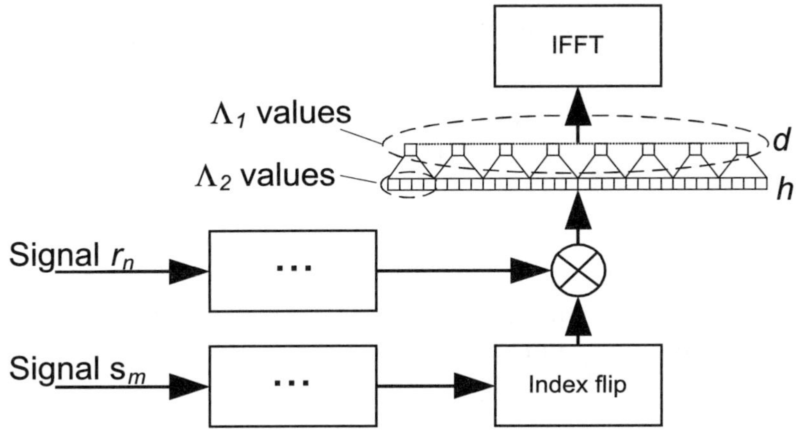

Figure 9.7 Reduction of the inverse-Fourier transform size. Note ". . ." stands for a forward FFT transform.

The inverse-Fourier transform of the block-averaged spectrum $\tilde{d}_\kappa$ evaluates to

$$\begin{aligned} d_m &= \frac{1}{\Lambda_1}\sum_{\kappa=0}^{\Lambda_1-1} \tilde{d}_\kappa \exp\left\{2\pi i \frac{m\kappa}{\Lambda_1}\right\} \\ &= \frac{1}{\Lambda_1}\sum_{\kappa=0}^{\Lambda_1-1} \frac{1}{\Lambda_2}\sum_{k=\kappa\Lambda_2}^{(\kappa+1)\Lambda_2-1} \tilde{h}_k \exp\left\{2\pi i \frac{m\kappa}{\Lambda_1}\right\} \\ &= \frac{1}{\Lambda}\sum_{\kappa=0}^{\Lambda_1-1}\sum_{k=\kappa\Lambda_2}^{(\kappa+1)\Lambda_2-1} \tilde{h}_k \exp\left\{2\pi i \frac{m\kappa(k)}{\Lambda_1}\right\} \\ &= \frac{1}{\Lambda}\sum_{k=0}^{\Lambda-1} \tilde{h}_k \exp\left\{2\pi i \frac{m\kappa(k)}{\Lambda_1}\right\} \end{aligned} \tag{9.24}$$

where the relationship of κ to k is given by

$$\kappa(k) = \left\lfloor \frac{k}{\Lambda_2} \right\rfloor \tag{9.25}$$

The symbol $\lfloor\ \rfloor$ denotes the largest integer smaller than the containing number (i.e., the **floor** operation). The inverse FFT can be further written as

$$\begin{aligned} d_m &= \frac{1}{\Lambda}\sum_{k=0}^{\Lambda-1} \tilde{h}_k \exp\left\{2\pi i m\left(\frac{\kappa(k)}{\Lambda_1} - \frac{k}{\Lambda} + \frac{k}{\Lambda}\right)\right\} \\ &= \frac{1}{\Lambda}\sum_{k=0}^{\Lambda-1} \tilde{h}_k^{block,m} \exp\left\{2\pi i \frac{mk}{\Lambda}\right\} \end{aligned} \tag{9.26}$$

with

$$\begin{aligned} \tilde{h}_k^{block,m} &= \tilde{h}_k \exp\left\{-\frac{2\pi i m}{\Lambda_1}\left(\frac{k}{\Lambda_2} - \kappa(k)\right)\right\} \\ &= \tilde{h}_k \exp\left\{-\frac{2\pi i m}{\Lambda_1}\operatorname{fract}\left(\frac{k}{\Lambda_2}\right)\right\} \end{aligned} \tag{9.27}$$

The function **fract** gives the fractional part (between zero and one) of a real number. In the above equations, the factors m/Λ_1 and **fract**(. . .) are both between zero and one. The difference of the block-averaged spectrum to the original spectrum depends on m and is a slowly varying complex sinusoidal given as

$$\tilde{b}_k^m = \exp\left\{-\frac{2\pi i m}{\Lambda_1}\operatorname{fract}\left(\frac{k}{\Lambda_2}\right)\right\} \tag{9.28}$$

Its inverse-Fourier transform depends on m and is formally defined as

$$
\begin{aligned}
b_n^m &= \mathrm{IFFT}_k\left\{\exp\left\{-\frac{2\pi i m}{\Lambda_1}\mathrm{fract}\left(\frac{k}{\Lambda_2}\right)\right\}\right\} \\
&\approx \delta_{n,0} \quad \text{for} \quad |\mathrm{m}| \ll \Lambda_1
\end{aligned}
\tag{9.29}
$$

The correlation value d_m obtained by block averaging is the mth value of the original correlation value h_m convoluted with b_n^m. For $m = 0$ or for small values of m, this is approximately a Kronecker delta, which implies that for small values of m, the correlation values obtained via block averaging equal the original correlation values. For larger values of m, deviations occur that cause a degradation of the obtained correlation peak. Sagiraju showed that this degradation is below 3 dB if $|m| < \Lambda_1/2$ and much less than 1 dB if $|m| < \Lambda_1/4$. Furthermore, a special weighting scheme can be employed, which is applied before the IFFT to reduce the degradation of the correlation peak [8]. As an example, Figure 9.8 shows the absolute value of b_n^m for typical settings that may occur for GPS C/A-code acquisition.

From Figure 9.8, one sees that, for higher values of m, the block-averaged correlation peak diminishes and receives contributions from side peaks of the original correlation function.

One could interpret the method more easily if the time domain and frequency domain are exchanged for the moment. Then one can think of the correlation function to be calculated as the spectrum and the spectrum as the waveform. The

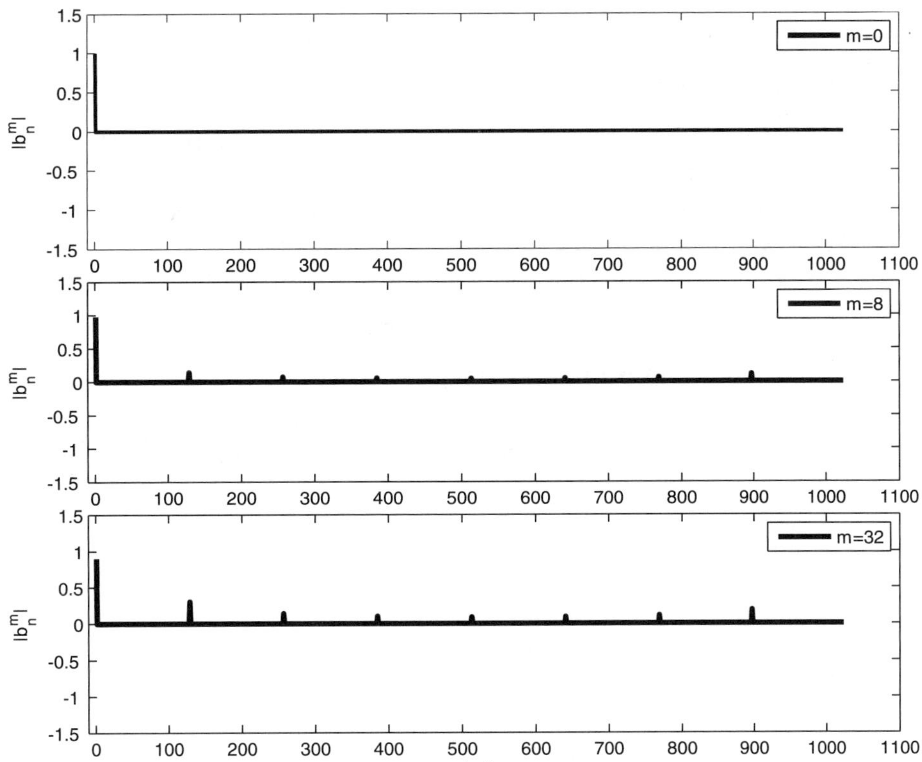

Figure 9.8 Exemplary values of b_n^m from (9.29) for $\Lambda_1 = 128$ and $\Lambda_2 = 8$.

"spectrum" has a sharp peak (the peak of the correlation function) that indicates a band-limited "waveform" and it is possible to filter the waveform and reduce its sample rate without significant information loss.

The advantage of this method is that the computational burden is reduced at least by a factor of Λ_2 and it may find its applications for assisted acquisition (if the code phase search range can be limited) and for frequency-domain tracking (where only correlation values around zero Doppler are needed).

9.5.6 Circular Correlation with Doppler Preprocessing

In case the signal r is periodic (as defined in Section 9.5.3.2) and if M is an integer multiple of the signal period, the circular correlation approach of Section 9.5.3.2 can be substantially simplified. This was first observed for the GPS C/A-code signal by Akopian [9] and processing load reductions of the order of the periodicity (i.e., maximum 20 in the case of the GPS C/A-code) can be obtained [10].

To apply this method, it is assumed that the value M can be written as

$$M = M_1 M_2 \quad M_1, M_2 \in \mathbb{N} \tag{9.30}$$

where M_1 is the period in samples of the signal r and M_2 is the number of periods contained in r. In the case of the GPS C/A-code signal, a period contains 1,023 PRN code chips; this period is represented (using resampling) by M_1 samples. An efficient choice for M_1 is 2^{11} or 2^{12} to allow FFT techniques. A typical GPS C/A-code value for M_2 is 8 to reduce the number of cases with a correlation over a data bit boundary. Time-domain index values are split up in the following way

$$n = n_2 M_1 + n_1 \quad n_2 \in \{0, \ldots, M/M_1 - 1\}, \quad n_1 \in \{0, \ldots, M_1 - 1\} \tag{9.31}$$

and frequency-domain index values as

$$k = k_2 + \frac{M}{M_1} k_1 \quad k_2 \in \{0, \ldots, M/M_1 - 1\}, \quad k_1 \in \{0, \ldots, M_1 - 1\} \tag{9.32}$$

An index of type k_1 is called *coarse* frequency index, an index of type k_2 is called *fine* frequency index.

With these settings, the periodicity of r is expressed as

$$r_{n_2 M_1 + n_1} = r_{n_1} \tag{9.33}$$

The Fourier transform of r simplifies to

$$\begin{aligned}
\tilde{r}_k = \tilde{r}_{k_1 M/M_1 + k_2} &= \sum_{n_1=0}^{M_1-1} \sum_{n_2=0}^{M/M_1-1} r_{n_2 M_1 + n_1} \exp\left\{-\frac{2\pi i}{M}(n_2 M_1 + n_1)\left(k_1 \frac{M}{M_1} + k_2\right)\right\} \\
&= \sum_{n_1=0}^{M_1-1} \sum_{n_2=0}^{M/M_1-1} r_{n_1} \exp\left\{-\frac{2\pi i}{M}\left(n_2 k_2 M_1 + n_1 k_1 \frac{M}{M_1} + n_1 k_2\right)\right\} \\
&= \frac{M}{M_1} \delta_{k_2,0} \sum_{n_1=0}^{M_1-1} r_{n_1} \exp\left\{-2\pi i\left(\frac{n_1 k_1}{M_1} + \frac{n_1 k_2}{M}\right)\right\} \\
&= \frac{M}{M_1} \delta_{k_2,0} \sum_{n_1=0}^{M_1-1} r_{n_1} \exp\left\{-2\pi i\left(\frac{n_1 k_1}{M_1}\right)\right\}
\end{aligned} \tag{9.34}$$

where (9.9) has been used to obtain the third line. As the whole expression of the third line is multiplied with a Kronecker delta, the value of k_2 can also be set to zero inside the exponential function. Overall, the equation states that the Fourier transform of a periodic signal has nonvanishing components only for a fine frequency index $k_2 = 0$. The spectral-shifted signal obtained by applying (9.19) to the periodic signal r is written as

$$\tilde{r}_{k+p} = \tilde{r}_{(k_1+p_1)M/M_1+k_2+p_2} = \frac{M}{M_1}\delta_{k_2+p_2,0}\sum_{n_1=0}^{M_1-1} r_{n_1}\exp\left\{-2\pi i\left(\frac{n_1(k_1+p_1)}{M_1}\right)\right\} \tag{9.35}$$

The spectral-shifted circular FFT of (9.21) evaluates for a periodic signal r to

$$\begin{aligned}
h_m(p) &= h_{m_2M_1+m_1}(p_1\,M/M_1+p_2)\\
&= \frac{1}{M}\sum_{k_1=0}^{M_1-1}\sum_{k_2=0}^{M/M_1-1}\tilde{s}_{-k_1M/M_1-k_2}\tilde{r}_{(k_1+p_1)M/M_1+k_2+p_2}\exp\left\{\frac{2\pi i}{M}(m_2M_1+m_1)\left(k_1\frac{M}{M_1}+k_2\right)\right\}
\end{aligned} \tag{9.36}$$

where only the first line of (9.21) has been rewritten using the index convention for periodic signals. Inserting (9.34) yields

$$\begin{aligned}
h_m(p) &= h_{m_2M_1+m_1}(p_1\,M/M_1+p_2)\\
&= \frac{1}{M}\sum_{k_1=0}^{M_1-1}\sum_{k_2=0}^{M/M_1-1}\tilde{s}_{-k_1M/M_1-k_2}\frac{M}{M_1}\delta_{k_2+p_2,0}\sum_{n_1=0}^{M_1-1} r_{n_1}\\
&\times\exp\left\{-2\pi i\left(\frac{n_1(k_1+p_1)}{M_1}-\frac{(m_2M_1+m_1)}{M}\left(k_1\frac{M}{M_1}+k_2\right)\right)\right\}\\
&= \frac{1}{M_1}\sum_{k_1=0}^{M_1-1}\tilde{s}_{-k_1M/M_1+p_2}\sum_{n_1=0}^{M_1-1} r_{n_1}\exp\left\{-2\pi i\left(\frac{n_1(k_1+p_1)}{M_1}-\frac{(m_2M_1+m_1)}{M}\left(k_1\frac{M}{M_1}-p_2\right)\right)\right\}\\
&= \frac{1}{M_1}\sum_{k_1=0}^{M_1-1}\tilde{s}_{-k_1M/M_1+p_2}\sum_{n_1=0}^{M_1-1} r_{n_1}\exp\left\{-2\pi i\left(\frac{n_1(k_1+p_1)}{M_1}-\frac{m_1k_1}{M_1}+\frac{p_2(m_2M_1+m_1)}{M}\right)\right\}
\end{aligned} \tag{9.37}$$

The Fourier transform of the signal s (which is, in general, not periodic with a periodicity of M_1) obtained with the index convention of periodic signals is given as

$$\begin{aligned}
\tilde{s}_{-k_1M/M_1+p_2} &= \sum_{a_1=0}^{M_1-1}\sum_{a_2=0}^{M/M_1-1} s_{a_2M_1+a_1}\exp\left\{-\frac{2\pi i}{M}(a_2M_1+a_1)\left(-k_1\frac{M}{M_1}+p_2\right)\right\}\\
&= \sum_{a_1=0}^{M_1-1}\sum_{a_2=0}^{M/M_1-1} s_{a_2M_1+a_1}\exp\left\{-2\pi i\left(-\frac{a_1k_1}{M_1}+\frac{(a_2M_1+a_1)p_2}{M}\right)\right\}
\end{aligned} \tag{9.38}$$

which is inserted into (9.37), yielding

$$
\begin{aligned}
h_m(p) &= h_{m_2 M_1 + m_1}(p_1 M/M_1 + p_2) \\
&= \frac{1}{M_1} \sum_{k_1=0}^{M_1-1} \sum_{a_1=0}^{M_1-1} \sum_{a_2=0}^{M/M_1-1} \sum_{n_1=0}^{M_1-1} s_{a_2 M_1 + a_1} r_{n_1} \\
&\times \exp\left\{-2\pi i \left(\frac{n_1(k_1+p_1)}{M_1} - \frac{m_1 k_1}{M_1} + \frac{p_2(m_2 M_1 + m_1)}{M} - \frac{a_1 k_1}{M_1} + \frac{(a_2 M_1 + a_1) p_2}{M} \right)\right\}
\end{aligned} \tag{9.39}
$$

Based on this formula, it is possible to rearrange the signal s in the form of a matrix with M_2 rows and M_1 columns. This matrix shall be denoted as $t(p_2)_{a1}$; its row index p_2 corresponds to a fine frequency and its column index a_1 corresponds to the sample index. The matrix $t(p_2)_{a1}$ is defined as

$$
t(p_2)_{a_1} = \sum_{a_2=0}^{M/M_1-1} s_{a_2 M_1 + a_1} \exp\left\{-2\pi i \left(\frac{(a_2 M_1 + a_1) p_2}{M} \right)\right\} \tag{9.40}
$$

The matrix $t(p_2)_{a1}$ averages the (in general, nonperiodic) signal s over all periods. Samples corresponding to the same index a_1 (but to different periods) are averaged. The averaging is repeated for all possible values of the fine-frequency index and the result is stored in a different row; for every row, different fine-frequency compensation factors are applied before averaging. The expression (9.39) can be further simplified as

$$
\begin{aligned}
h_m(p) &= h_{m_2 M_1 + m_1}(p_1 M/M_1 + p_2) \\
&= \frac{1}{M_1} \sum_{k_1=0}^{M_1-1} \sum_{a_1=0}^{M_1-1} \sum_{n_1=0}^{M_1-1} t(p_2)_{a_1} r_{n_1} \\
&\times \exp\left\{-2\pi i \left(\frac{n_1(k_1+p_1)}{M_1} - \frac{m_1 k_1}{M_1} + \frac{p_2(m_2 M_1 + m_1)}{M} - \frac{a_1 k_1}{M_1} \right)\right\} \\
&= \frac{1}{M_1} \sum_{k_1=0}^{M_1-1} \sum_{n_1=0}^{M_1-1} \tilde{t}(p_2)_{-k_1} r_{n_1} \exp\left\{-2\pi i \left(\frac{n_1(k_1+p_1)}{M_1} - \frac{m_1 k_1}{M_1} + \frac{p_2(m_2 M_1 + m_1)}{M} \right)\right\} \\
&= \frac{1}{M_1} \sum_{k_1=0}^{M_1-1} \tilde{t}(p_2)_{-k_1} \tilde{r}_{k_1+p_1} \exp\left\{-2\pi i \left(-\frac{m_1 k_1}{M_1} + \frac{p_2(m_2 M_1 + m_1)}{M} \right)\right\} \\
&= \frac{1}{M_1} \mathrm{IFFT}\{\mathrm{FLIP}\{\mathrm{FFT}\{t(p_2)\}\}\mathrm{SHIFT}_{p_1}\{\mathrm{FFT}\{r_{n_1}\}\}\} \exp\left\{-2\pi i \frac{p_2(m_2 M_1 + m_1)}{M}\right\}
\end{aligned} \tag{9.41}
$$

resulting in a frequency-domain correlation scheme depicted in Figure 9.9. The important point in this equation is that all FFTs are of the order M_1 instead of M.

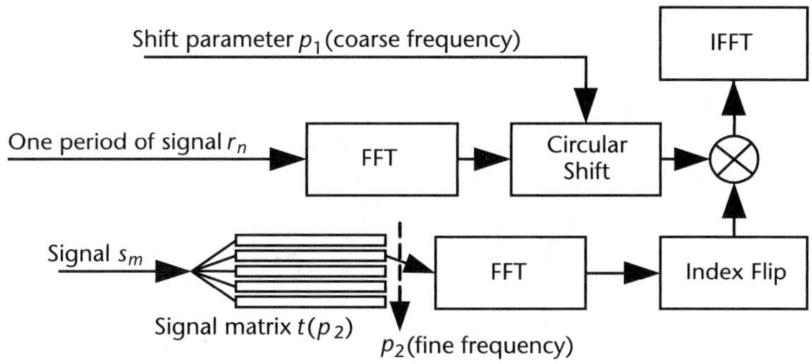

Figure 9.9 Circular correlation with Doppler preprocessing.

The matrix $t(p_2)_{a1}$ is independent of p_1. After the FFTs of the matrix $t(p_2)_{a1}$ and of one period of the signal r are computed, it suffices to calculate one IFFT of size M_1 to search one Doppler bin. Recall that, within the circular-correlation approach of Section 9.5.3.2, an IFFT of size M is required to search one Doppler bin.

A particular simple case occurs if a Doppler bin with a fine frequency of $p_2 = 0$ is searched. In this case, the computation of the $p_2 = 0$ row of the matrix $t(p_2)_{a1}$ simplifies to an averaging of samples.

9.5.7 Handling Secondary Codes

The algorithm presented in Section 9.5.6 proves to be a very efficient realization of the circular-correlation method. However, its prerequisite (a periodic signal r) is fulfilled only for a few GNSS signals. One example is the GPS C/A-code where the PRN code repeats itself 20 times within one data bit. Other examples are pilot signal components. However, most of the modernized GNSS signals (e.g., GPS L5 and most of the Galileo signals) use tiered codes, which are given by a convolution of a long primary and a short secondary code. This structure is being used for data as well as for pilot components. The combined tiered code can be considered as a very long code sequence, but it would be more convenient if the primary code alone could be used for acquisition. The combined tiered code is eventually too long to be acquired directly. In other words, although it is possible to consider a tiered-code pilot signal as a (long) periodic signal, it is more practical to consider it as a "nearly periodic" signal with the shorter period of the primary code.

This section shows how the method of Section 9.5.6 can be adapted to cope with tiered codes and to apply the techniques for a nearly periodic signal. The presented solution is not as efficient as that in Section 9.5.6 but can be considerably more efficient than a full-length code acquisition, especially in a case when the secondary code-phase search space can be limited (e.g., if aiding time information is available to the GNSS receiver).

To proceed we assume that r is a tiered code signal; that is, it can be written as

$$r_n = u_n v_n \tag{9.42}$$

where the primary code signal v fulfills the identity

$$u_{n_2 M_1 + n_1} = u_{n_1} \tag{9.43}$$

and for the secondary code signal v

$$v_{n_2 M_1 + n_1} = v_{n_2 M_1} \tag{9.44}$$

holds. The same index notation as in Section 9.5.6 has been assumed. The signal u in this section corresponds to the periodic signal r of Section 9.5.6.

Equation (9.42) can be rewritten as

$$r_n = u_n \frac{1}{M} \sum_{k=0}^{M-1} \tilde{v}_k \exp\left\{2\pi i \frac{nk}{M}\right\} \tag{9.45}$$

where $\tilde{v}_k$ is the Fourier transform of the secondary code signal v.

Inserting this equation into the formula for circular correlation with spectral shifting (9.21) yields

$$\begin{aligned} h_m(p) &= \sum_{n=0}^{M-1} s_n r_{n+m} \exp\left\{-2\pi i \frac{(n+m)p}{M}\right\} \\ &= \sum_{n=0}^{M-1} s_n u_{n+m} \frac{1}{M} \sum_{k=0}^{M-1} \tilde{v}_k \exp\left\{2\pi i \frac{(n+m)k}{M}\right\} \exp\left\{-2\pi i \frac{(n+m)p}{M}\right\} \\ &= \frac{1}{M} \sum_{n=0}^{M-1} s_n u_{n+m} \sum_{k=0}^{M-1} \tilde{v}_k \exp\left\{2\pi i \frac{(n+m)(k-p)}{M}\right\} \\ &= \frac{1}{M} \sum_{k=0}^{M-1} \tilde{v}_k \sum_{n=0}^{M-1} s_n u_{n+m} \exp\left\{2\pi i \frac{(n+m)(k-p)}{M}\right\} \\ &= \frac{1}{M} \sum_{k=0}^{M-1} \tilde{v}_k h_m^u(p-k) \end{aligned} \tag{9.46}$$

with

$$h_m^u(p) = \sum_{n=0}^{M-1} s_n u_{n+m} \exp\left\{-2\pi i \frac{(n+m)p}{M}\right\} \tag{9.47}$$

being the correlation function of the signal s with the strictly periodic signal u. To obtain $h_m{}^u(p)$, the method of Section 9.5.6 can be used. The correlation function including the secondary code $h_m(p)$ is the properly weighted sum of the correlation functions obtained for the periodic signal $h_m{}^u(p)$ at different frequency offsets. The evaluation of (9.46) is numerically not efficient because it contains a sum over the full spectral width (i.e., k ranges from 0 to $M - 1$). Further simplifications can be obtained if we analyze $\tilde{v}_k$. Using the index notation of Section 9.5.6, it evaluates to

$$\begin{aligned} \tilde{v}_k = \tilde{v}_{k_1 M/M_1 + k_2} &= \sum_{a_1=0}^{M_1-1} \sum_{a_2=0}^{M/M_1-1} v_{a_2 M_1} \exp\left\{-\frac{2\pi i}{M}(a_2 M_1 + a_1)\left(k_1 \frac{M}{M_1} + k_2\right)\right\} \\ &= \sum_{a_2=0}^{M/M_1-1} v_{a_2 M_1} \exp\left\{-\frac{2\pi i a_2 k_2 M_1}{M}\right\} \sum_{a_1=0}^{M_1-1} \exp\left\{-\frac{2\pi i a_1}{M}\left(k_1 \frac{M}{M_1} + k_2\right)\right\} \end{aligned} \tag{9.48}$$

Defining the factor $\beta(k)$ as

$$\beta(k) = \frac{1}{M_1} \sum_{a_1=0}^{M_1-1} \exp\left\{-\frac{2\pi i a_1 k}{M}\right\}$$
$$\Leftrightarrow \beta(k_1\,M/M_1 + k_2) = \frac{1}{M_1} \sum_{a_1=0}^{M_1-1} \exp\left\{-\frac{2\pi i a_1}{M}\left(k_1 \frac{M}{M_1} + k_2\right)\right\} \tag{9.49}$$

and simplifying it to

$$\beta(k) = \frac{1}{M_1} \sum_{a_1=0}^{M_1-1} \exp\left\{-\frac{2\pi i a_1 k}{M}\right\} = \frac{1}{M_1} \frac{1-\exp\left\{-\frac{2\pi i M_1 k}{M}\right\}}{1-\exp\left\{-\frac{2\pi i k}{M}\right\}} \tag{9.50}$$

yields

$$\tilde{v}_{k_1 M/M_1 + k_2} = \tilde{v}'_{k_2} M_1 \beta(k_1\,M/M_1 + k_2) \tag{9.51}$$

Here, $\tilde{v}'_{k_2}$ denotes the Fourier transform of the secondary code just accounting for the index k_2 (actually a short Fourier transform). Overall, the correlation function is then given as

$$h_m(p) = \frac{M_1}{M} \sum_{k=0}^{M-1} \tilde{v}'_{k_2(k)} \beta(k) h_m^u(p-k) \approx \frac{M_1}{M} \sum_{k=-BM_2}^{BM_2-1} \tilde{v}'_{k_2(k)} \beta(k) h_m^u(p-k) \tag{9.52}$$

where the summation can be limited by choosing, for example, $B = 1 \ldots 3$. This limitation is motivated beause the function $\beta(k)$ exhibits a narrow peak, as shown in Figure 9.10.

The extent to which the frequency bins can be further limited has to be determined on a case-by-case basis. A particular simple case occurs if the Fourier transform $\tilde{v}'_{k_2}$ exhibits a dominant peak. Then a narrow limitation of the summation range could be performed. Defining

$$w_{m_1}(p) = w_{m_1}(p_1\,M/M_1 + p_2) = \frac{1}{M_1} \sum_{l_1=0}^{M_1-1} \tilde{t}(p_2)_{-l_1} \tilde{r}_{l_1+p_1} \exp\left\{\frac{2\pi i m_1 l_1}{M_1}\right\} \tag{9.53}$$

yields

$$h_m(p) \approx \frac{1}{M} \sum_{k_1=-B}^{B-1} \sum_{k_2=0}^{M_2-1} \tilde{v}^{*}_{k_2} \beta(k_1\,M/M_1 + k_2) \times$$
$$\times w_{m_1}(p - k_1\,M/M_1 - k_2) \exp\left\{-2\pi i \frac{p_2(m_2 M_1 + m_1)}{M}\right\} \tag{9.54}$$

and a block diagram of the resulting acquisition scheme is shown in Figure 9.11. It generalizes the scheme of Figure 9.9 for arbitrary secondary codes.

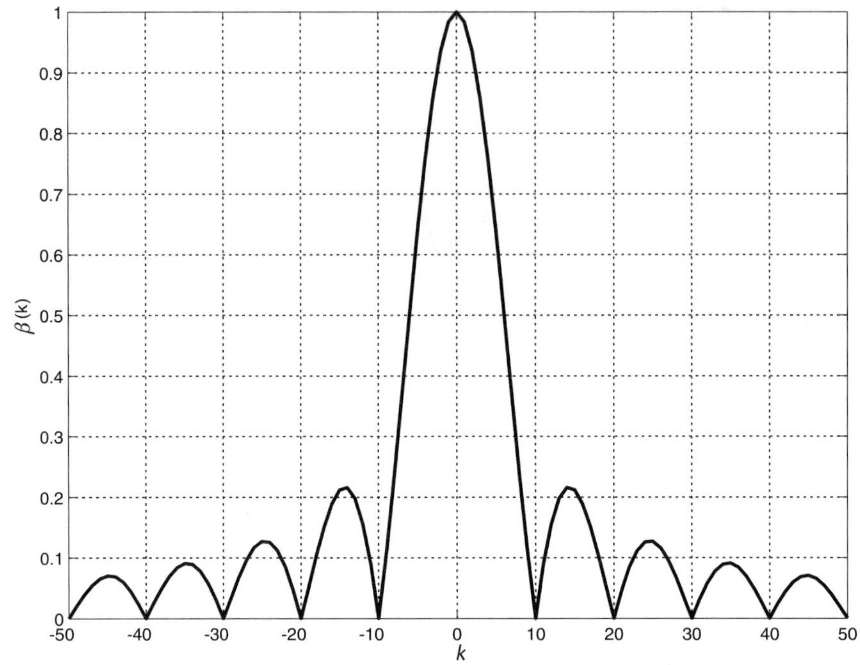

Figure 9.10 The magnitude of $\beta(k)$ plotted as a function of k for $M_1 = 10{,}230$, $M_2 = 10$.

If the secondary code is constant, then $\tilde{v}^*_{k_2}\beta(k_1\, M/M_1 + k_2) = \delta_{k_1,0}\delta_{k_2,0}$ and the summation over k extends only over one term (see Section 9.5.6). Another interesting case occurs if the v contains D periods of the secondary code. Then the summation over k includes only $2BM_2\,/\,D$ terms.

Overall, the presented method allows applying the Doppler-preprocessing method of Section 9.5.3.2 also for tiered-code signals. The Doppler-preprocessing method gives a processing-performance gain of about M_2 and avoids the computa-

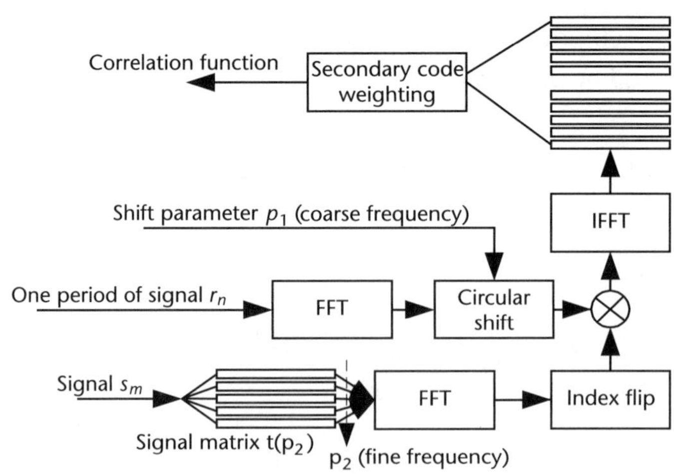

Figure 9.11 Circular correlation with Doppler preprocessing for tiered codes.

tion of full-length (i.e., M) FFTs. For a tiered code with a single period, a summation of the order of M_2 terms is required, thus the processing gain is partly lost. Nevertheless, the method allows for working with shorter FFTs, which is of special importance if only a limited portion of the correlation function needs to be computed (i.e., if the code-phase search range can be limited).

9.5.8 Asymptotic Computational Performance

The asymptotic performance of an acquisition algorithm is given as the number of required operations in the limit of an infinite long coherent integration time. We assume that a fixed Doppler range shall be tested; thus, the number of Doppler bins increases linearely with T_{coh}. Time-domain correlation shows a third-order performance, the application of the convolution theorem a second-order performance and the secondary code scheme a first-order performance. The asymptotic computational performance is summarized in Table 9.6.

It should be noted that the secondary code scheme of Figure 9.11 is the most efficient and realizes a coherent correlation over multiple periods of the secondary code. It requires a pilot signal.

9.5.9 Reported FFT Performance Values

For a successful implementation of a frequency-domain correlator, an efficient FFT implementation is required. The computational complexity (i.e., the number of mathematical operations, either multiplication or addition) of a complex FFT of size N is approximately given as [5]

$$\#Ops_{FFT} = 5N \log_2 N \tag{9.55}$$

More specifically, there are $N \log_2 N$ complex additions required, as well as $(N/2) \log_2 N - 3/2\, N + 2$ complex multiplications, see Section 3.5.1 of Diniz, da Silva, and Netto's book [7].

An FFT performance test for a complex FFT with 32-bit floating-point values and with 16-bit fixed-point values gave typical performance values of around 2.4–2.8 GOPS (giga operations per second; either floating-point or integer) for the test system described in Table 9.1 and as shown in Figure 9.12. As the FFT library, the well-performing Intel library was used [2]. An N = 8,192 FFT can thus be calculated within around 200 μs. The performance depends on the FFT size and a relative performance reduction can be seen if the FFT size exceeds a value of 2^{16} = 65,536. Presumably, for larger FFTs the CPU L2 cache (see Table 9.1) is too small. Interestingly the fixed-point 16-bit FFT performs with a similar speed as the 32-bit

Table 9.6 Asymtotic Computational Performance of Different Acquisition Schemes

Scheme	*Assumption*	*Number of Operations*
Time-domain correlation, Figure 9.3	None	$O(T_{coh}^3)$
Spectral shifting, Figure 9.6	Periodic reference signal	$O(T_{coh}^2 \log T_{coh})$
Secondary code, Figure 9.11	Periodic tiered code, periodic secondary code	$O(T_{coh})$

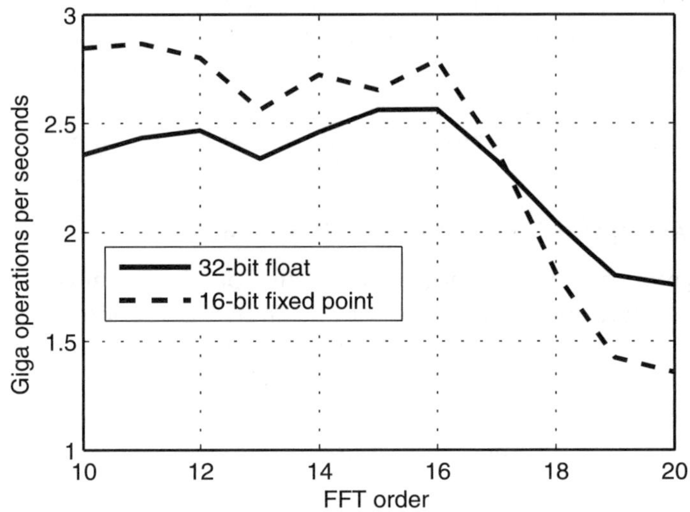

Figure 9.12 Computational (complex) FFT performance for the system of Table 9.1.

floating-point FFT. Fixed-point FFTs require well-scaled signals to avoid rounding effects and are thus less convenient to use. Fixed-point FFTs are, however, common in GNSS ASIC acquisition engines or on GNSS-software receivers running on DSPs/embedded processors.

Reported FFT performance values for other computer systems can be found, for example, in the publication by Frigo and Johnson [5]. A 3.0 GHz Intel Core 2 Duo, for example, can achieve, for a complex 32-bit floating-point FFT with N = 65,536, a performance of 12.5 GOPS, being five times faster than the test system of Table 9.1. For the Intel Core 2 Duo, the FFT performance significantly reduces if $N > 262{,}144$.

By contrast, a Sony Playstation 3 can perform FFTs up to N = 2,097,152 with a performance of around 15 GOPS (values for longer FFTs are not reported) using the FFTW library [5]. The Playstation utilizes the Cell Broadband Engine Architecture that consists of one power processor element (PPE) running the operating system and eight synergistic processor elements (SPE), which are autonomous processors storing their data and program in their own local memory [11]. The PPE and the SPE are interconnected by a high-speed element interconnect bus (EIB). The SPEs may run up to a speed of 4 GHz (laboratory results) and during each clock cycle they may perform four multiplications and four additions (vector dot operation or multiply-and-add command) on two times four 32-bit floating-point numbers. A peak performance of 256 GFLOPS per cell processor is the result. The Playstation utilizes only six of the eight SPEs and runs at a speed of 3.2 GHz resulting in a peak performance of 154 GFLOP. Comparing this value to the FFT performance listed in Frigo and Johnson's Web page shows that this library exploits about 10% of the maximum performance of a Playstation. A more efficient FFT implementation has been documented in the work by Chow, yielding 46.8 GFLOPS (GOPS) on an 8-SPE cell processor running at 3.2 GHz for an FFT length of N = 16,777,216 [12].

When comparing the Playstation and the Pentium M system, it should be noted that the Pentium M system performs well at low FFT sizes but breaks down at large

FFT sizes, and the Playstation behaves exactly the opposite. The reason lies in the fact that, for cache-based systems (like the Pentium processors), the FFT performance breaks down if the data does not fit into the cache, whereas for multicore systems like the cell, the data-transfer overhead is simply too large for short FFTs. The FFT characteristics of both systems are compared in Table 9.7. It should be noted that the values reported therein should not be taken literally, especially because reported FFT-performance values are highly implementation-dependent and because power consumption values have only indicative character as they are not based on real measurements [13].

Two things should, however, be noted. First, the cell gives significantly more GFLOPS per consumed electrical power, which may be explained by the fact that this system is newer (see also Section 9.7). Second, the ratio between the reported peak-performance values (which is, in both cases, derived from the performance of the multiply-and-add command) and the achievable FFT performance is, in both cases, around 0.2–0.3.

9.5.10 Discussion and Number of Correlators

In the following discussion, the frequency-domain and time-domain correlations shall be compared in terms of operations (i.e., multiplications and additions) required to achieve the same result. The figure *number of equivalent correlators* will be introduced. Frequency-domain techniques gain performance with an increasing number of correlators. Their use in signal acquisition is common practice in GNSS SDRs. For the following example, we assume a conventional frequency technique, namely zero padding, as described in Section 9.5.3.1.

If M correlation values of two real-valued signals of length N shall be calculated in the time domain, then for each sum a number of N multiplications and N additions have to be performed, resulting in

$$\#Ops_{time} = 2MN \tag{9.56}$$

operations. Here, we assume that the first signal is of a fixed length and it is not periodic. For each correlation value, the second signal is taken from a vector of size $M + N - 1$, each time shifted by one sample. We assume that $M \leq N$ and that $M + N - 1$ is an integer power of two.

Table 9.7 Comparison Between Two CPU Architectures for FFT

Parameter	*Test System, Table 9.1*	*Cell Broadband Engine*
Number of cores	1	1+8
CPU clock speed	2.26 GHz	3.2 GHz
Peak performance 32-bit float	9 GFLOP	205 GFLOPS (SPEs only)
Peak Performance 16-bit int	18 GOPS	—
CPU power consumption	5–27	50–80
CPU chip technology	90 nm	90 nm
Max. reported FFT performance	2.8 GOPS, 2.6 GFLOP	46.8 GFLOPS
FFT GFLOP/watt	0.096 GFLOP/W	0.61 GFLOP/W
FFT performance ratio (float)	0.29	0.23
Preferred FFT length	Short	Long

To achieve the same result with frequency-domain techniques, the following steps have to be performed (see Figure 9.4):

1. Zero pad the first signal to a length of $M + N - 1$;
2. Forward FFT of the real-valued zero-padded first signal;
3. Forward FFT of the real-valued second signal;
4. Frequency-domain multiplication;
5. Inverse FFT of the frequency-domain product.

A real-valued FFT needs half the operations as a complex valued FFT, thus (2), (3), and (5) need all together

$$\#Ops_{FFT,1} = 10(M + N - 1)\log_2(M + N - 1) \tag{9.57}$$

operations. The spectral multiplication needs

$$\#Ops_{FFT,2} = 6(M + N - 1) \tag{9.58}$$

operations, yielding an overall number of operations of

$$\#Ops_{FFT} = \#Ops_{FFT,1} + \#Ops_{FFT,2} \approx (M + N)(10\log_2(M + N) + 6) \tag{9.59}$$

The last expression shows that for large N, the number of operations increases with N **log** N, whereas in the time domain the increase is quadratic (assuming M proportional to N).

In the following example, the *break-even* number of correlators M_{be} shall be determined, for which both the time domain and frequency domain require the same amount of operations. It is obtained by solving the equation

$$\begin{aligned} & 2M_{be}N = (M_{be} + N)(10\log_2(M_{be} + N) + 6) \approx 10N\log_2 N \\ & \Leftrightarrow M_{be} \approx 5\log_2 N \end{aligned} \tag{9.60}$$

where we assume that the number of correlators is much smaller than the number of samples. For example, if a GPS C/A-code signal with a coherent integration time of 1 ms and a sample rate of 20.48 MHz is considered, then frequency-domain techniques become more efficient if more than 71 correlators need to be calculated, which corresponds to a code-phase range of 3.5 chips.

The ratio between the time span represented by N samples divided by the CPU time of the FFT implementation to complete the correlation defines the correlation efficiency of an FFT-based correlator algorithm. The effective number of correlators #*Corr* is defined as the product of correlation efficiency multiplied with N. The effective number of correlators corresponds to the number of hardware correlators that would give the exact same result within the same time. Note that hardware correlators are understood to work intrinsically in real time (and not faster). The definition of #*Corr* assumes that all available correlation values from the frequency domain correlation are exploited (i.e., $M = N$ is assumed for the most effective use of the frequency-domain correlation). The effective number of correlators is thus

$$\#Corr = N\frac{T_{coh}}{T_{CPU}} = \frac{NT_{coh}f_{OPS}}{2N(10\log_2(2N)+6)} = \frac{T_{coh}f_{OPS}}{2(10\log_2 N+16)} \tag{9.61}$$

Here T_{coh} denotes the length of the signal in seconds (usually the coherent integration time given as the number of samples N divided by the sample rate). T_{CPU} is the time of the CPU to perform the computation in seconds, f_{OPS} is the number of FFT operations the CPU can perform within one second. The longer T_{coh} is, the more efficient the FFT. For example, for a 20-ms signal, $N = 2^{15}$, and $f_{OPS} = 2.5$ GOPS, the effective number of correlators evaluates to 150,602. If Doppler preprocessing can be applied (see Section 9.5.6), the number of effective correlators further increases (e.g., by a factor of approximately 10–20 for the case of the GPS C/A-code).

9.6 Reality Check for Signal Tracking

Based on the clock-tick measurements of the preceding sections, an estimate on how many channels the system of Table 9.1 can track in real time shall be presented. Acquisition is not considered here. It is assumed that the system runs a GPS L1 C/A-code receiver using a sample rate of 23.104 MHz with a 1.5-bit sign/magnitude representation. Let N denote the number of channels in the following Table 9.8, which gives an overview of how many clock ticks are required to process one IF sample.

Each sample has to be converted to 16 bits, then, for each channel, four reference signals have to be generated (i.e., the I/Q component of the D- and P-correlator signals) and then those signals are correlated with the received sample. It is assumed that all data is read from the main memory and not from the CPU caches.

The CPU clock frequency shown in Table 9.1 is 2.26 GHz and thus, at maximum, 97.8 clock ticks are available to process one IF sample. If more clock ticks would be consumed, then the software does not run in real time anymore, simply because it gets too slow. Solving the equation for the total number of clock ticks of Table 9.8 for N yields a value of $N = 3.07$, which would limit the laptop to track only three channels in real time.

It is clear that a direct implementation of a GNSS SDR on the laptop would not be feasible, even though optimized assembler routines are used. Optimization at the algorithmic level is therefore required.

From Table 9.8 it is evident that most time is spent in generating the reference signals. As mentioned above, a possible solution to reduce the efforts for reference signal generation is to reuse already-generated reference signals for multiple

Table 9.8 Clock Ticks Per Sample for a GPS L1 Receiver

Operation	*Clock Ticks*	*Multiplicity Factor*	*Sum*
Bit conversion	5.48	1	5.48
Reference-signal generation	5.64	$4N$	$22.56N$
Correlation	1.88	$4N$	$7.52N$
Total of clock ticks			$5.48 + 30.08N$

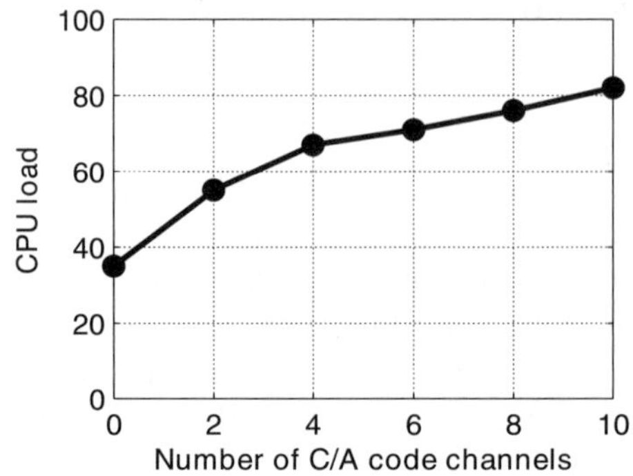

Figure 9.13 CPU load of the system in Table 9.1 for a different number of channels running ipexSR.

correlations. This approach was used in the software receiver, ipexSR [14]. The reuse implies that the correlation point does not smoothly follow the true signal parameters. However, Section 4.3.2.10 demonstrated that the correlation point is allowed to vary within the linearity region centered on the true code-phase and Doppler values. This technique requires assumptions on the characteristics of the Doppler frequency and code-rate variations.

Figure 9.13 shows the CPU load actually used by the ipex software receiver when running a different number of C/A-code channels in real time. The CPU-load measurement was taken from the Windows Task Manager and corresponds to the ipexSR process value (and not to the total CPU load read from the task manager, which also would include other processes running on the same computer).

It is observed that the "idle" load of the receiver is about 35%, which is needed to keep the USB data transfer alive, do the necessary preprocessing of the incoming signals (e.g., bit conversion), and to update the GUI. This value also includes the navigation solution or writing output files. When channels are activated, the CPU load increases but the rate of increase decreases with the number of channels. In fact, for 4 to 10 channels we observe an increase of CPU load of 2.5% per channel, which equals to additional 2.4 CPU clock ticks per channel and IF sample (note 100% corresponds to 97.8 clock ticks per IF sample). Table 9.4 shows that 4.69 clock ticks are necessary to correlate 8 + 8 16-bit values if the signals are in the CPU cache. Equivalently, 2.35 clock ticks are needed to correlate one IF sample with four reference signals if the reference signals need not to be generated and are stored within the CPU cache. The conclusion is that the ipexSR shows a nice *channel scaling* behavior in the sense that, for a larger number of channels, the software runs at the theoretical achievable limit of available processor power with respect to the presented algorithmic implementation (i.e., IF samples and reference signals are available within the CPU caches when it comes to correlation).

9.7 Power Consumption

The test system of Table 9.1 was equipped with an 80-Wh rechargeable battery. The CPU, a Pentium M 780, has a specified thermal design power of 27W. Most of the

CPU processing power is needed to run the receiver core algorithms discussed in this chapter. In a typical high-end GPS C/A-code configuration, tracking all satellites in view, the average CPU load is about 80%. The L1 USB front end consumes less than 2.5W, yielding an (theoretical) overall receiver power consumption of below 30W, ignoring the power consumption of hard disks, screens, and other peripherals.

A real test showed that the run time of the system is around two hours (display switched off), depending on the chosen sample rate as listed in Table 9.9. One recognizes that reducing the sample rate by a factor of four increases the runtime by 35%. This indicates that the receiver power consumption is less affected by the actual processing done and that most of the power is needed to keep the system alive (e.g., to power the hard disk, mainboard, and chip set). Furthermore, the CPU cannot reduce its clock rate (to save power) even when the low sample rate is used because it is constantly working. In the high-sample-rate mode, the system seems to consume a power of 42W.

The runtime of the presented system is not particularly high. Although the system is not optimized for GNSS receiver usage (e.g., the hard disk is always powered on) it might be interesting to know when technology evolution can allow using a laptop as a GNSS receiver replacement. To be more specific, we would like to estimate when this software allows 10 hrs of continuous operation without external power having a performance of a geodetic GPS L1/L2 receiver.

The answer can only be found on a qualitative basis founded on Moore's law. It predicts that the number of gates doubles every 18 months. On the other hand, it is observed that dissipated power by CPUs is approximately constant. Thus, it is expected that for a given algorithm, the required power halves every 18 months. As we need an increase in power of 526% (= 600 minutes / 114 minutes), we estimate that this will be available in 3.6 years. Furthermore, increasing the number of frequency bands from L1 to L2 requires doubling the processing power, which needs 1.5 additional years. Thus, we could estimate that in 2011, laptop systems could be available that would allow for running a L1/L2 GNSS SDR continuously for 10 hours. Although these arguments are very heuristic, they indicate that a high-end portable GNSS SDR could be realized by COTS components in the not-so-distant future. The discussion also indicates that a laptop may not be the best hardware platform for a GNSS SDR in terms of power consumption. A dedicated hardware platform development would bring a significant increase in power efficiency.

For example, an embedded system like that described in the work by Toradex delivers about 2.15 GOPS with a power consumption of around 800 mW [15]. The board is based on an Intel PXA320 processor running at 806 MHz. A multiply-and-add command accumulating 2 × 4 16-bit integers (thus performing 8 operations) needs 3 clock ticks to be executed [16]. The ratio between peak performance and

Table 9.9 Run Time of Test System Table 9.1 Without External Power

	Run Time
Sample rate = 5.776 MHz	154 min
Sample rate = 23.104 MHz	114 min

power consumption is about six times better than for the test system of Table 9.1. The extent to which this can be attributed to the processor architecture or to the general system design is, however, unclear. The cell processor alone (without other components like chip set, memory, and so forth) has a similar ratio between peak performance and power consumption like the embedded system (see Table 9.7) but works with 32-bit floating point numbers.

9.8 Discussion

The presented algorithms form the core of a GNSS SDR. For signal tracking, efficient implementations have been sought and, for acquisition, frequency-domain techniques are used. Tracking and acquisition behave differently. Whereas acquisition is performed on-demand, tracking algorithms run continuously and define the ultimate real-time behavior of a GNSS SDR.

From the discussion of Section 9.6, it is obvious that the presented reference assembler implementation of the core (tracking) algorithms is not efficient enough to allow a direct coding of a multichannel high-end SDR under the given premises (like multibit signal representation and a high sample rate of, e.g., 23.104 MHz). Although the implementation itself is optimized and uses only a few assembler instructions, signal generation and bit conversion execute rather slowly on the given CPU architecture, mostly because the required assembler command (table lookup) is not well supported by the CPU. By contrast, correlation and FFT execute sufficiently fast. Modern general-purpose processors partly support the implementation of those algorithms because they provide special vector instructions, and due to the multicore architecture they natively facilitate processing of GNSS signals in parallel. So far, there is no support for instructions working with a small (e.g., 2–4) number of bits, which would be particularly advantageous for processing GNSS-signal samples. Indeed, the processors either support at minimum 8- or 16-bit instructions causing mainly a memory-bus bandwidth problem. Some algorithms, however, (e.g., vector multiplication) can be efficiently implemented with 1-bit data. Signal-processing algorithms benefit only partly from a large number of bits: correlation benefits to some extent (e.g., interference cancellation), but FFT algorithms do benefit, allowing longer FFT lengths.

The most efficient way to speed up a GNSS SDR to allow real-time operation is higher-level optimization. Two key issues have been identified. First, massive parallel correlators should be implemented via FFT techniques. Second, reuse of reference signals should be done to avoid reference-signal generation via resampling. Both techniques in unison are the key ingredients for a real-time multifrequency GNSS receiver.

It is essential that a GNSS SDR respects the CPU memory architecture and tries to exploit the various CPU caches. A GNSS SDR processes a large amount of data and reading those data from main memory would cause a drastic performance reduction. Working with a 1-bit or 2-bit signal representation, as proposed by Ledvina, definitely reduces the memory bus load and also allows efficient signal correlation [4]. The main drawback of this method is, however, the cumbersome access of signal samples because the CPU does not support direct bit access and, of course, the increased implementation losses of the receiver due to the low number of bits.

A software radio requires a suitable hardware platform. Modern embedded (or UMPC) processors with high computational power only need a few watts in operation. This advantageous benefit is lost if the software radio runs a conventional PC whose other components (graphic chip, hard disk, or display) need much more power and are not needed for a software radio. For example, a laptop like that described in Table 9.1 is definitely not the best platform to run a GNSS SDR. An embedded system like [15], the proposed RTK system of Chapter 10, or a system built around the cell processor yields around six times more operations per watt than the laptop of Table 9.1.

References

[1] SiSoftware Ltd., "SiSoftware Sandra Lite (Win32 x86), 2008.1.12.34," http://www.sisoftware.eu/, 2007.

[2] Intel Corp., "Intel Integrated Performance Primitives v5.0 for Windows on Intel Pentium Processors," http://www.intel.com, 2007.

[3] Fog, A., "Instruction Tables, Lists of Instruction Latencies, Throughputs and Microoperation Breakdowns for Intel and AMD CPU's," http://www.agner.org/optimize, 2008.

[4] Ledvina, B., et al., "Performance Tests of a 12-Channel Real-Time GPS L1 Software Receiver," *Proc. 16th Int. Technical Meeting of the Satellite Division of the Institute of Navigation (ION-GPS) 2003*, Portland, September 9–12, 2003, pp. 679–688.

[5] Frigo, M., and S. G. Johnson, "FFTW: Fastest Fourier Transform in the West," http://www.fftw.org, 2007.

[6] Cooley, J. W., and J. W. Tukey, "An Algorithm for the Machine Computation of Complex Fourier Series," *Math. of Comput.*, Vol. 19, No. 1965, pp. 297–301.

[7] Diniz, P. S. R., E. A. B. da Silva, and S. L. Netto, *Digital Signal Processing, System Analysis and Design*, Cambridge, U.K.: Cambridge University Press, 2002.

[8] Sagiraju, P. K., S. Agaian, and D. Akopian, "Reduced Complexity Acquisition of GPS Signals for Software Embedded Applications," *IEE Proc. Radar, Sonar, and Navigation*, Vol. 153, No. 1, 2006, pp. 69–78.

[9] Akopian, D., "A Fast Satellite Acquisition Method," *Proc. 14th Int. Technical Meeting of the Satellite Division of the Institute of Navigation (ION-GPS) 2001*, Salt Lake City, UT, September 11–14, 2001, pp. 2871–2881.

[10] Sagiraju, P. K., G. V. S. Raju, and D. Akopian, "Fast Acquisition Implementation for High Sensitivity Global Positioning Systems Receivers Based on Joint and Reduced Space Search," *IET Radar, Sonar & Navigation*, Vol. 2, No. 5, 2008, pp. 376–387.

[11] Hofstee, H. P., "Introduction to the Cell Broadband Engine," http://www-05.ibm.com/e-business/uk/innovation/special/satin/nonflash/pdf/2053_IBM_CellIntro.pdf, 2005.

[12] Chow, A. C., G. C. Fussum, and D. A. Brokenshire, "A Programming Example: Large FFT on the Cell Broadband Engine," http://www-01.ibm.com/chips/techlib/techlib.nsf/techdocs/0AA2394A505EF0FB872570AB005BF0F1/$file/GSPx_FFT_paper_legal_0115.pdf, 2005.

[13] Wang, D.T., "ISSCC 2005: The CELL Microprocessor," http://www.realworldtech.com/page.cfm?ArticleID=RWT021005084318&p=2, 2005.

[14] Anghileri, M., et al., "Performance Evaluation of a Multi-Frequency GPS/Galileo/SBAS Software Receiver," *Proc. 20th Int. Technical Meeting of the Satellite Division of the Institute of Navigation (ION-GNSS) 2007*, Fort Worth, September 25–28, 2007, pp. 2749–2761.

[15] Toradex AG, "Colibri XScale(R) PXA320 Datasheet," Rev. 1.3, http://www.toradex.com/downloads/Colibri_PXA320_Datasheet_Rev_1.3.pdf, 2007.

[16] Intel Corp., "Intel XScale® Technology: Intel® Wireless MMX™ 2 Coprocessor, Programmers Reference Manual," Rev. 1.5, http://download.intel.com/design/intelxscale/31451001.pdf, 2006.

CHAPTER 10

GNSS SDR RTK System Concept

In this chapter, an innovative receiver concept will be outlined. It exploits the inherent software-receiver advantages to realize a *real-time kinematic* (RTK) positioning system. RTK positioning relies on a roving receiver that is aided by a reference station providing correction data to the rover in real time. It aims at a 2-cm accuracy level, required to perform geodetic measurements, in forest areas.

The proposed RTK concept exploits the following advantages of software radio technology:

- Incorporation of aiding data at correlator level (plus conventional code and phase pseudorange correction), thereby increasing carrier tracking stability significantly;
- Use of a high performance computing platform to realize high sensitivity acquisition, sophisticated multipath mitigation, and carrier-phase processing algorithms;
- Optional incorporation of low-cost pseudolites, whose signals penetrate areas of high signal attenuation.

The proposed RTK system should thereby outperform available hardware-based receivers, where it is difficult to realize the above-mentioned techniques and of which the author has no knowledge at this time. The development costs for this system are significantly reduced by using SDR technology, which is important to this niche-market application.

The RTK system relies on two main technology enablers (a low power platform and a cost-efficient high data-rate communication link). In this chapter, these enablers and the system design will be presented and the key algorithms will be given.

10.1 Technology Enablers

The ever-increasing mobile Internet applications drive the development of low-power CPUs, increase the data rate of mobile phone networks, and cut data transfer costs. A summary of recent developments will be given later.

10.1.1 Ultra-Mobile PCs

Since 2007, personal computers with very small form factors and low power consumption have become increasingly popular. These PCs are termed *ultra-mobile PCs*, *mobile Internet devices*, or *netbooks*. The application domain of those devices

Table 10.1 Thermal Design Power of UMPC Components

Unit	*Solution*	*Thermal Design Power*
CPU	Intel Atom Z530	2.2W
Chipset	Intel System Controller Hub US15W	2.3W
Memory (DDR2)	—	< 1W
Hard disk (mechanical)	—	< 3W
Hard disk (solid state drive)	DuraDrive AT SATA 35	< 1W

ranges from embedded solutions or help provisions for inexpensive laptops for developing countries or for educational purposes.

These devices rely on recently developed components minimizing the power consumption. An exemplary selection of devices is listed in Table 10.1 [1, 2]. Integrating these components into a single module yields a computing platform with 50%–100% computational performance as in the laptop test system of Table 9.1 [3]. However, the power consumption is drastically reduced. It should, however, be noted that an integrated module like the one by LiPPERT supports only a simple graphical display.

Both CPUs—the Pentium M 780 [4] and the Atom Z530 [1]—have a similar clock rate (2.26 MHz/1.6 MHz) and support the same assembler instructions (the Atom additionally supports SSE3 instructions). Both are single-core CPUs, but the Atom supports hyper-threading, thus simulating two cores. The Pentium M was introduced in March 2003, the Atom Z530 in April 2008. Their ratio of the thermal design power is 27W/2.2W = 12.3.

On the other hand, Moore's law for power consumption, as discussed in Section 9.7, predicts a power ratio of $2^{61/18} = 10.5$, which agrees with the observation. Here, 61 is the number of months that passed between the market introduction of the two CPUs and Moore's law predicts halving of the power consumption every 18 months.

Overall, a clear tendency toward low-power-consumption computing platforms is observed, which is well-suited for the implementation of a GNSS SDR.

10.1.2 Cost-Effective High-Rate Data Links

The data link between the reference station (network) and the roving receiver is a critical element in a RTK system. A potential failure and/or latency may significantly disturb the system performance. In the late eighties, when the RTCM 2.x standards were introduced, the maximum possible data rate enforced considerable constraints on the amount, type, and format of the data to be exchanged between the reference and the rover receiver [5]. For example, radio modems or, later in the early 1990s, GSM-based modems were used where the data rate was 9,600 bit/s. It should also be mentioned that transferring the RTK data stream involved significant costs for the user.

Due to the still-increasing popularity of mobile Internet applications, strong efforts have been put into the development of modernized mobile-phone networks. Currently, mobile-phone terminals allow much higher data rates. A GSM-based EDGE class-10 phone allows a downlink rate of 220 kbit/s and an uplink rate of

110 kbit/s. The availability of the EDGE-based data transfer is high. For example, T-Mobile offers EDGE over most of Germany and other providers upgraded their GSM-based networks with EDGE in 2008. Higher data rates can be achieved with UMTS (downlink 384 kbit/s, uplink 64 kbit/s) or HSDPA/HSUPA (downlink 7.2 Mbit/s, uplink 3.6 Mbit/s), but the availability of these services, especially in rural areas, is questionable. Also, other services like WiMAX (IEEE 802.16) will offer high data rates in the future. A second important observation is the introduction of flat rates for data services. A monthly fixed price (e.g., 10 EUR per month in the German O2 network) covers all data transfers up to a volume of 200 MByte in 2008.

Overall, a clear tendency is observed that data-link constraints are, at least by a factor of 10, less stringent than 10 years ago; this additional available bandwidth should be used to transfer more and better correction data between the reference station and the rover.

10.2 System Overview

This section gives an overview of the proposed system installation, target applications, used signals, and a possible test bed.

10.2.1 Setup

The core elements of the system are the reference and the rover GNSS receiver. The antenna of the reference receiver is installed at a fixed location with known coordinates. The roving receiver moves dynamically in the vicinity of the reference receiver. The rover and reference receiver are connected by a data link. Both receivers process signals received from the same GNSS satellites. A schematic of the proposed system is shown in Figure 10.1.

The setup is basically identical to a conventional differential GPS system with one reference station. The important difference lies in the fact that the rover receiver

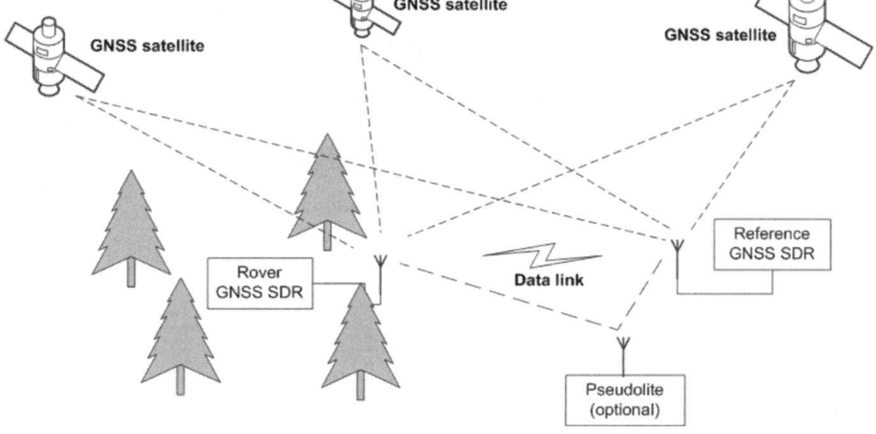

Figure 10.1 RTK system overview.

Table 10.2 Target System Key Parameters

Parameter	*Value Rover*	*Value Ref. Sta.*
Time-to-first fix	10 sec	—
Forest penetration loss (GNSS)	< 25 dB	0 dB
Forest penetration loss (pseudolite)	< 50 dB	< 25 dB
Minimum C/N_0	20 dBHz	45 dBHz
Horizontal accuracy	2 cm (2DRMS)	—
Maximum baseline length	500m	—
Maximum user speed	2 m/s	—

is allowed to work in a degraded signal environment, for example within a forest. When a conventional RTK system is not able to process GNSS carrier-phase measurements anymore, the proposed system continues to work.

The test system can be enhanced by placing pseudolites around the area of operation. The carrier phase of the pseudolite signals is processed simultaneously and is identical to the GNSS signals and further stabilizes the position solution. The pseudolite signals have the advantage that the broadcast signal power can be adjusted to overcome high signal-attenuation effects. On the contrary, the pseudolite signals are greatly affected by the multipath, because the pseudolite signals travel near the Earth's surface and a building or trees can cause multiple reflections. Therefore, pseudolite-code pseudorange measurements are of reduced precision, but the carrier phase is still useful (carrier-phase multipath errors are maximally one quarter of the carrier wavelength). The key performance parameters are summarized in Table 10.2.

10.2.2 Sample Applications

The proposed system targets precision surveying applications, which are carried out today using terrestrial means. This includes applications like cadastral surveying, identification, and monitoring of tree growth or tracking of people who are moving under canopies. Typical accuracy requirements are listed in Table 10.3 and point out that the use of carrier-phase measurements is inevitable because code measurements affected by a multipath are typically distorted at the decimeter level, making it impossible to achieve submeter positioning accuracy.

10.2.3 Test Installation and Used Signals

We chose as a test area the Galileo Test Bed GATE [6], which is a few-square-kilometer test range at the southern border of Germany equipped with six pseudolites that are normally used to simulate Galileo satellites [7]. The pseudolites can be operated in a base mode, where they act as normal pseudolites. Within this area, GPS, (true) Galileo, and GLONASS signals can be received as well, rendering a nice range to validate the proposed RTK system.

Table 10.3 Required Accuracy for Different Applications

Application	*Horizontal 2DRMS*
Cadastral surveying	2 cm
Tree location	20 cm
People tracking	1m

Table 10.4 Signal Parameters

Parameter	*Value*
GNSS signal	GPS C/A
Pseudolite signal	Galileo E1 OS-like (optionally pulsed)
Pseudolite PRN chip length (data/pilot)	4,092/4,092
Pseudolite chip rate	1.023 Mchip/s

For this discussion, we assume that the system will process GPS C/A-code measurements and the pseudolites will broadcast a BOC(1,1) Galileo E1 OS-like signal. The extension of the proposed system to other and more frequencies is possible, but will not be discussed here. Signal parameters are summarized in Table 10.4. The use of only one signal carrier frequency implies that the baseline length (i.e., the distance between reference receiver and rover) is limited to a few kilometers, because ionospheric effects are not corrected. Nevertheless, the reader shall be reminded that this is a niche-market application, providing the user with high accuracy in an area that, so far, is not covered. Therefore, it is expected that the limited range is acceptable. Additionally, it should be mentioned that the range of a pseudolite signal is also limited, because the pseudolite signals require a free line of sight (at least with respect to topographic variations).

10.3 Key Algorithms and Components

The proposed system concept follows a conventional RTK system design, which is not be reviewed here. Instead the reader is referred to well-known GNSS textbooks [8–10]. The roving system determines its position based on double-difference carrier-phase measurements and fixes the carrier-phase integer ambiguities based on integer LSQ adjustments, also using code pseudorange measurements. It is widely recognized that the ability of an RTK system to correctly resolve double-difference carrier-phase ambiguities is essential for centimeter-accurate positioning. The ability of the system to correctly resolve the carrier-phase ambiguities strongly depends on the number of received and processed signals [11]. Therefore, it is required that both receivers track all in-view signals and continuously output code-pseudorange, Doppler, and carrier-phase measurements.

In the following section, the innovative aspects of the proposed system are described; these are required to achieve the high navigational accuracy in a degraded signal environment. They are based on SDR technology and include a high-sensitivity acquisition engine, assisted tracking, and low-cost pseudolites.

10.4 High-Sensitivity Acquisition Engine

The roving system works in a degraded signal environment; the GNSS signals are especially attenuated by the canopy. Therefore, the roving receiver requires a high-sensitivity acquisition engine. Conventional RTK GNSS receivers are typically not equipped with such an engine because conventional carrier-phase measurements in a degraded signal environment have many cycle slips, or the carrier phase cannot be tracked at all. Therefore, there is little need to acquire low-power signals. In contrast, most mass-market GPS receivers are equipped with a high-sensitivity engine, which is based on massive parallel correlators or on frequency-domain techniques [12].

For the proposed RTK system, a special acquisition unit is designed. It is based on several key components:

- A relatively stable oscillator (e.g., a TCXO) used in both receivers, whose short-term stability can be predicted with an accuracy of several hertz;
- Exploiting the knowledge of an approximate initial position (rover distance < 500m from the reference station);
- Provision of aiding data in the form of satellite ephemeris, clock corrections, and broadcast navigation data bits;
- Time synchronization of a rover and reference station with an accuracy of better than 30 ms using the data link via, for example, the NTP protocol [13].

10.4.1 Doppler Search Space

A stable reference clock allows limitation of the Doppler search range during the signal-acquisition phase. Figure 10.2 shows the clock drift of a TCXO measured over 10 days [14]. This oscillator is integrated in a USB front end of a software receiver and the drift is computed by using GPS measurements on a fixed-site location [15]. The drift is with respect to the GPS time scale. The TCXO has a nominal frequency tolerance of ±2,000 ppb and a frequency stability over temperature (–30°C to –85°C) of ±500 ppb. In fact, the strong variations shown in Figure 10.2 are caused by temperature variations, caused by heating the room where the front end was located.

Figure 10.2 demonstrates that the frequency is stable within a few hertz accuracy when no temperature changes occur. Temperature changes may cause a frequency change of ±100 ppb/°C, but are unlikely to occur [14]. In Figure 10.2, a temperature change of 10°C causes a frequency change of 60 ppb, corresponding to a Doppler shift of 94.52 Hz at the GPS L1 frequency.

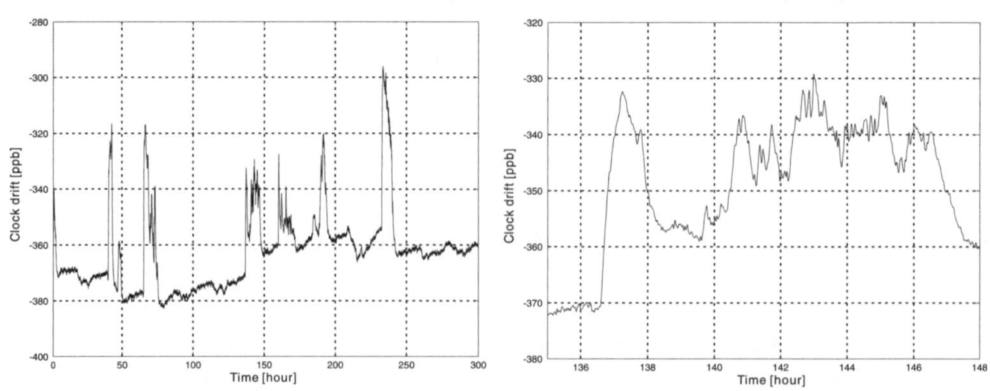

Figure 10.2 Long-term clock drift for a TCXO [14].

The absolute value of the clock drift of the roving receiver can be calibrated if at least one satellite signal is acquired (or tracked), because satellite/pseudolite parameters are known and the user dynamics are limited to less than 2 m/s (corresponding to a Doppler uncertainty of ±10.51 Hz at GPS L1). This allows for the calculation of a predicted Doppler value, and the difference between the predicted minus the observed Doppler value can be attributed to the clock drift.

An uncertainty in the user position causes a Doppler variation, which shall be assessed below. This uncertainty is important for the satellite signals (but not for the pseudolite signals) because the satellite moves with the speed of several km/s.

The range rate r is defined as a timely variation of the geometric distance between the transmitter located at $\mathbf{x}_{sat}$ and the receiver at $\mathbf{x}_{rec}$. The transmitter and the receiver have a velocity of $\mathbf{v}_{sat}$ and $\mathbf{v}_{rec}$. The range rate r is given as the velocity difference projected into the line of sight

$$r = \frac{1}{\left|\mathbf{x}_{sat} - \mathbf{x}_{rec}\right|}(\mathbf{x}_{sat} - \mathbf{x}_{rec})\cdot(\mathbf{v}_{sat} - \mathbf{v}_{rec}) \tag{10.1}$$

If we assume that the receiver position changes by an amount of $\delta\mathbf{x}_{rec}$, the range rate changes will be in an amount of δr

$$r + \delta r = \frac{1}{\left|\mathbf{x}_{sat} - \mathbf{x}_{rec} - \delta\mathbf{x}_{rec}\right|}(\mathbf{x}_{sat} - \mathbf{x}_{rec} - \delta\mathbf{x}_{rec})\cdot(\mathbf{v}_{sat} - \mathbf{v}_{rec}) \tag{10.2}$$

A first-order Taylor series expansion in the position change of the geometric distance yields

$$\left|\mathbf{x}_{sat} - \mathbf{x}_{rec} - \delta\mathbf{x}_{rec}\right|^{-1} \approx \left|\mathbf{x}_{sat} - \mathbf{x}_{rec}\right|^{-1}\left(1 - \frac{(\mathbf{x}_{sat} - \mathbf{x}_{rec})\cdot\delta\mathbf{x}_{rec}}{\left|\mathbf{x}_{sat} - \mathbf{x}_{rec}\right|^{2}}\right) \tag{10.3}$$

which is used to obtain a first order approximation of the range rate change as

$$\begin{aligned}
r + \delta r &\approx \left|\mathbf{x}_{sat} - \mathbf{x}_{rec}\right|^{-1}\left(1 - \frac{(\mathbf{x}_{sat} - \mathbf{x}_{rec})\cdot\delta\mathbf{x}_{rec}}{\left|\mathbf{x}_{sat} - \mathbf{x}_{rec}\right|^{2}}\right)(\mathbf{x}_{sat} - \mathbf{x}_{rec} - \delta\mathbf{x}_{rec})\cdot(\mathbf{v}_{sat} - \mathbf{v}_{rec}) \\
&\approx \left|\mathbf{x}_{sat} - \mathbf{x}_{rec}\right|^{-1}(\mathbf{x}_{sat} - \mathbf{x}_{rec})\cdot(\mathbf{v}_{sat} - \mathbf{v}_{rec}) - \left|\mathbf{x}_{sat} - \mathbf{x}_{rec}\right|^{-3}((\mathbf{x}_{sat} - \mathbf{x}_{rec})\cdot\delta\mathbf{x}_{rec}) \\
&\quad \times((\mathbf{x}_{sat} - \mathbf{x}_{rec})\cdot(\mathbf{v}_{sat} - \mathbf{v}_{rec})) - \left|\mathbf{x}_{sat} - \mathbf{x}_{rec}\right|^{-1}\delta\mathbf{x}_{rec}\cdot(\mathbf{v}_{sat} - \mathbf{v}_{rec}) \\
&= r - \left|\mathbf{x}_{sat} - \mathbf{x}_{rec}\right|^{-3}((\mathbf{x}_{sat} - \mathbf{x}_{rec})\cdot\delta\mathbf{x}_{rec})((\mathbf{x}_{sat} - \mathbf{x}_{rec})\cdot(\mathbf{v}_{sat} - \mathbf{v}_{rec})) \\
&\quad - \left|\mathbf{x}_{sat} - \mathbf{x}_{rec}\right|^{-1}\delta\mathbf{x}_{rec}\cdot(\mathbf{v}_{sat} - \mathbf{v}_{rec})
\end{aligned} \tag{10.4}$$

The maximum absolute value of the range rate change is bounded by assuming the in-products in (10.4) and their maximum or minimum values. Furthermore, the receiver velocity can be ignored with respect to the satellite velocity, finally yielding

$$|\delta r| \leq |\mathbf{x}_{sat} - \mathbf{x}_{rec}|^{-1} |\delta \mathbf{x}_{rec}||\mathbf{v}_{sat} - \mathbf{v}_{rec}| + |\mathbf{x}_{sat} - \mathbf{x}_{rec}|^{-1} |\delta \mathbf{x}_{rec}||\mathbf{v}_{sat} - \mathbf{v}_{rec}|$$

$$\approx 2|\mathbf{x}_{sat} - \mathbf{x}_{rec}|^{-1} |\delta \mathbf{x}_{rec}||\mathbf{v}_{sat}| \quad (10.5)$$

Assuming a maximum position error of 500m, a satellite–receiver distance of 20,000 km, and a satellite velocity of 5 km/s yields a maximum range rate error of 0.25 m/s that translates on GPS L1 to a Doppler error of 1.2 Hz.

Overall, we expect that after an initial acquisition of one signal (or other frequency calibration of the rover receiver clock), the Doppler search range is limited to be within ±24 Hz at GPS L1 for the rest of the signals to be acquired.

10.4.2 Correlation Method

We propose as acquisition method to use a single coherent integration as discussed in Section 5.7.1. There are a number of arguments that favor this solution over a noncoherent detection:

- The method achieves a higher sensitivity with the same total integration time.
- Mitigation of signal interference is improved by taking into account the navigation data bits. Aiding data is available for GPS C/A-code signals to perform a data wipe-off.
- Due to the long coherent integration time, the obtained frequency estimate is precise, allowing an easy handover to tracking.
- Efficient implementations exist for time-assisted acquisition.
- The Doppler search space is narrow; thus, the number of Doppler bins is limited and the computational demands are reasonable.

For a coherent integration time of 1 second, 24 Doppler bins have to be searched (after coarse clock calibration). If we assume a code-phase resolution of 0.25 μs and a time search range of 60 ms, 240,000 code phase bins need to be searched. Assuming all bins are independent (see Section 5.8), 5.76 million bins need to be tested. If we require a system false-detection probability of 1%, this translates to a single-bin false-detection probability of 1.7×10^{-9}.

Assuming further a detection probability of 0.9, the clairvoyant sensitivity is shown in Table 10.5. Because an FFT-based acquisition scheme will be used, the maximum Doppler error is $(2T_{coh})^{-1}$ and, as described in Section 5.8.5, the Doppler loss is 3.9 dB in this case. The sum of further losses caused by the finite code-phase resolution (see Section 5.8.5), by nonlinear oscillator jitter (see Section 10.4.3), and by line-of-sight dynamics (see Section 10.4.4) is assumed to be less than 1 dB.

10.4.3 Clock Stability

The receiver oscillator has to be stable during the coherent integration time so that the summation of the samples is not corrupted by carrier-phase variations. The effect of clock stability on coherent integration has been investigated by Sıçramaz Ayaz and López-Risueño [16, 17], and both papers indicate that the TCXO stability

Table 10.5 Acquisition Sensitivity Budget

Parameter	*Value*
Coherent integration time T_{coh}	1,000 ms
Single bin sensitivity	14.6 dBHz
Max. Doppler losses	3.9 dB
Other losses (code-phase, nonlinear carrier-phase)	1 dB
Total sensitivity	19.5 dBHz

is sufficient to support a coherent integration time of 1 second. In the next section, the key results will be summarized and verified by a Monte Carlo simulation.

The clock stability may be expressed as the power spectral density of frequency fluctuations $S_y(f)$ in [Hz^{-1}] [16]

$$S_y(f) = h_{-2}f^{-2} + h_{-1}f^{-1} + h_0 \tag{10.6}$$

We use y to denote the instantaneous clock frequency divided by the nominal frequency, to be distinguished from the frequency argument f of the power spectral density. The parameters h_{-2}, ..., h_0 are explained in Table 10.6. A more convenient way to express the clock stability is the Allan variance, which is the variance of the instantaneous frequency difference over a given period τ. It is related to the spectral density via [16]

$$\sigma_y^2(\tau) = \frac{1}{2}\left\langle (y(t+\tau) - y(t))^2 \right\rangle = h_{-2}\frac{(2\pi)^2}{6}\tau + h_{-1}2\log 2 + h_0\frac{1}{2\tau} \tag{10.7}$$

It is natural to assume that, during the coherent integration time, carrier-phase fluctuations caused by the clock jitter should be much smaller than half of the carrier wavelength. Fluctuations in y multiplied by the nominal carrier frequency f_{RF} yield frequency fluctuations. The latter fluctuations need to be multiplied by the coherent integration time to get carrier-phase fluctuations in cycles. Overall, one would expect that the following guideline should hold

$$T_{coh}\sigma_y(T_{coh})f_{RF} \ll 0.5 \tag{10.8}$$

However, López-Risueño has shown that this guideline is too conservative and should be replaced by the requirement

$$10 f_{RF}\sqrt{\frac{T_{coh}}{f_s}}\sigma_y(1/f_s) \ll 0.5 \tag{10.9}$$

where f_s denotes the sampling frequency in hertz.

A more precise estimate of the influence of the clock stability on the coherent integration can be obtained via simulation. Therefore, specific parameters of a TCXO, which are listed in Table 10.6, have to assumed.

Winkel presented a method to simulate a stochastic process of a power spectral density given by (10.6) [18]. The inputs of this simulation are the clock parameters and the output is the instantaneous clock error in seconds. Multiplying it by the speed of light gives the clock error in meters, denoted as $\Phi(t)$. As pointed out by

Table 10.6 TCXO Parameters [16, 18]

Parameter	*Value*
White frequency noise h_0	1×10^{-21} Hz^{-1}
Flicker frequency noise h_{-1}	1×10^{-20}
Random walk frequency noise h_{-2}	2×10^{-20} Hz

Sıçramaz Ayaz, only the nonlinear variations reduce the correlation result because a constant term is irrelevant for acquisition and a linear term is absorbed by slightly adjusting the Doppler frequency [17]. If $\Delta\Phi(t)$ denotes the detrended (constant and linear contribution removed) clock error in meters, the loss due to clock jitter is given as

$$L_{clock} = 1/T_{coh} \left| \int_{t=0}^{T_{coh}} \exp\left\{ i2\pi \frac{\Delta\Phi(t)}{\lambda} \right\} \right| \tag{10.10}$$

In Figure 10.3, the detrended clock jitter, expressed in millimeters, is shown over a time interval of 1 second. The result of three Monte Carlo simulations is shown. The clock jitter is not larger than a few millimeters, which is much smaller than the carrier wavelength λ (0.1903m at GPS L1). Calculating the loss via (10.10) for a number of 5,000 Monte Carlo runs and plotting them as a histogram in Figure 10.4 demonstrates that the clock-jitter loss for these particular TCXO parameters is negligible.

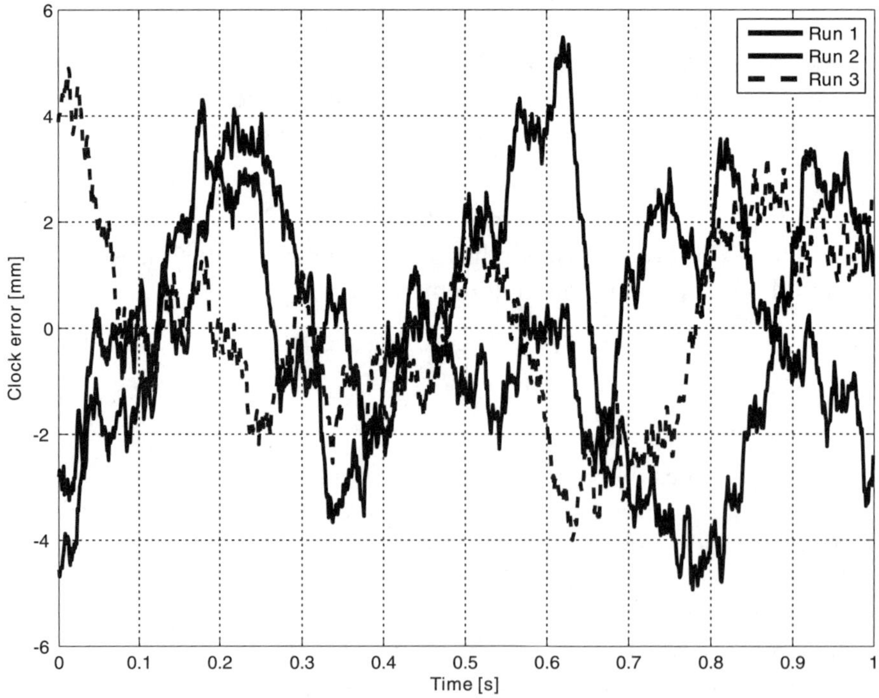

Figure 10.3 Three Monte Carlo simulations of the TCXO clock error (detrended).

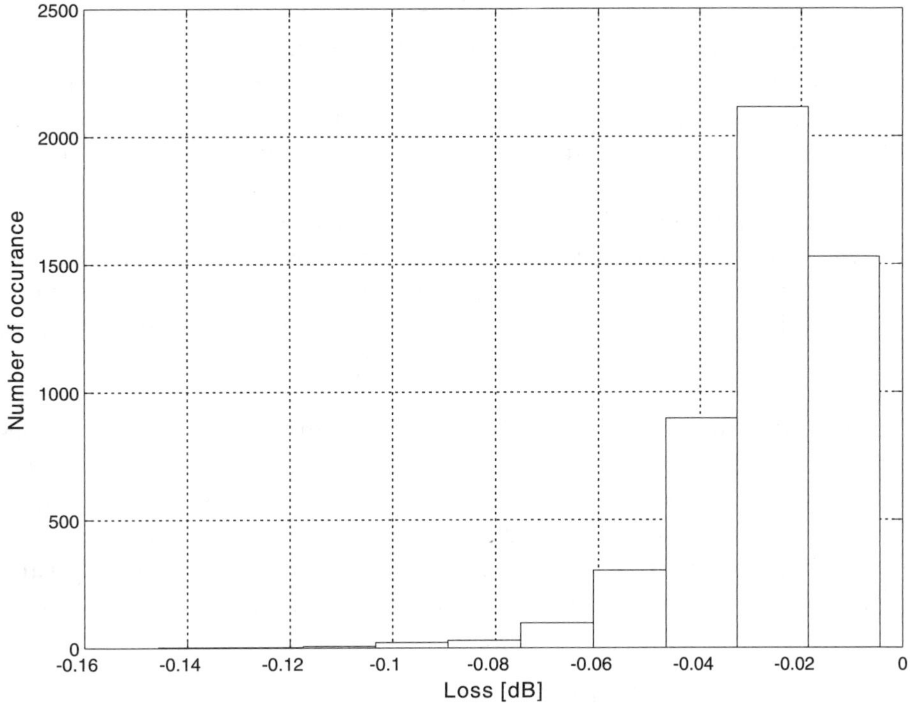

Figure 10.4 Oscillator-jitter loss (10.10) for a TCXO for a coherent integration of 1 second.

10.4.4 Line-of-Sight Dynamics

Similar to the clock jitter, nonlinear variations of the geometric distance between transmitter and receiver during the coherent integration interval cause a reduced correlation amplitude. Distance variations may occur due to transmitter and/or receiver motion. Now we discuss why correlation losses caused by line-of-sight dynamics can be ignored.

For the proposed system, it is reasonable to assume that, during an acquisition phase, the user is not moving or is moving in a gentle way, thereby not corrupting the acquisition engine. After the signals are tracked, shorter integration times will be used and user dynamics are allowed to increase. Line-of-sight accelerations caused by the satellite are below 0.2 m/s^2 for GPS satellites. This causes detrended line-of-sight variations below 2 cm over an interval of 1 second. The associated correlation loss is less than 0.3 dB.

10.4.5 Flow Diagram and FFT Algorithms

Acquisition is a time-consuming process and it is advisable to shrink the search space to the smallest possible size. For a reduced search space, efficient FFT algorithms are available. A reduced Doppler search space requires less search time simply because a smaller number of IFFT operations have to be performed (see Section 9.5.4). A reduced code-phase search space allows the application of techniques presented in Section 9.5.5, thereby reducing the IFFT order but keeping the same coherent integration time.

Overall, the acquisition flow diagram shown in Figure 10.5 is proposed. When the rover receiver is started up for the first time, one signal needs to be acquired to determine the coarse clock drift value. This step may require a longer time, but is needed only rarely for frontend calibration. After coarse clock calibration, the rover receiver can do a warm start. During a warm start, only one GPS signal needs to be acquired. The Doppler search range is narrow, but the time search range still covers the ±30 ms uncertainty. Nevertheless, the methods in Section 9.5.5 can be used. After the warm start is completed, the time search range can be narrowed down to a few microseconds, thereby allowing a much more efficient use of the techniques of Section 9.5.5. After four signals have been acquired, the receiver switches into the vector-hold tracking and all-in-view signals are tracked.

The acquisition algorithm for the pseudolite signal depends on the broadcast signal structure. If Galileo-like signals are broadcast, the same techniques as for GPS signals are proposed to be used. This includes data wipe-off and additionally allows the warm start with pseudolite signals to occur. In case the pseudolites broadcast only a pilot signal (which is actually the preferred method), the use of FFT techniques as described in Section 9.5.6 is possible. They are especially valuable during the cold-start acquisition, when a larger Doppler space needs to be searched.

10.4.6 Acquisition Time

In the next section, an estimate of the time required by the GNSS SDR to obtain a first position fix shall be estimated. The estimation is based on a number of assumptions summarized in Table 10.7. We assume that the clock drift has already been calibrated in a previous cold start run. Then the warm start Doppler search range is ±12 Hz. Furthermore, the timing uncertainty of the receiver is ±30 ms. We use a coherent integration time of 1 second, thus the total correlation time is 1,060 ms as can be seen in Figure 10.6.

The reduced code-phase search-space technique discussed in Section 9.5.5 is applied. The forward FFT of the local and received signal is created with a large FFT of 4 million samples. The inverse FFT is created with a much smaller number (0.13 million samples). Signal samples are stored as complex float values, each sample thus occupying eight bytes (real plus imaginary part).

For FFT performance calculations, we refer to Section 9.5.9 and, especially, to Figure 9.12. The time needed to calculate the reference signal is based on Section

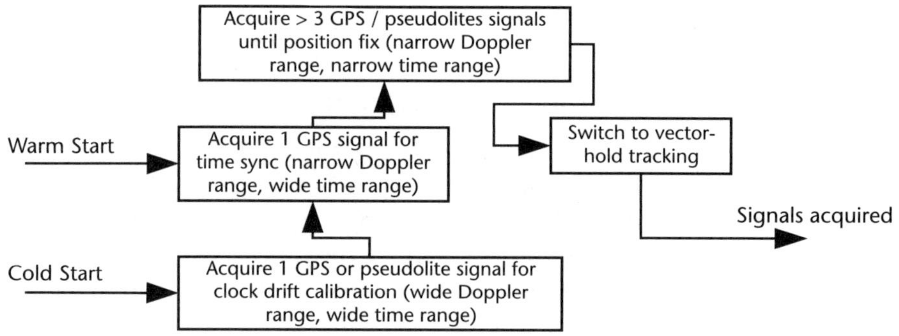

Figure 10.5 Acquisition flow diagram.

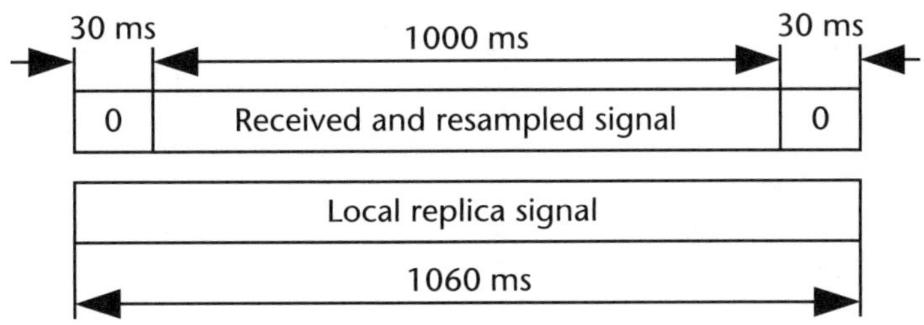

Figure 10.6 Timing of received and local signal.

9.3 and for all other operations it is assumed that their performance is determined by the memory bandwidth. This is a realistic assumption because the actual operation to be performed is almost trivial; only getting and storing the values from and into memory requires significant time. For example, to multiply the signal and replica spectrum in the frequency domain, 3 (= two input vectors plus one output vector) × 8 (= a complex number is represented as two float numbers each 4 bytes long) × 4,194,304 (= FFT length) MBytes of data needs to be transferred, which takes 37.5 ms for a memory bandwidth of 2.5 GByte/s.

Tables 10.8 and 10.9 give the time needed to acquire the first and following signals. Assuming that each signal acquisition is successful, four signals are acquired after 7.4 seconds. This allows the receiver to compute a position and stop the acquisition process. Note that for Table 10.9 the timing uncertainty is at the microsecond level. Thus, the IFFT computation time is negligible.

10.5 Assisted Tracking

The assistance data provided by the reference station to the rover receiver includes information to aid tracking. The assistance data link provides:

- Code and phase pseudorange measurements of the reference receiver;
- Broadcast navigation data bits;
- Prompt correlator values of the reference receiver.

Table 10.7 Summary of Warm-Start Acquisition Parameters

Parameter	*Value*
Signal	GPS C/A
Sample rate	3.96 MHz
Code-phase resolution	0.26 chip
Total correlation time	1,060 sec
Code-phase search window	60 ms
Forward FFT order *N*/size	2^{22} = 4,194,304
Inverse FFT order *N*/size	2^{17} = 131,072
Decimation factor before IFFT	2^5 = 32
Number of Doppler bins	24
Memory bandwidth	2.5 GByte/s
FFT performance for *N* =17 / 22	2.4/1 GFLOP

Table 10.8 Warm-Start Acquisition Execution Time (First Signal)

Parameter	*Number of Runs*	*Time per Run*	*Total Time*
Signal generation	1	10 ms	10 ms
Resampling	1	13 ms	13 ms
Forward FFT	2	461 ms	922 ms
Frequency multiplication	24	38 ms	912 ms
Decimation	24	13 ms	312 ms
Inverse FFT	24	5 ms	120 ms
Peak search	24	0.4 ms	10 ms
Total			2,299 ms

First, the broadcast navigation data bits allow the rover receiver to remove them and to extend the coherent integration time during tracking. More specifically, the dependency on the broadcast navigation data bits is removed by multiplying the rover correlation values with the data-bit values obtained through the assistance link (data wipe-off process). The assistance data also allows limiting the region of Doppler and code-phase values for single channels (vector-hold tracking). Double-difference correlator values are formed for high carrier-phase tracking stability. Vector-hold tracking and double-difference correlators are described in the following section.

10.5.1 Vector-Hold Tracking

In Section 4.3.3, vector tracking and independent channel tracking are discussed and it is shown that estimated pseudoranges are independent of the correlation point, as long as the correlation point is within the linear region of the discriminators.

For the proposed system, conventional DLL/FLL tracking will fulfill the requirement to track signals down to 20 dBHz [19]. DLL/FLL parameters, which allow tracking of GPS C/A-code signals even down to 10–12 dBHz, are shown in Table 10.10 [20].

Tracking problems occur if a signal is further attenuated; the channel may lose its lock. Loss-of-lock detection is done via signal-power estimation. If the signal power estimate falls beyond a certain tracking threshold, its function for navigation is discontinued and the *vector-hold* tracking method is applied.

Vector-hold tracking is a hybrid tracking method between independent channel tracking and vector tracking. Vector-hold tracking uses a similar DLL/FLL loop architecture as independent channel tracking, but continuously compares the in-

Table 10.9 Warm-Start Acquisition Execution Time (After the First Signal Has Been Acquired)

Parameter	*Number of Runs*	*Time per Run*	*Total Time*
Signal generation	1	10 ms	10 ms
Resampling	1	13 ms	13 ms
Forward FFT	1	461 ms	461 ms
Frequency multiplication	24	38 ms	912 ms
Decimation	24	13 ms	312 ms
Inverse FFT	24	—	—
Peak search	24	—	—
Total			1,708 ms

Table 10.10 DLL/FLL Parameters for GPS

Parameter	*DLL Value*	*FLL Value*
Order	2	1
Bandwidth	0.5 Hz	0.25 Hz
Coherent integration time	20 ms	10 ms

stantaneous code-phase/Doppler values against nominal values derived from the receiver's PVT estimate. The nominal DLL code-phase value is derived from the nominal pseudorange and the nominal Doppler value is derived from the pseudorange rate value. Vector-hold tracking requires a valid PVT solution, transmitter positions, and clock corrections. If the code-phase or Doppler value of a single tracking channel deviates from the nominal value more than a certain threshold, the receiver forces the channel to stay within the threshold by adjusting the code-phase and/or Doppler values of the NCO. The value of the threshold equals the size of the linear region of the code-phase and frequency discriminators as discussed in Section 4.3.2.10.

The main advantage of vector-hold tracking compared to vector tracking is that truly independent measurements are obtained for each channel above the signal-power threshold, making it easier to attribute measurement errors (e.g., multipath) to single channels, whereas in full vector tracking, all error sources are mixed together in the final position estimate. On the other hand, weak-signal tracking is aided by the strong signals, thereby periods of signal outages are bridged. Vector-hold tracking does not improve the tracking stability if all signals are attenuated identically.

10.5.2 Double-Difference Correlator

Using single- or double-difference code/phase pseudorange measurements is a commonly used method to eliminate common mode errors in GNSS positioning [9]. Forming receiver single-difference observations between two receivers removes the observation dependency on the satellite clock error and satellite hardware biases. Forming single-difference observations between two satellites removes the observation dependency on the receiver clock error and receiver hardware biases. Double differencing combines a receiver plus a satellite single difference and is a commonly used method for carrier-phase-based positioning. Double-differencing allows the carrier-phase ambiguities to be constrained to their integer values within the positioning algorithm [9].

For the proposed positioning system, double-differencing will already be done at the correlator level to increase the coherent integration time and to increase the carrier-phase tracking stability. Double-difference correlators are formed at the rover receiver and are used to track the (double-difference) carrier phase, which is then used for positioning.

10.5.2.1 Tracking Scheme

Forming double-difference correlators implies that two receivers track signals from two transmitters, as shown in Figure 10.7.

For convenience, the indices k and l shall be used to distinguish observables from different transmitters, whereas the indices m and n shall be used to distinguish different receivers; that is,

$$\begin{aligned} k,l &\rightarrow sat \\ m,n &\rightarrow rec \end{aligned} \tag{10.11}$$

Both receivers track both transmitted signals. As discussed in the previous section, the vector-hold method might be used for stability tracking results. Vector-hold tracking does not track the carrier phase; instead, a new double-difference carrier-phase tracking method will be described now.

In Section 4.3.3.4, the phase-tracking process is described as a piggyback module attached to frequency and code tracking, doing phase unwrapping. The scheme is depicted in Figure 10.8 and each tracking channel outputs the P-correlator value C_P (see Section 7.3.1) with the nominal receiver time t_{rec} and the nominal transmit time t_{sent} of the signal. The nominal receiver time is derived from the receiver internal time scale, whereas the nominal transmit time is based on the transmitter time scale (eventually derived from the broadcast navigation message). Furthermore, the channel outputs, for each coherent integration period, the phase of the replica signal φ_0 at the midpoint of the integration interval. In the case when a PLL is used, φ_0 would be an estimate of the received carrier phase φ.

The carrier-phase value φ received at the nominal receiver time $t_{rec}^{k,m}$ and broadcast at the nominal transmit time t_{sent}^{k} shall be denoted for the transmitter k and the receiver m as

$$\varphi = \varphi^{k,m}(t_{rec}^{k,m}, t_{sent}^{k}) \tag{10.12}$$

10.5.2.2 Synchronized Measurements

It is common practice to subtract the nominal carrier-phase increase (i.e., the increase that occurs due to the nonzero working frequency f_{nom}, usually the nominal

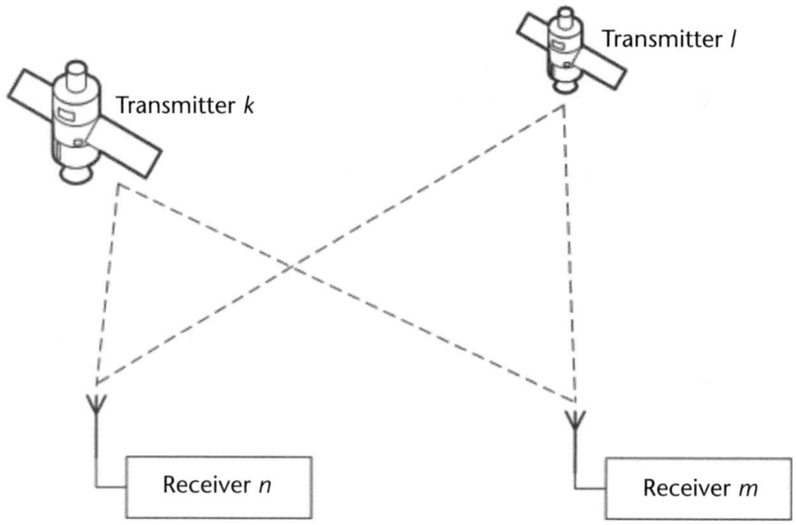

Figure 10.7 Double-difference scheme.

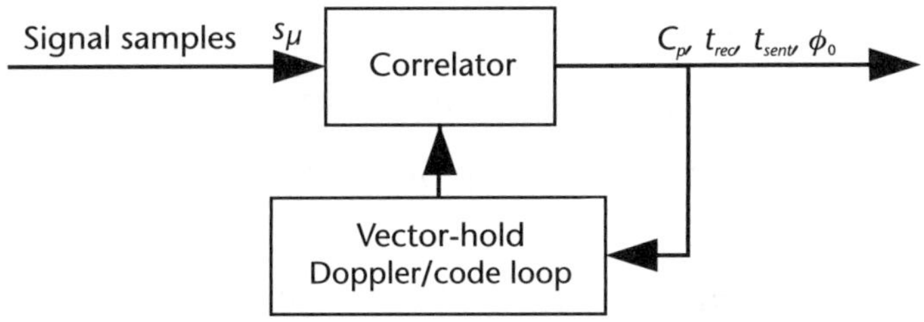

Figure 10.8 Data input for double-difference carrier-phase tracking.

IF), to obtain numerically smaller values. Furthermore, most positioning algorithms require that the carrier phase of different transmitters be measured at the same epoch, which is denoted as $t^m_{rec;sync}$ in the nominal receiver time scale. The reduced and synchronized carrier phase $\tilde{\varphi}^{k,m}$ is defined as

$$\tilde{\varphi}^{k,m}(t^m_{rec;sync}, t^k_{sent}) = \varphi^{k,m}(t^m_{rec;sync}, t^k_{sent}) - 2\pi f_{nom} t^m_{rec;sync} \tag{10.13}$$

and is the value found in RINEX observation files [21].

The times of the measurement epochs (in nominal receiver time) $t^{k,m}_{rec}$, when the estimates are actually available, differ from the synchronized receiver time $t^m_{rec;sync}$ by an inter- or extrapolation time interval $\delta t^{k,m}_{rec}$

$$t^{k,m}_{rec} = t^m_{rec;sync} + \delta t^{k,m}_{rec} \tag{10.14}$$

and forming reduced carrier phases requires an inter- or extrapolation

$$\begin{aligned}\tilde{\varphi}^{k,m}(t^m_{rec;sync}, t^k_{sent}) &= \varphi^{k,m}(t^{k,m}_{rec} - \delta t^{k,m}_{rec}, t^k_{sent}) - 2\pi f_{nom} t^m_{rec;sync} \\ &\approx \varphi^{k,m}(t^{k,m}_{rec}, t^k_{sent}) - 2\pi f_{nom} t^m_{rec;sync} - (2\pi f_{nom} + \omega^{k,m})\delta t^{k,m}_{rec}\end{aligned} \tag{10.15}$$

where $\omega^{k,m}$ is the Doppler estimate at the epoch $t^{k,m}_{rec}$. The accuracy of the reduced carrier phase therefore depends on the accuracy of the estimated Doppler. This inter- or extrapolation can be done precisely if PLL tracking is used (and the PLL is locked), because, in that case, Doppler estimation errors are small and the Doppler inter- or extrapolation errors are disregarded. The use of double-difference correlators allows carrier-phase tracking at much lower signal-power values and Doppler inter- or extrapolation errors need to be accounted for.

The non reduced and unsynchronized double-difference carrier phase is defined as

$$\nabla\Delta\varphi = \varphi^{k,m}(t^{k,m}_{rec}, t^k_{sent}) - \varphi^{l,m}(t^{l,m}_{rec}, t^l_{sent}) - \varphi^{k,n}(t^{k,n}_{rec}, t^k_{sent}) + \varphi^{l,n}(t^{l,n}_{rec}, t^l_{sent}) \tag{10.16}$$

and the synchronized and reduced double-difference carrier phase (normally used for carrier phase positioning) as

$$\nabla\Delta\tilde{\varphi} = \tilde{\varphi}^{k,m}(t^m_{rec;sync}, t^k_{sent}) - \tilde{\varphi}^{l,m}(t^m_{rec;sync}, t^l_{sent}) - \tilde{\varphi}^{k,n}(t^n_{rec;sync}, t^k_{sent}) + \tilde{\varphi}^{l,n}(t^n_{rec;sync}, t^l_{sent}) \tag{10.17}$$

Both are related via Doppler based inter- or extrapolation

$$\begin{aligned}\nabla\Delta\tilde{\varphi} &= \nabla\Delta\varphi - \delta t_{rec}^{k,m}\omega^{k,m} + \delta t_{rec}^{l,m}\omega^{l,m} + \delta t_{rec}^{k,n}\omega^{k,n} - \delta t_{rec}^{l,n}\omega^{l,n} \\ &\quad + 2\pi f_{nom}\left(\delta t_{rec}^{l,m} - \delta t_{rec}^{l,n} + \delta t_{rec}^{k,n} - \delta t_{rec}^{k,m}\right) \\ &= \nabla\Delta\varphi - \delta t_{rec}^{k,m}\omega^{k,m} + \delta t_{rec}^{l,m}\omega^{l,m} + \delta t_{rec}^{k,n}\omega^{k,n} - \delta t_{rec}^{l,n}\omega^{l,n} + 2\pi f_{nom}\nabla\Delta\delta t_{rec}\end{aligned} \tag{10.18}$$

The value of the last term $2\ \pi_{f\ nom}\ \nabla\Delta\delta t_{rec}$ can be calculated exactly because it is completely based on internal receiver time readings. Based on the channel output of Figure 10.8, a carrier-phase-related exponential $P^{k,m}$ is obtained from the P-correlator ${C_P}^{k,m}$ and the midpoint phase ${\varphi_0}^{k,m}$ of the reference signal (see Section 7.3.1)

$$P^{k,m} = C_P^{k,m}\exp\{i\varphi_0^{k,m}\} \propto \exp\{i\varphi^{k,m}\} \tag{10.19}$$

where the symbol $\propto$ denotes proportionality. The unsynchronized double-difference prompt correlator is defined as

$$\nabla\Delta P = P^{k,m}\,\overline{P^{l,m}\,P^{k,n}}\,P^{l,n} \propto \exp\{i\nabla\Delta\varphi\} \tag{10.20}$$

yielding an unsynchronized carrier-phase estimate as

$$\nabla\Delta\varphi = \text{angle}\{\nabla\Delta P\} \tag{10.21}$$

A synchronized double-difference prompt correlator is given as

$$\nabla\Delta\tilde{P} = \nabla\Delta P\exp\{2\pi i(-\delta t_{rec}^{k,m}\omega^{k,m} + \delta t_{rec}^{l,m}\omega^{l,m} + \delta t_{rec}^{k,n}\omega^{k,n} - \delta t_{rec}^{l,n}\omega^{l,n} + f_{nom}\nabla\Delta\delta t_{rec})\} \tag{10.22}$$

and a synchronized carrier-phase estimate as

$$\nabla\Delta\tilde{\varphi} = \text{angle}\{\nabla\Delta\tilde{P}\} \tag{10.23}$$

10.5.2.3 Unwrapping

The double-difference correlator principle relies on first forming the double-difference correlator via (10.20) and synchronizing the correlator by applying (10.22). Then those correlator values are used to set up a carrier-phase unwrapping algorithm. The advantage of (10.22) over undifferenced carrier-phase tracking is that common mode errors like receiver and satellite clock errors are eliminated and tracking-loop bandwidths can be chosen to be much smaller. Also, possibly present navigation data bits can be ignored when forming (10.20).

For simplicity, a conventional PLL is used to perform the double-difference phase unwrapping process. A PLL *unwraps* and *filters* the carrier-phase estimates simultaneously. PLL equations for various loop orders are readily available [19]. The PLL noise performance is characterized by the tracking loop bandwidth B_{PLL}. The reader should keep in mind that other unwrapping algorithms and/or filter algorithms can also be used.

10.5.2.4 Stability Condition

Based on (10.23), an estimate of the reduced double-difference carrier phase is obtained. This carrier-phase estimate needs to be unwrapped to include the full carrier cycle count in the estimate. For the unwrapping to work in a stable manner, noise in the estimated wrapped phase needs to be limited (i.e., the 3-sigma noise needs to be less than $\pi/2$ to avoid a 2π ambiguity (a cycle-slip) during unwrapping). The related double-difference carrier-phase tracking-loop stability is determined by thermal noise, interpolation noise, and, eventually, atmospheric noise. Ignoring the last contribution, the stability equation adapted from [19] is

$$\sigma_{\nabla\Delta\tilde{\varphi}}^2 = \sigma_{th}^2 + \sigma_{in}^2 < \left(\frac{\pi}{6}\right)^2 \tag{10.24}$$

The closed-loop thermal-noise variance σ_{th} in radians is considered to be independent from the interpolation noise σ_{in} in radians. Both variances are discussed below.

We assume that a double-difference carrier-phase tracking loop with a bandwidth of B_{PLL} is used. We assume that two strong signals are tracked (by the reference receiver) with two weak signals (by the rover receiver). The weak signals dominate the thermal-noise error budget. Both weak signals are assumed to have identical power.

The closed-loop thermal noise is related to the carrier-phase discriminator noise of Sections 4.3.2.8 and 8.1.3 [19] and

$$\sigma_{th}^2 = 2T_{coh}B_{PLL}\,\mathrm{var}\langle\nabla\Delta\varphi\rangle \underset{\substack{\text{2 strong, 2 weak}\\ \text{signals}}}{\approx} 4T_{coh}B_{PLL}\,\mathrm{var}\langle\varphi\rangle \tag{10.25}$$

The variance of the double-difference noise is twice the variance of the weak-signal undifferenced noise. Contributions from the strong signals are disregarded. Based on (4.88) we obtain

$$\sigma_{th}^2 = \frac{2B_{PLL}}{C/N_0}\left(1 + \frac{1}{2T_{coh}C/N_0}\right) \tag{10.26}$$

The interpolation noise is treated in a similar manner as the thermal noise. The interpolation noise is caused by frequency-tracking errors and is then filtered by the double-difference PLL. The frequency-tracking-error variance σ_ω is described in Sections 4.3.2.7 and 8.1.2. By simple error propagation based on (10.18) and (10.25), the closed-loop interpolation noise variance is given by

$$\sigma_{in}^2 = 4T_{coh}B_{PLL}\left\langle\delta t^2\right\rangle_{Time}\sigma_\omega^2 \tag{10.27}$$

where $\left\langle\delta t^2\right\rangle_{Time}$ is the RMS interpolation time. Again, we assume that the closed-loop interpolation noise error budget is dominated by the two weak signals (i.e., the Doppler of the strong signals is sufficiently well known).

Using, for example, (4.84) to describe the Doppler-discriminator noise and converting the Doppler-discriminator noise into a closed-loop frequency-tracking error (assuming a frequency-loop bandwidth B_{FLL}), we obtain

$$\sigma_{\hat{\omega}}^2 = 2T_{coh}B_{FLL}\operatorname{var}\langle\hat{\omega}\rangle_{\mathrm{N}} = \frac{12B_{FLL}}{\chi_{freq}T_{coh}{}^2C/N_0}\left(1+\frac{1}{T_{coh}C/N_0}\right) \tag{10.28}$$

Assuming that the mean inter- or extrapolation time equals half the coherent integration time (i.e., $\left\langle \delta t^2 \right\rangle_{Time} = T_{coh}^2/4$) the closed-loop interpolation noise is consequently

$$\begin{aligned}\sigma_{in}^2 &= 4T_{coh}B_{PLL}\langle\delta t\rangle_{Time}^2\frac{12B_{FLL}}{\chi_{freq}T_{coh}{}^2C/N_0}\left(1+\frac{1}{T_{coh}C/N_0}\right)\\ &= \frac{12T_{coh}B_{PLL}B_{FLL}}{\chi_{freq}C/N_0}\left(1+\frac{1}{T_{coh}C/N_0}\right)\end{aligned} \tag{10.29}$$

The closed-loop double-difference reduced and synchronized carrier-phase variance is the sum of thermal noise and interpolation noise and needs to fulfill the stability condition

$$\begin{aligned}\sigma_{\nabla\Delta\tilde{\varphi}}^2 &= \frac{2B_{PLL}}{C/N_0}\left(1+\frac{1}{2T_{coh}C/N_0}\right)+\frac{12T_{coh}B_{PLL}B_{FLL}}{\chi_{freq}C/N_0}\left(1+\frac{1}{T_{coh}C/N_0}\right)\\ &= \frac{2B_{PLL}}{C/N_0}\left(1+\frac{1}{2T_{coh}C/N_0}+\frac{6T_{coh}B_{FLL}}{\chi_{freq}}+\frac{6B_{FLL}}{\chi_{freq}C/N_0}\right)<\frac{\pi^2}{36}\end{aligned} \tag{10.30}$$

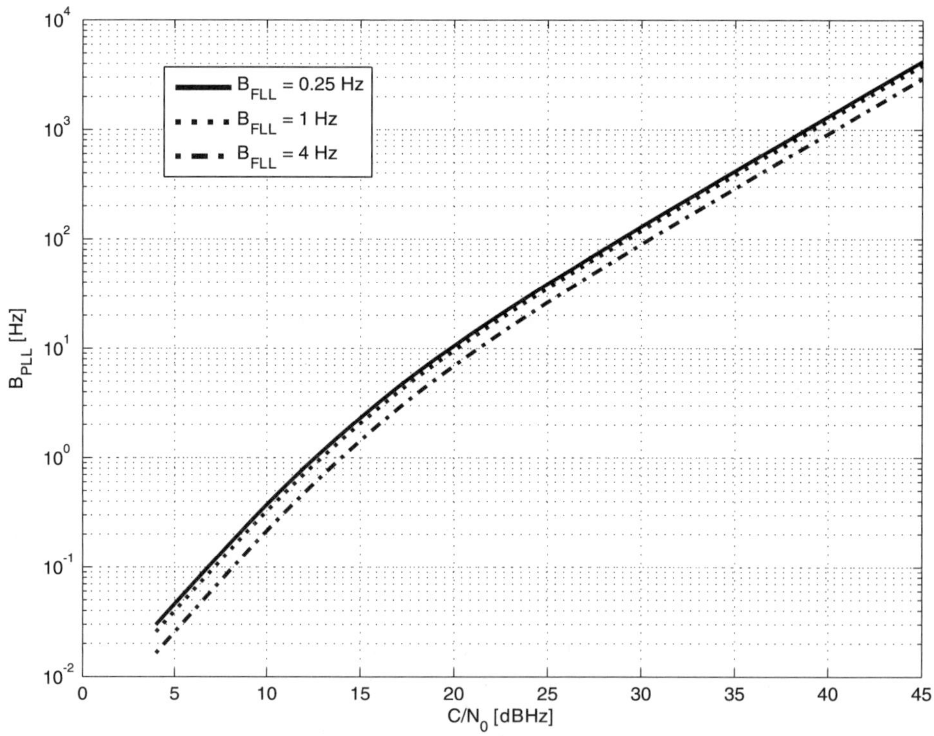

Figure 10.9 Maximum B_{PLL} plotted as a function of the weak signal power and B_{FLL} (T_{coh} = 20 ms, χ_{freq} = 1).

Equation (10.30) allows calculating the maximum allowable B_{PLL} value for a given value of B_{FLL}. An exemplary evaluation is shown in Figure 10.9. The figure shows that B_{FLL} = 1 Hz and B_{PLL} = 10 Hz allow cycle-slip-free tracking of 20-dBHz signals. For a slowly moving user, B_{PLL} = 1 Hz allows tracking signals down to 13 dBHz.

10.6 Low-Cost Pseudolites

The proposed positioning system can be enhanced by pseudolites if the GNSS signal attenuation is too strong to allow precise positioning. The pseudolite signals are designed to overcome an additional penetration loss of 25 dB (see Table 10.2). A key requirement on the pseudolites is that they are of low complexity (and thus of low cost) and that they can be installed easily. The basic idea is to set up the pseudolites at points with known coordinates. After switching them on, they start broadcasting their signal autonomously and independent of other pseudolites or the GPS signals. Neither calibration nor any other complex startup procedure is required.

Pseudolite signals match well with a software radio, because the signal processing as described for example in Chapter 9 works with 16 bits, which represents a dynamic amplitude range of 96 dB. Multibit ADCs are readily available; for example, the ADC14L020 provides 14-bit resolution at 20 MHz with a power consumption of 150 mW [22]. Overall, a software radio can cope well with strong signal-power variations caused by the near–far effect and the timely variable attenuation of a pseudolite signal. Furthermore, the proposed positioning system is based on differential methods. Thus, neither the pseudolite clock error needs to be determined, nor is there a need to broadcast a navigation data message by the pseudolite. Because of the differential positioning method, pseudolite hardware biases are partly eliminated. Furthermore, the use of the carrier phase makes the system robost against distortions of the signal-autocorrelation function caused by amplitude-dependent front-end effects. A low complexity pseudolite can be used as discussed by Manandhar [23]. In the latter work, the pseudolites act as proximity sensors and are not used for ranging.

In Table 10.11, a possible pseudolite configuration is presented, which is adapted from the existing transmitters placed in the GATE area [7]. However, in contrast to the real GATE transmitters, only a pilot signal is broadcast. The broadcast signal is not synchronized to any time frame and a comparably low quality oscillator is used. Therefore the pseudolite can, in principle, be realized with a small FPGA for IF signal generation plus an RF upconverter reducing components' costs to a minimum. The pseudolite can operate autonomously or, eventually, a low data-rate data link allows control of the broadcast signal power.

The major disadvantage of pseudolite signals in comparison to satellite-borne signals is the near-far effect. Signals received from nearby transmitters are of much higher power than signals received from more distant transmitters. A worst-case situation is depicted in Figure 10.10.

If we assume that the minimum distance from the rover to the pseudolites can be limited to be larger than 50m, geometric signal-power variations on the order of 26 dB are the consequence (assuming that the inverse signal power is proportional

Table 10.11 Pseudolite Parameters

Parameter	*Value*
Modulation scheme	BOC(1,1)
PRN-code length	4,092
Data message	No
Sync. to GPS time	No
Oscillator	TCXO
Pulsing (optional)	on: 4 ms, off: 90–95 ms
Signal-power control	Optional
Pseudolite placing	At point with known coordinates
Number of pseudolites	Environment specific (1.6)

to the squared distance). Attenuation factors and signal-power levels derived from Table 10.2 and Figure 10.10 are summarized in Table 10.12.

In the worst case, the rover receives a weak signal of 20 dBHz simultaneously with a strong signal of 96.4 dBHz [= 20 dBHz (minimum C/N_0) + 50 dB (penetration loss reserve) + 26.4 dB (geometric signal power variation reserve)], which causes significant cross-correlation effects. The near-far effect is expressed more for the roving receiver, because the reference receiver is never near the pseudolites and it receives all signals without strong attenuation. Two possible solutions to surmount this interference are discussed in the next section. One is based on continuous-time signals, the other one on pulsed signals.

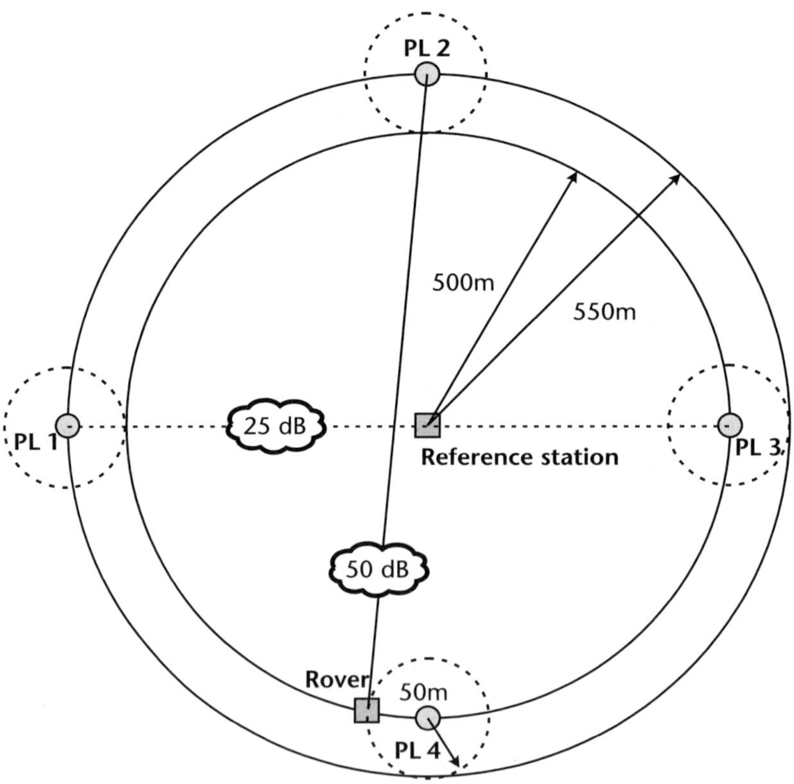

Figure 10.10 Worst-case near–far effect for pseudolite signals.

Table 10.12 Pseudolite Near–Far Parameters Without Dynamic Pseudolite Power Adjustment

Parameter	*Value Rover*	*Value Ref. Sta.*
Minimum distance pseudolite-receiver	50m	500m
Maximum distance pseudolite-receiver	1,050m	550m
Geometric signal-power variation	26.4 dB	0.8 dB
Minimum received signal power	20 dBHz	45 dBHz
Maximum received signal power	96.4 dBHz	70.8 dBHz
Worst-case signal-power difference between two different signals	76.4 dB	25.8 dBHz

10.6.1 Continuous-Time Signals

If the pseudolite signals were broadcast continuously in time, the high interference caused by the extreme signal-power variation would make tracking of weak signals impossible. For the presented pseudolite signal, the cross-correlation protection is of a similar magnitude as between two GPS C/A-code signals [24]. Two methods are proposed to overcome the interference problem [25]. The GPS signals have a cross-correlation protection of around –21 dB against Galileo or GPS signals, and the Galileo signals have a protection of around –23 dB against Galileo or GPS signals.

10.6.1.1 Signal-Power Adaption

The transmitted signal power can be dynamically adjusted based on actual measurements of rover and reference station received power levels. The pseudolite signals can be adjusted in a way that the rover receives signals approximately at a level not much greater than 41 dBHz. In this case the PRN-code cross-correlation is high enough to avoid any incorrect locks on false cross-correlation peaks. Dynamic signal-power adaptation is realized in GATE within the extended base mode [7].

10.6.1.2 Interference Cancellation

Interference-cancellation schemes can be employed in the receiver to reduce the effect of a strong signal. Signal-cancellation schemes require generation of a strong signal replica in the channel tracking the weak signal and are touched upon in Sections 4.2.3 and 5.7.3. Interference cancellation might be necessary at the reference station, if the dynamic signal-power adaptation requires that a pseudolite transmits with full power and simultaneously the reference station receives a low power signal with 45 dBHz. In this case, the high-power pseudolite signal has 76.4 dBHz and the difference at the reference station between the high-power and low-power signal is 31.4 dB. This value is around 10 dB above the cross-correlation protection. Cancellation schemes are comparably easy to implement in a software receiver, but require a significantly greater computational load.

10.6.2 Pulsed Signals

Using pulsed-pseudolite signals is common practice to reduce the near–far effect. This method is based on first detecting a pulse from a strong pseudolite and blanking it out for the channels tracking other signals [25]. Two conditions need to be fulfilled when pseudolite signals are present:

- The system must be able to detect strong pulses and mitigate the effect on weaker signals. The degradation caused by the pulsed signals needs to be acceptable.
- The pulse repetition rate must be high enough to allow continuous-phase tracking.

In the next section it shall be demonstrated that the proposed pulsing scheme is able to fulfill both requirements.

10.6.2.1 Pulse Repetition Rate

First, the pulse repetition rate is, according to Table 10.11, 100 ms. Correlator values are therefore available with a rate of 10 Hz, which allows a maximum tracking-loop bandwidth of 4 Hz [26]. This is sufficient for the DLL/FLL/PLL parameters, which were introduced above.

10.6.2.2 Pulse Detection

Pulse detection can basically be performed by two means: by using a properly chosen detection algorithm as described in Chapter 5 or by predicting the pulse arrival time if sufficient information is available to the signal-processing unit.

For pulse detection, the use of an energy detector is advisable (see section 5.5) because of its simple implementation. Its detection performance is less than that of a matched filter, but undetected weak pulse signals do not degrade the other signals. This will be demonstrated later. An energy detector working with 1-ms batches assuming $p_f = 10^{-5}$ and $p_d = 0.9$ can detect pulsed signals with a C/N_0 value (during transmission) of 53 dBHz in the background of white noise. It should be noted that this corresponds to an average C/N_0 value of 39 dBHz (a pulsing ratio of 25 corresponds to 14 dB). Figure 10.11 shows a simulated energy detector output for two different pseudolite signal-power levels. A detailed list of simulation parameters can be found in Table 10.13. On the right side of Figure 10.11, cases are clearly visible in which two or more pulses overlap.

Overall, pulses can be detected with sufficient sensitivity and signal samples affected by pulsed signals can be identified.

Once both receivers have computed a first positioning solution (for example using the GPS signals), the reference receiver is able to communicate pulse transmission periods to the roving receiver. In this case, the roving receiver can precisely predict the pulsed signal arrival time. Consequently, an energy detector to detect pulses is required only during signal acquisition but not during tracking.

10.6.2.3 Mitigation

Once a pulse has been detected, its effect has to be mitigated. As has been mentioned above, the advantage of a pulsed signal is the use of pulse blanking, which is

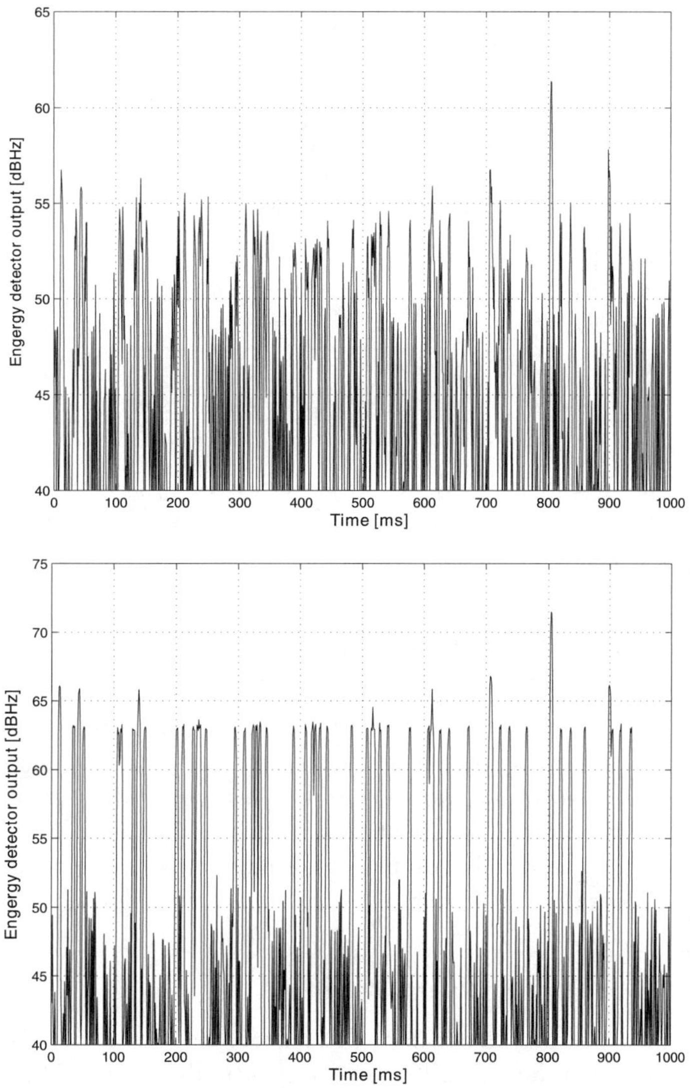

Figure 10.11 Energy detector (1 ms) output receiving thermal noise plus one weak GPS C/A-code signal (20 dBHz) plus 6 pseudolite signals. Mean single pseudolite C/N_0: left = 39 dBHz, right = 49 dBHz.

a comparably simple to implement mitigation scheme. The fundamental assumption of pulse blanking is that the affected signal samples do not carry any useful information and can therefore be ignored. Pulse blanking is applied for all channels, but obviously not for the channel tracking the respective pseudolite whose pulse has been detected.

When the correlator values are formed, the integration interval must exclude the affected samples. This can be achieved either by properly chosen integration indices or by setting the affected signal samples to zero.

Pulse blanking reduces the GPS C/A-code signal-tracking performance, because fewer discriminator values are available to steer the tracking loops. If pulse blanking occurs with a rate much higher than typical loop bandwidths, pulse blanking reduces the effective received signal power by

Table 10.13 Simulation Parameters for Pulsed Signal Detection

Parameter	*Value*
Number pseudolite signals	6
Pseudolite repetition rate	PRN1: 99 ms, PRN2: 98 ms, . . . , PRN6: 94 ms
Single pseudolite mean power	39, 49 dBHz
Single pseudolite peak power	53, 63 dBHz
Number of GPS signals	1 (PRN 1)
GPS signal power	20 dBHz
Sample rate	4.19 MHz
Sample type	Complex
Duration	1 sec

$$L_{PB} = -10\log_{10}\frac{T_{total} - T_{blan}}{T_{total}} \tag{10.31}$$

where L_{PB} is the pulse blanking loss in decibels, T_{blank} is the duration of pulse blanking in seconds within the repetition rate T_{total} in seconds. For an assumed number of six pseudolite signals of Table 10.13, the pulse blanking loss is therefore around 1.2 dB.

10.6.2.4 Navigation Data-Bit Decoding

Pulse blanking also reduces the GPS C/A data-bit estimation performance. In the worst case, a consecutive train of pulses may completely hide a broadcast navigation data bit, making it impossible to retrieve its value. This situation is, however, unlikely to occur and is not repeatable in time. Only a detailed simulation could clarify the percentage of affected data bits (which is, however, beyond the scope of this text). Eventually the proposed navigation system has to rely on external ephemeris data.

10.6.2.5 Overlapping Pseudolite Pulses

If two or more pulses overlap, the respective pulsed signal-correlator values are degraded due to the blanking and are eventually useless. The correlation values are reduced by

$$L_{over} = -10\log_{10}(1-\eta) \tag{10.32}$$

where L_{over} is the overlapping loss in decibels and η is the overlapping ratio (0, no overlap; 1, total overlap). For a total overlap, the respective correlator values are useless. In order to avoid timely constant situations of totally overlapping pulses, the pulse-repetition periods are chosen to be non-equal and are chosen to be

$$T_{R,k} = T_{R,nom} - kT_{R,delta} \tag{10.33}$$

$T_{R,k}$ is the pulse-repetition time for the pseudolite k in seconds. $T_{R,nom}$ is the nominal repetition rate and $T_{R,delta}$ is the repetition rate decrement between two pseudolites. Partial overlapping of two pseudolites pulses with an index difference of Δk occurs approximately every $T_{over;rep}$ seconds,

$$T_{over;rep} = \frac{{T_{R,nom}}^2}{\Delta k T_{R,delta}} \tag{10.34}$$

and overlapping lasts for $T_{over;dur}$ seconds,

$$T_{over;dur} = \frac{T_{R,nom}T_P}{\Delta k T_{R,delta}} \tag{10.35}$$

where T_P is the pulse duration in seconds. For the chosen settings, overlapping between two pseudolites with an index difference of 1 occurs every 10 seconds with a duration of 0.4 second (3 correlator values are partly degraded, 1 is strongly degraded). Overlapping between two pseudolites with an index difference of 4 occurs every 2.5 seconds with a duration of 0.1 second (one correlator value). One would expect that periods of overlapping can be bridged by the tracking loops, which have a loop bandwidth of less than 1 Hz. However, a final conclusion can only be drawn after a detailed simulation has been performed.

10.6.2.6 Undetected Pulses

The most severe signal degradation due to the presence of undetected pseudolite signals is expected to occur when a weak GPS C/A-code signal is acquired. This case is investigated by a numerical simulation. The GPS C/A-code signal power is 20 dBHz and 6 pseudolite signals are present each with a mean C/N_0 power of 39 or 59 dBHz (corresponding to a peak C/N_0 during transmission period of 53 or 73 dBHz). We assume that the pulses have not been detected and are present during the signal-acquisition process. This is likely to occur for the 39-dBHz pseudolite signals, because this signal power value is near the energy detector sensitivity. The

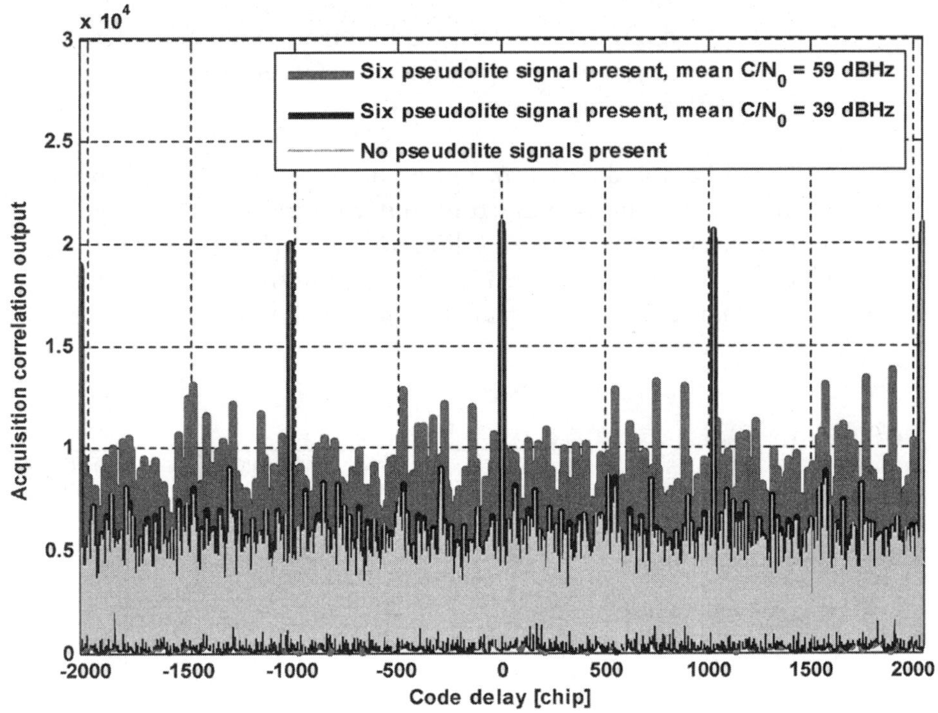

Figure 10.12 GPS C/A-code acquisition results, with and without the presence of pseudolite signals.

second case is only included for illustration purposes but not expected to occur. The acquisition process is detailed in Section 10.4. A coherent integration time of 1 second is used to acquire the GPS C/A-code signal. All signals are simulated at baseband and all signals (including the GPS signal) have an identical Doppler of 0 Hz. This scenario represents the worst case for cross-correlations [27].

The simulation is repeated three times, each time using identical noise samples. In the first case, no pseudolite signals are present, the second and third cases include the 6 pseudolite signals. The result is shown in Figure 10.12.

The basic finding from this simulation is that undetected pseudolite pulses cause little or no effect on the weak GPS C/A-code acquisition. Three kinds of cross-correlation protection schemes actually apply for this case: different PRN codes, different modulation schemes and the data wipe-off processes. Together, cross-correlations between pseudolite signals and the GPS signals are effectively mitigated without using any further cancellation schemes.

10.7 RTK Engine

Carrier-phase-based RTK positioning is used to determine a precise position estimate for the proposed navigation system. RTK algorithms can be found in many textbooks [8–10]. A modern RTK algorithm is described by Tiberius and automatically treats observations on different carrier frequencies in an optimal way [11].

A schematic for the proposed algorithm is shown in Figure 10.13. This algorithm needs the following input data:

- Code pseudoranges, carrier pseudoranges, Doppler values and prompt correlator values (from reference and rover receiver);
- Transmitter positions.

It should be noted that transmitter clock corrections are only required if the transmitters are moving to allow transmitter-position computation at the true transmission time. Thus, for the pseudolites no clock corrections are required.

Figure 10.13 contains a feed-back link from the position solution to the rover receiver signal processing. This link is required to do vector-hold tracking described in Section 10.5.1.

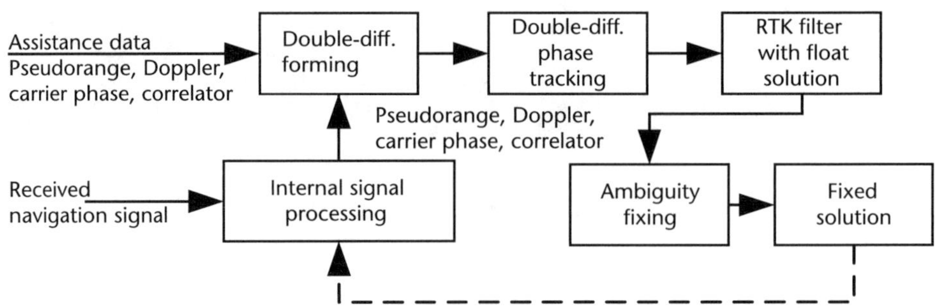

Figure 10.13 Rover double-difference carrier-phase tracking and RTK positioning.

The rover receiver gets all reference station data and forms double-difference code pseudoranges and double-difference correlator values. The double-difference correlator values are used to estimate the double-difference carrier phase as described in Section 10.5.2. Code and phase observations are processed in a Kalman-filter to estimate the *float* position with the *float* carrier-phase ambiguity. Based on the ambiguity covariance matrix, the Z-transform is used to de-correlate the ambiguities and to possibly constrain the ambiguities to their integer values [11]. Proper verification is needed to avoid false ambiguity fixing. If the ambiguities are fixed, the position is recalculated using the integer ambiguity values to obtain the *fixed* solution. During the double-difference phase tracking process, cycle-slip detection and eventually correction are necessary to avoid gross errors in the fixed ambiguities.

References

[1] Intel Corp., "Intel® Atom™ processor Z5xx Series, Data Sheet," Rev 1, Doc. No. 319535-001US, http://download.intel.com/design/chipsets/embedded/datashts/319535.pdf, 2008.

[2] Intel Corp., "Intel® System Controller Hub US15W for Embedded Computing," Doc. No. 319545-002, http://download.intel.com/design/chipsets/embedded/prodbrf/319545.pdf, 2008.

[3] LiPPERT, t. e. c., "CoreExpress™-ECO, Next Generation Computer-On-Module, Specifications," http://www.lippertembedded.com, 2008.

[4] Intel Corp., "Intel® Pentium® M Processor with 2-MB L2 Cache and 533-MHz Front Side Bus, Data Sheet," Rev 2, Doc. No. 305262-002, http://download.intel.com/design/mobile/datashts/30526202.pdf, 2005.

[5] Radio Technical Commission For Maritime Services, S. C. No. 1, "RTCM Recommended Standards For Differential GNSS (Global Navigation Satellite Systems) Service, Version 2.3," RTCM Paper 136-2001/SC104-STD, 2001.

[6] IFEN GmbH, "GATE Home, Hompage of the Galileo Test and Development Environment," http://www.gate-testbed.com, 2008.

[7] Heinrichs, G., et al., "First Outdoor Positioning Results with Real Galileo Signals by Using the German Galileo Test and Development Environment," *Proc. 20th Int. Technical Meeting of the Satellite Division of the Institute of Navigation (ION-GNSS) 2007*, Fort Worth, TX, September 25–28, 2007, pp. 1576–1587.

[8] Hofmann-Wellenhof, B., H. Lichtenegger, and E. Wasle, *GNSS: Global Navigation Satellite Systems: GPS, GLONASS, Galileo & More*, Vienna: Springer, 2008.

[9] Leick, A., *GPS Satellite Surveying*, New York: Wiley, 2004.

[10] Teunissen, P. J. G., and A. Kleusberg, *GPS for Geodesy*, Berlin: Springer, 1996.

[11] Tiberius, C., et al, "0.99999999 Confidence Ambiguity Resolution with GPS and Galileo," *GPS Solutions*, Vol. 6, No. 1–2, 2002, pp. 96–99.

[12] SiRF Technology, Inc., http://www.sirf.com, 2008.

[13] Wikipedia, "Network Time Protocol (NTP)," http://en.wikipedia.org/wiki/Network_Time_Protocol, 2008.

[14] Rakon Ltd., "IT5305BE, 23.104 MHz, SMD GPS TCXO," http://www.rakon.com, 2008.

[15] IFEN GmbH, "NavX®-NSR: GPS/Galileo Navigation Software Receiver," Brochure, http://www.ifen.com, 2007.

[16] López-Risueño, G., et al., "User Clock Impact on High Sensitivity GNSS Receivers," *Proc. ENC-GNSS 2008*, Toulouse, April 22–25, 2008.

[17] Sıçramaz Ayaz, A., T. Pany, and B. Eissfeller, "Performance of Assisted Acquisition of the L2CL Code in a Multi-Frequency Software Receiver," *Proc. 20th Int. Technical Meeting of the Satellite Division of the Institute of Navigation (ION-GNSS) 2007,* Fort Worth, TX, September 25–28, 2007, pp. 1830–1838.

[18] Winkel, J. Ó., *Modeling and Simulating GNSS Signal Structures and Receivers*, University of Federal Armed Forces Munich, Werner-Heisenberg-Weg 39, D-85577 Neubiberg, http://www.unibw.de/unibib/digibib/ediss/bauv, 2003.

[19] Kaplan, E. D., and C. J. Hegarty, (eds.), *Understanding GPS: Principles and Applications,* 2nd ed., Norwood, MA: Artech House, 2006.

[20] Anghileri, M., et al., "Performance Evaluation of a Multi-Frequency GPS/Galileo/SBAS Software Receiver," *Proc. 20th Int. Technical Meeting of the Satellite Division of the Institute of Navigation (ION-GNSS) 2007*, Fort Worth, TX, September 25–28, 2007, pp. 2749–2761.

[21] Gurtner, W., "RINEX: The Receiver Independent Exchange Format Version 2," ftp://igscb.jpl.nasa.gov/igscb/data/format/rinex2.txt, 1998.

[22] National Semiconductor, "ADC14L020 - 14-Bit, 20 MSPS, 150 mW A/D Converter from the PowerWise® Family," http://www.national.com, 2008.

[23] Manandhar, D., et al., "Development of Ultimate Seamless Positioning System Based on QZSS IMES," *Proc. 21st Int. Technical Meeting of the Satellite Division of the Institute of Navigation (ION-GNSS) 2008,* Savannah, September 16–19, 2008, pp. 1698–1705.

[24] Ávila Rodríguez, J. Á., *On Generalized Signal Waveforms for Satellite Navigation*, University of Federal Armed Forces Munich, Werner-Heisenberg-Weg 39, D-85577 Neubiberg, http://www.unibw.de/unibib/digibib/ediss/bauv, 2008.

[25] Elrod, B. D., and A. J. van Dierendonck, "Pseudolites," in *Global Positioning System: Theory and Applications, Volume II,* pp. 51–79, Parkinson, B. W., and J. J. Spilker, (eds.), Washington, D.C.: American Institute of Aeronautics and Astronautics Inc., 1996.

[26] Kazemi, P. L., "Optimum Digital Filters for GNSS Tracking Loops," *Proc. Int. Technical Meeting of the Satellite Division of the Institute of Navigation (ION-GNSS) 2008*, Savannah, GA, September 16–19, 2008, pp. 2304–2313.

[27] Spilker, J. J., Jr., "GPS Signal Structure and Theoretical Performance," in *Global Positioning System: Theory and Applications, Volume I,* pp. 57–120, Parkinson, B. W., and J. J. Spilker, (eds.), Washington, D.C.: American Institute of Aeronautics and Astronautics Inc., 1996.

CHAPTER 11

Exemplary Source Code

This book is accompanied by a few small computer programs to illustrate algorithms and methods presented herein.* The selection of the methods reflects the author's opinion of their relevance for a software receiver implementation. It includes the enhanced FFT acquisition algorithm of Section 9.5.7 and the multipath-estimating discriminator of Section 8.3. The latter program can be run with different signals and user dynamics, thereby allowing an easy adoption for a software receiver. Implicitly, the core algorithms of Chapter 9 are used. The simulation framework is MATLAB and the core algorithms are included as pieces of inline assembler code.

Additionally, exemplary code is provided to calculate the TCRLB (see Section 4.7.1) and to calculate the code phase, the Doppler, and the carrier-discriminator noise for arbitrary navigation signals and tracking schemes (see Section 8.1).

11.1 Intended Use

The software should allow better understanding of the material presented in the main text and allow easier use for the reader. The source code makes precise reference to sections and equation numbers of the main text. The reader should open the MATLAB scripts and the C files and work through the source code with the help of the book. The reader should then truly understand the methods and be able to modify them.

11.2 Setup

Running the computer programs requires only a few small steps, which are described in the following. The software can be downloaded from http://www.artechhouse.com/static/reslib/pany/pany1.html to a directory in your PC. In the following, it is assumed that the software is copied to a folder named "c:\navsigproc."

11.2.1 Required Software

The software was tested with MATLAB 7.2 (R2006a) and Microsoft Visual C++ 2005. It should run (with minor modification) with any later version. The C++ compiler is required to compile the MATLAB mex files containing the C and assembler code.

* Software available at www.artechhouse.com.

11.2.2 Preparing the Simulation

After copying the files and starting MATLAB, switch to the working directory "c:\navsigproc" and run the MATLAB script "prepareSimulation.m." This script will configure the C++ compiler and then compile all mex files.

11.2.3 Signal Selection and Simulation Parameters

Before running any of the routines listed in the next sections, open the script "init.m" from within MATLAB and adjust the simulation parameters according to your needs. Then, run "init.m." It might be useful to change the navigation signal type (see Section 1.9), the used sample rate, or the user dynamics.

11.3 Routines

The provided routines are all located in the chosen directory, for example, "c:\navsigproc," and will be described in Sections 11.3.1 through 11.3.4.

11.3.1 True Cramér-Rao Lower Bound

The script "calcTcrlb.m" computes the two code-phase TCRLBs of Section 4.6 and the conventional code-phase CRLB of Section 4.3.2.6 (with and without squaring loss) for the navigation signal specified in "init.m." After you start the script, the four bounds will be plotted as a function of C/N_0.

11.3.2 Discriminator Noise Analysis

The script "calcDiscNoise.m" should help carry out performance analyses of the waveform-based tracking scheme with the noncoherent discriminators of Section 8.1. Before starting "calcDiscNoise.m," you may set the considered navigation signal and the used D- and P-correlator reference signals. The script will visualize the code-phase error, the Doppler error, and the carrier-phase error as a function of the C/N_0 value. In addition, a code multipath error envelope will be plotted assuming one line-of-sight signal and one multipath signal with a constant attenuation and a variable geometric delay.

11.3.3 FFT Acquisition

The script "FftAcquisition.m" implements two FFT acquisition schemes and compares the results. After start-up, a window will appear visualizing the correlation function as a function of the code phase.

The script implements the circular shift method of Section 9.5.4 and the secondary code acquisition scheme of Section 9.5.7 (note that the Doppler preprocessing method of Section 9.5.6 is a special case of Section 9.5.7). An artificial tiered-code input signal is used and primary and secondary code can be changed in the script. The input signal is a code-phase-shifted version of the reference signal. The input signal is *not* related to the settings in "init.m." The Doppler search bin as well as the the Doppler summation range B can be modified. The latter parameter allows the computational burden to be reduced at the cost of reduced sensitivity.

11.3.4 Simplified Vector Tracking with Multipath Mitigation and Spectral Whitening

The script "track.m" includes a bit-true simulation of a received and quantized navigation signal and processes this signal with the multipath-estimating discriminator of Section 8.3. The signal is first generated by starting the script "generate-Signal.m." After starting "track.m," code-phase, Doppler, and carrier-phase errors are plotted as a function of time. Additionally, the estimated number of multipath signal components are displayed.

You may specify the used navigation signal type, the user dynamics, and the multipath settings in the script "init.m." Furthermore, the position and type of the correlators can be modified within "track.m." The underlying P- and D-correlator reference signals can be changed in "init.m." as well as the number of ADC bits, sample rate, and bandwidth. The spectral characteristics of the received noise can be selected and the method of spectral whitening (see Section 6.4) can be switched on or off.

The script "track.m" also demonstrates how the estimated errors have to be combined with predicted-pseudorange values. This is the basis for vector tracking, where the prediction is based on the receiver's position estimate.

Appendix

This appendix summarizes a few methods that are used to derive detection and estimation algorithms in the main text.

A.1 Complex Least-Squares Adjustment

This section extends a standard least-squares adjustment [1], which is modified to include complex-valued random variables as parameters and as observations. By using complex-valued random variables, we obtain a more compact notation for the least-squares adjustment. Complex-valued random variables appear for I- and Q-components of the IF signal, for the correlator outputs, or for the complex-valued signal amplitude. In that case, the I- and Q-components share the same stochastic and functional model allowing the use of the compact complex notation.

A.1.1 Definitions

In this section, lower case letters are used to indicate all random variables. If we have two i.i.d. real-valued random variables x and y, then we define a new complex-valued random variable z as

$$z = x + iy \tag{A.1}$$

The mean value <...> and the variance var <...> of z is given by

$$\begin{aligned} \langle z \rangle &= \langle x \rangle + i \langle y \rangle \\ \operatorname{var}\langle z \rangle &= \langle \bar{z} z \rangle - \langle \bar{z} \rangle \langle z \rangle = \langle x^2 + y^2 \rangle - \langle x \rangle^2 - \langle y \rangle^2 = 2\operatorname{var}\langle x \rangle = 2\operatorname{var}\langle y \rangle \end{aligned} \tag{A.2}$$

This syntax can be extended to two vectors of real valued i.i.d. random variables $\mathbf{x}$ and $\mathbf{y}$, namely

$$\mathbf{z} = \mathbf{x} + i\mathbf{y} \tag{A.3}$$

With Q, a symmetric matrix and the hermitian conjugate z^* of z, we obtain

$$\begin{aligned} \mathbf{z}^* Q \mathbf{z} &= (\mathbf{x}^T - i\mathbf{y}^T) Q (\mathbf{x} + i\mathbf{y}) \\ &= \mathbf{x}^T Q \mathbf{x} + \mathbf{y}^T Q \mathbf{y} + i(\mathbf{x}^T Q \mathbf{y} - \mathbf{y}^T Q \mathbf{x}) = \mathbf{x}^T Q \mathbf{x} + \mathbf{y}^T Q \mathbf{y} \end{aligned} \tag{A.4}$$

because the scalar value $\mathbf{x}^T Q \mathbf{y}$ can be written as

$$\mathbf{x}^T Q \mathbf{y} = (\mathbf{x}^T Q \mathbf{y})^T = \mathbf{y}^T Q^T \mathbf{x} = \mathbf{y}^T Q \mathbf{x} \tag{A.5}$$

Furthermore, we see

$$\left\langle \mathbf{z}^*\mathbf{z} \right\rangle = \left\langle (\mathbf{x}^T - i\mathbf{y}^T)(\mathbf{x} + i\mathbf{y}) \right\rangle = \left\langle \mathbf{x}^T\mathbf{x} \right\rangle + \left\langle \mathbf{y}^T\mathbf{y} \right\rangle = 2\left\langle \mathbf{x}^T\mathbf{x} \right\rangle = 2\left\langle \mathbf{y}^T\mathbf{y} \right\rangle \tag{A.6}$$

Therefore, if we have two vectors of real-valued random variables whose respective covariance matrices are identical, then they can be combined to one complex-valued vector of random variables, whose covariance matrix is real-valued and two times the covariance matrix of one of the real-valued vectors.

A.1.2 Probability Density Function

If the two vectors of i.i.d. random variables $\mathbf{x}$ and $\mathbf{y}$ are of length L, are Gaussian with mean $\mu_\mathbf{x}$ and $\mu_\mathbf{y}$, and have a common covariance matrix Q, the probability density function can be written as

$$p(\mathbf{x}, \mathbf{y}) = p(\mathbf{x})p(\mathbf{y}) \tag{A.7}$$

with

$$p(\mathbf{x}) = \frac{1}{(2\pi)^{L/2} \det(Q)^{1/2}} \exp\left\{ -\frac{1}{2}(\mathbf{x} - \mu_\mathbf{x})^T Q^{-1} (\mathbf{x} - \mu_\mathbf{x}) \right\} \tag{A.8}$$

and similarly for $\mathbf{y}$. Therefore

$$p(\mathbf{x}, \mathbf{y})$$

$$= \frac{1}{(2\pi)^L \det(Q)} \exp\left\{ -\frac{1}{2}\left[(\mathbf{x} - \mu_\mathbf{x})^T Q^{-1} (\mathbf{x} - \mu_\mathbf{x}) + (\mathbf{y} - \mu_\mathbf{y})^T Q^{-1} (\mathbf{y} - \mu_\mathbf{y}) \right] \right\} \tag{A.9}$$

$$= \frac{1}{(2\pi)^L \det(Q)} \exp\left\{ -\frac{1}{2}\left[(\mathbf{z} - \mu_\mathbf{z})^* Q^{-1} (\mathbf{z} - \mu_\mathbf{z}) \right] \right\}$$

with

$$\mu_\mathbf{z} = \mu_\mathbf{x} + i\mu_\mathbf{y} \tag{A.10}$$

See also Theorem 15.1 in Kay's book for further information [2]. There the matrix C=2Q is used yielding

$$p(\mathbf{z}) = \frac{1}{\pi^L \det(C)} \exp\left\{ -\left[(\mathbf{z} - \mu_\mathbf{z})^* C^{-1} (\mathbf{z} - \mu_\mathbf{z}) \right] \right\} \tag{A.11}$$

We also obtain

$$\left\langle \mathbf{z}\mathbf{z}^* \right\rangle = C = 2Q \tag{A.12}$$

A.1.3 The Adjustment

The complex LSQ adjustment procedure is based on the following function model

$$\mathbf{l} = \mathbf{l}(\mathbf{x}) \tag{A.13}$$

where $\mathbf{x}$ is the vector of parameters and $\mathbf{l}$ is the deterministic part of the measurements. Both are assumed to be complex-valued.

We assume the following linearized observation model

$$\mathbf{k} = \mathbf{l}(\mathbf{x}_0) + A\Delta\mathbf{x} + \mathbf{v} \tag{A.14}$$

where the vector of parameters is split into the a priori values (initial guesses) $\mathbf{x}_0$ and the corrections $\Delta\mathbf{x}$,

$$\mathbf{x} = \mathbf{x}_0 + \Delta\mathbf{x} \tag{A.15}$$

The observations are denoted as $\mathbf{k}$ and the measurement errors as $\mathbf{v}$.

The observations $\mathbf{k}$ and the measurement errors $\mathbf{v}$ are assumed to be complex-valued random variables. The a priori values $\mathbf{x}_0$ are assumed to be known deterministic quantities and the corrections $\Delta\mathbf{x}$ are deterministic but unknown quantities. Both $\mathbf{x}_0$ and $\Delta\mathbf{x}$ are generally complex-valued. The measurement errors are assumed to have zero mean and a real symmetric covariance matrix $Q_\mathbf{v}$

$$\langle \mathbf{v} \rangle = 0, \quad \langle \mathbf{v}\mathbf{v}^* \rangle = Q_\mathbf{v} \Rightarrow \mathrm{var}\langle \mathbf{k} \rangle = \left\langle \left(\mathbf{k} - \langle \mathbf{k} \rangle\right)\left(\mathbf{k}^* - \langle \mathbf{k}^* \rangle\right) \right\rangle = Q_\mathbf{v} \tag{A.16}$$

The design matrix A is given by

$$A_{i,j} = \frac{\partial l_i(\mathbf{x})}{\partial x_j}\Big|_{\mathbf{x}=\mathbf{x}_0} \tag{A.17}$$

The reduced observations are defined as

$$\mathbf{k}' = \mathbf{k} - \mathbf{l}(\mathbf{x}_0) \quad \Rightarrow \quad \langle \mathbf{k}' \rangle = A\Delta\mathbf{x}, \quad \mathrm{var}\langle \mathbf{k}' \rangle = Q_\mathbf{v} \tag{A.18}$$

and the LSQ principle is formulated as

$$\chi/2 = (\mathbf{k}' - A\Delta\mathbf{x})^* Q_\mathbf{v}^{-1} (\mathbf{k}' - A\Delta\mathbf{x}) \rightarrow \min_{\Delta\mathbf{x}} \tag{A.19}$$

The derivative of χ with respect to $\Delta\mathbf{x}^*$ yields

$$\frac{\partial}{\partial \Delta\mathbf{x}^*}\left[(\mathbf{k}' - A\Delta\mathbf{x})^* Q_\mathbf{v}^{-1}(\mathbf{k}' - A\Delta\mathbf{x})\right] = -A^* Q_\mathbf{v}^{-1}(\mathbf{k}' - A\Delta\mathbf{x}) = 0 \tag{A.20}$$

Note, that this is a set of complex-valued equations. The derivative with respect to $\Delta\mathbf{x}$ yields the same set of equations (but complex-conjugated). When using the complex notation, the variables $\Delta\mathbf{x}$ and $\Delta\mathbf{x}^*$ are considered to be independent (deterministic) variables; for example,

$$\frac{\partial \Delta\mathbf{x}}{\partial \Delta\mathbf{x}^*} = 0 \tag{A.21}$$

To prove this statement consider a sufficiently continuous real-valued function $f(x,y)$ of two real-valued unknowns. A necessary condition that it achieves a minimum or maximum value, is that its derivatives with respect to x and y are zero. Substituting x and y with two independent complex-valued variables defined by

$$\begin{pmatrix} z \\ z^* \end{pmatrix} = \begin{pmatrix} 1 & i \\ 1 & -i \end{pmatrix} \begin{pmatrix} x \\ y \end{pmatrix} \tag{A.22}$$

results in

$$\begin{pmatrix} \dfrac{\partial f}{\partial x} \\ \dfrac{\partial f}{\partial y} \end{pmatrix} = \begin{pmatrix} 1 & 1 \\ i & -i \end{pmatrix} \begin{pmatrix} \dfrac{\partial f}{\partial z} \\ \dfrac{\partial f}{\partial \tilde{z}^*} \end{pmatrix} \tag{A.23}$$

and thus

$$\begin{pmatrix} \dfrac{\partial f}{\partial x} \\ \dfrac{\partial f}{\partial y} \end{pmatrix} = 0 \quad \Leftrightarrow \quad \begin{pmatrix} \dfrac{\partial f}{\partial z} \\ \dfrac{\partial f}{\partial \tilde{z}^*} \end{pmatrix} = 0 \tag{A.24}$$

Therefore looking for the function's extreme values with respect to z and $\tilde{z}^*$ is equal to looking for the extreme values with respect to x and y. We require that the function f needs to be holomorphic (complex differentiable) in z and $\tilde{z}^*$ as its first derivative exists. We require that z and $\tilde{z}^*$ are complex-conjugate with respect to each other, thereby solving the original LSQ problem. This implies that the two equations on the right-hand side of (A.24) are identical, since

$$f\left(z + \tilde{z}^*, i\left(z - \tilde{z}^*\right)\right) = \overline{f\left(z + \tilde{z}^*, i\left(z - \tilde{z}^*\right)\right)} \tag{A.25}$$

The solution of (A.20) is the well known LSQ estimator equation, but now in complex notation

$$\Delta\hat{\mathbf{x}} = N^{-1}A^*Q_v{}^{-1}\mathbf{k}' \tag{A.26}$$

with the normal matrix N

$$N = A^*Q_v{}^{-1}A \tag{A.27}$$

The normal matrix and its inverse are hermitian ($N^* = N$) and thus its eigenvalues are real-valued. The estimated corrections are themselves random variables (because they depend on the observations) and from (A.18) we obtain for the expected value

$$\left\langle \Delta\hat{\mathbf{x}} \right\rangle = N^{-1}A^*Q_v{}^{-1}\left\langle \mathbf{k}' \right\rangle = N^{-1}A^*Q_v{}^{-1}A\Delta\mathbf{x} = \Delta\mathbf{x}\,, \tag{A.28}$$

and from (A.18) for the variance

$$\begin{aligned} &\left\langle \left(\Delta\hat{\mathbf{x}} - \left\langle \Delta\hat{\mathbf{x}} \right\rangle\right)\left(\Delta\hat{\mathbf{x}}^* - \left\langle \Delta\hat{\mathbf{x}}^* \right\rangle\right) \right\rangle \\ &= \left\langle N^{-1}A^*Q_v{}^{-1}(\mathbf{k}'^* - \left\langle \mathbf{k}'^* \right\rangle)(\mathbf{k}' - \left\langle \mathbf{k}' \right\rangle)Q_v{}^{-1}AN^{-1} \right\rangle \\ &= N^{-1}A^*Q_v{}^{-1}Q_vQ_v{}^{-1}AN^{-1} = N^{-1}A^*Q_v{}^{-1}AN^{-1} = N^{-1} \end{aligned} \tag{A.29}$$

A.1.4 Real- and Complex-Valued Estimated Parameters

The estimated corrections (and thus the estimated parameters themselves) are a vector of complex-valued random variables $\Delta\hat{\mathbf{x}}$. Choosing complex-valued parameters may be natural in the case of the signal amplitude (i.e., I- and Q-component) but is less obvious for parameters such as the code-phase or Doppler frequency. In the latter case, it may be necessary to force the parameters to be real-valued by imposing boundary conditions (e.g., $x_k - \overline{x}_k = 0$). This is most easily achieved after the setup of the LSQ adjustment by decomposing the complex-valued normal equation in its real and imaginary part. As a caveat, decomposing increases the dimension of the equation system by a factor of two. In case of an *uncorrelated* complex-valued variable (uncorrelated with respect to the other variables), the following observation is very useful to circumvent the decomposition.

Suppose the complex-valued estimated parameter x_i has a variance of

$$\mathrm{var}\langle \Delta x_i \rangle = (\mathrm{N}^{-1})_{ii} \tag{A.30}$$

As N^{-1} is hermitian, its diagonal elements are real-valued, thus the variance of the real part is

$$\begin{aligned} \mathrm{var}\langle \Delta x_{i,re} \rangle &= \mathrm{var}\left\langle \frac{1}{2}(\Delta x_i + \Delta \overline{x}_i) \right\rangle \\ &= \frac{1}{4}(\mathrm{var}\langle \Delta x_i \rangle + \mathrm{var}\langle \Delta \overline{x}_i \rangle) = \frac{1}{2}\mathrm{var}\langle \Delta x_i \rangle = \frac{(\mathrm{N}^{-1})_{ii}}{2} \end{aligned} \tag{A.31}$$

with the same being true for the imaginary part and

$$\begin{aligned} &\left\langle \left(\Delta x_{i,re} - \langle \Delta x_{i,re} \rangle\right)\left(\Delta x_{i,im} - \langle \Delta x_{i,im} \rangle\right) \right\rangle = \\ &= \frac{1}{4i}\left\langle \left(\Delta x_i + \Delta \overline{x}_i - \langle \Delta x_i \rangle - \langle \Delta \overline{x}_i \rangle\right)\left(\Delta x_i - \Delta \overline{x}_i - \langle \Delta x_i \rangle + \langle \Delta \overline{x}_i \rangle\right) \right\rangle \\ &= \frac{1}{4i}\left(\left\langle \Delta x_i^2 - \Delta \overline{x}_i^2 \right\rangle + \left(\langle \Delta x_i \rangle^2 - \langle \Delta \overline{x}_i \rangle^2\right)\right) = 0 \end{aligned} \tag{A.32}$$

because Δx_i and $\Delta \overline{x}_i$ are independent but identically distributed random variables. Thus, the covariance matrix of the real and imaginary component is given by

$$\mathrm{var}\left\langle \begin{pmatrix} \Delta x_{i,re} \\ \Delta x_{i,im} \end{pmatrix} \right\rangle = \frac{1}{2}\begin{pmatrix} \left(\mathrm{N}^{-1}\right)_{ii} & 0 \\ 0 & \left(\mathrm{N}^{-1}\right)_{ii} \end{pmatrix} \tag{A.33}$$

In that case forcing the imaginary part to be zero does not change the real part, which greatly simplifies the whole adjustment procedure. The imaginary component can simply be discarded. Note, however, that the same is generally not true if Δx_i correlates with other parameters.

A.1.5 A Posteriori Variance of Unit Weight

To calculate the a posteriori variance of unit weight, we expand the expression (A.20) after doing the LSQ adjustment into

$$
\begin{aligned}
&\left\langle \chi/2 \right\rangle = \left\langle (\mathbf{k}' - A\Delta\hat{\mathbf{x}})^* Q_{\mathbf{v}}^{-1}(\mathbf{k}' - A\Delta\hat{\mathbf{x}}) \right\rangle = Tr\left\langle (\mathbf{k}' - A\Delta\hat{\mathbf{x}})^* Q_{\mathbf{v}}^{-1}(\mathbf{k}' - A\Delta\hat{\mathbf{x}}) \right\rangle \\
&= Tr\left\langle (\mathbf{k}' - AN^{-1}A^* Q_{\mathbf{v}}^{-1}\mathbf{k}')^* Q_{\mathbf{v}}^{-1}(\mathbf{k}' - AN^{-1}A^* Q_{\mathbf{v}}^{-1}\mathbf{k}') \right\rangle \\
&= Tr\left\langle \mathbf{k}'^*(\mathbf{1} - Q_{\mathbf{v}}^{-1}AN^{-1}A^*)Q_{\mathbf{v}}^{-1}(\mathbf{1} - AN^{-1}A^* Q_{\mathbf{v}}^{-1})\mathbf{k}' \right\rangle \\
&= Tr\left\langle (\mathbf{1} - Q_{\mathbf{v}}^{-1}AN^{-1}A^*)Q_{\mathbf{v}}^{-1}(\mathbf{1} - AN^{-1}A^* Q_{\mathbf{v}}^{-1})\mathbf{k}'\mathbf{k}'^* \right\rangle \\
&= Tr\left\{ \left(Q_{\mathbf{v}}^{-1} - Q_{\mathbf{v}}^{-1}AN^{-1}A^* Q_{\mathbf{v}}^{-1} - Q_{\mathbf{v}}^{-1}AN^{-1}A^* Q_{\mathbf{v}}^{-1} + Q_{\mathbf{v}}^{-1}AN^{-1}A^* Q_{\mathbf{v}}^{-1}AN^{-1}A^* Q_{\mathbf{v}}^{-1} \right) \left\langle \mathbf{k}'\mathbf{k}'^* \right\rangle \right\} \\
&= Tr\left\{ \left(Q_{\mathbf{v}}^{-1} - Q_{\mathbf{v}}^{-1}AN^{-1}A^* Q_{\mathbf{v}}^{-1} - Q_{\mathbf{v}}^{-1}AN^{-1}A^* Q_{\mathbf{v}}^{-1} + Q_{\mathbf{v}}^{-1}AN^{-1}A^* Q_{\mathbf{v}}^{-1} \right) \left\langle \mathbf{k}'\mathbf{k}'^* \right\rangle \right\} \\
&= Tr\left\{ \left(Q_{\mathbf{v}}^{-1} - Q_{\mathbf{v}}^{-1}AN^{-1}A^* Q_{\mathbf{v}}^{-1} \right) \left\langle \mathbf{k}'\mathbf{k}'^* \right\rangle \right\}
\end{aligned}
\tag{A.34}
$$

The expected value for the reduced observations is obtained from

$$
Q_{\mathbf{v}} = \operatorname{var}\left\langle \mathbf{k}' \right\rangle = \left\langle \left(\mathbf{k}' - \left\langle \mathbf{k}' \right\rangle\right)\left(\mathbf{k}'^* - \left\langle \mathbf{k}'^* \right\rangle\right) \right\rangle = \left\langle \mathbf{k}'\mathbf{k}'^* \right\rangle - \left\langle \mathbf{k}' \right\rangle \left\langle \mathbf{k}'^* \right\rangle \tag{A.35}
$$

yielding

$$
\left\langle \mathbf{k}'\mathbf{k}'^* \right\rangle = Q_{\mathbf{v}} + \left\langle \mathbf{k}' \right\rangle \left\langle \mathbf{k}'^* \right\rangle = Q_{\mathbf{v}} + A\Delta\mathbf{x}\Delta\mathbf{x}^* A^* \tag{A.36}
$$

Inserting (A.36) into (A.34) we obtain

$$
\begin{aligned}
&\left\langle \chi/2 \right\rangle = Tr\left\{ \left(Q_{\mathbf{v}}^{-1} - Q_{\mathbf{v}}^{-1}AN^{-1}A^* Q_{\mathbf{v}}^{-1} \right)\left(Q_{\mathbf{v}} + A\Delta\mathbf{x}\Delta\mathbf{x}^* A^* \right) \right\} \\
&= Tr\left\{ \left(\mathbf{1} - Q_{\mathbf{v}}^{-1}AN^{-1}A^* + Q_{\mathbf{v}}^{-1}A\Delta\mathbf{x}\Delta\mathbf{x}^* A^* - Q_{\mathbf{v}}^{-1}AN^{-1}A^* Q_{\mathbf{v}}^{-1}A\Delta\mathbf{x}\Delta\mathbf{x}^* A^* \right) \right\} \\
&= Tr\left\{ \left(\mathbf{1} - Q_{\mathbf{v}}^{-1}AN^{-1}A^* + A^* Q_{\mathbf{v}}^{-1}A\Delta\mathbf{x}\Delta\mathbf{x}^* - A^* Q_{\mathbf{v}}^{-1}AN^{-1}A^* Q_{\mathbf{v}}^{-1}A\Delta\mathbf{x}\Delta\mathbf{x}^* \right) \right\} \\
&= Tr\left\{ \left(\mathbf{1} - Q_{\mathbf{v}}^{-1}AN^{-1}A^* + A^* Q_{\mathbf{v}}^{-1}A\Delta\mathbf{x}\Delta\mathbf{x}^* - A^* Q_{\mathbf{v}}^{-1}A\Delta\mathbf{x}\Delta\mathbf{x}^* \right) \right\} \\
&= Tr\left\{ \left(\mathbf{1} - Q_{\mathbf{v}}^{-1}AN^{-1}A^* \right) \right\} = Tr\left\{ \left(\mathbf{1}_r - \mathbf{1}_u \right) \right\} = r - u
\end{aligned}
\tag{A.37}
$$

where $\mathbf{1}_h$ denotes a h times h unit matrix. The symbol r denotes the number of complex-valued observations, the symbol u the number of complex-valued estimated corrections

The adjusted observations are given as

$$
\mathbf{k}'_a = \mathbf{k}' - A\Delta\hat{\mathbf{x}} = \mathbf{k}' - AN^{-1}A^* Q_{\mathbf{v}}^{-1}\mathbf{k}' = (\mathbf{1} - AN^{-1}A^* Q_{\mathbf{v}}^{-1})\mathbf{k}' \tag{A.38}
$$

with the stochastic properties

$$\langle \mathbf{k}'_a \rangle = 0, \quad \langle \mathbf{k}'_a \mathbf{k}'^*_a \rangle = (\mathbf{1} - AN^{-1}A^* Q_v^{-1}) Q_v (\mathbf{1} - AN^{-1}A^* Q_v^{-1})^* \tag{A.39}$$

We analyze the covariance matrix of the adjusted observations and obtain

$$\begin{aligned} \langle \mathbf{k}'_a \mathbf{k}'^*_a \rangle &= (\mathbf{1} - AN^{-1}A^* Q_v^{-1}) Q_v (\mathbf{1} - Q_v^{-1} AN^{-1}A^*) \\ &= Q_v - AN^{-1}A^* - AN^{-1}A^* + AN^{-1}A^* Q_v^{-1} AN^{-1}A^* \\ &= Q_v - AN^{-1}A^* \end{aligned} \tag{A.40}$$

Based on the adjusted observation the vector $\mathbf{k}''$ of complex-valued random variables is defined as

$$\mathbf{k}'' = \sqrt{Q_v^{-1}}\, \mathbf{k}'_a \quad \text{with} \quad \langle \mathbf{k}'' \rangle = 0 \tag{A.41}$$

and we obtain for χ

$$\chi/2 = \mathbf{k}''^* \mathbf{k}'' \tag{A.42}$$

For the covariance matrix of $\mathbf{k}''$ we obtain the matrix M given as

$$M = \langle \mathbf{k}'' \mathbf{k}''^* \rangle = \sqrt{Q_v^{-1}} (Q_v - AN^{-1}A^*) \sqrt{Q_v^{-1}} = \mathbf{1} - \sqrt{Q_v^{-1}} AN^{-1}A^* \sqrt{Q_v^{-1}} \tag{A.43}$$

The matrix is idempotent as the following relation holds

$$\begin{aligned} MM &= \left(\mathbf{1} - \sqrt{Q_v^{-1}} AN^{-1}A^* \sqrt{Q_v^{-1}}\right)\left(\mathbf{1} - \sqrt{Q_v^{-1}} AN^{-1}A^* \sqrt{Q_v^{-1}}\right) \\ &= \mathbf{1} - \sqrt{Q_v^{-1}} AN^{-1}A^* \sqrt{Q_v^{-1}} - \sqrt{Q_v^{-1}} AN^{-1}A^* \sqrt{Q_v^{-1}} + \sqrt{Q_v^{-1}} AN^{-1}A^* \sqrt{Q_v^{-1}} \sqrt{Q_v^{-1}} AN^{-1}A^* \sqrt{Q_v^{-1}} \\ &= \mathbf{1} - \sqrt{Q_v^{-1}} AN^{-1}A^* \sqrt{Q_v^{-1}} = M \end{aligned} \tag{A.44}$$

Furthermore, it is hermitian because

$$M^* = \mathbf{1} - \sqrt{Q_v^{-1}} AN^{-1}A^* \sqrt{Q_v^{-1}} = M \tag{A.45}$$

Since M is hermitian it only has real eigenvalues and it can be diagonalized. This is written as

$$M = D\Lambda D^* \quad \Leftrightarrow \quad D^* M D = \Lambda \tag{A.46}$$

with an unitary matrix D

$$DD^* = \mathbf{1} \tag{A.47}$$

and a diagonal matrix Λ. Since M is hermitian and idempotent, its eigenvalues can either be one or zero.

The vector of complex-valued random variables $\mathbf{k}'''$ is defined as

$$\mathbf{k}''' = D^*\mathbf{k}'' \tag{A.48}$$

For the covariance matrix we obtain

$$\left\langle \mathbf{k}'''\mathbf{k}'''^* \right\rangle = D^*MD = \Lambda \tag{A.49}$$

where the Λ is a diagonal matrix containing the eigenvalues of M, and the main diagonal is populated either by zeros or ones. Thus the elements of $\mathbf{k}'''$ are uncorrelated complex-valued random variables whose variance is either zero or one. If we express the expected value of χ as a function of $\mathbf{k}'''$, we obtain

$$\left\langle \chi/_2 \right\rangle = \left\langle \mathbf{k}''^*\mathbf{k}'' \right\rangle = \left\langle \mathbf{k}'''^* D^* D\mathbf{k}''' \right\rangle = \left\langle \mathbf{k}'''^*\mathbf{k}''' \right\rangle = \sum_{j=1}^{r} \Lambda_{jj} \tag{A.50}$$

We know from (A.37) that this equation also evaluates to r–u and we conclude that the vector $\mathbf{k}'''$ includes exactly r–u random variables with unit variance. The other u elements are random variables of zero variance.

So far, we made no assumption on the actual distribution of our measurement errors. However, to obtain a quality measure for the LSQ adjustment, we assume that the measurement errors $\mathbf{v}$ are Gaussian. Because the operations to this point involving random variables were linear, also the random variables contained in $\mathbf{k}'''$ are either complex Gaussian variables with unit variance or they vanish. Thus, for χ we obtain [see (A.2)]

$$\chi = 2\mathbf{k}'''^*\mathbf{k}''' = 2\sum_{j=1}^{r}\left(k_j'''\overline{k_j'''}\right) = \sum_{j=1}^{r-u}\left(2\left(k_{j,re}'''\right)^2 + 2\left(k_{j,im}'''\right)^2\right) \tag{A.51}$$

The minimum cost χ is the sum of $2(r$–$u)$ real-valued random variables, each having unit weight. Therefore χ is distributed according to the chi-squared distribution with $2(r$–$u)$ degrees of freedom, which is formally written as

$$\chi \sim \chi^2_{2(r-u)} \tag{A.52}$$

Compared to the LSQ adjustment with real-valued observations and real-valued estimated parameters, we observe that the degree of freedom, the effective number of parameters and the effective number of observations, doubles.

A.1.6 Example

To illustrate the complex LSQ adjustment method, let's assume we have the following simplified model for a correlator output $\mathbf{l}$.

$$l_i = aR(\tau + \tau_i) \tag{A.53}$$

Here l_i is the complex valued correlator output for a correlator with an offset of τ_i with respect to the prompt correlator value. The received code phase is symbolized as τ. The symbol a denotes the complex-valued signal amplitude.

$$a = |a| \exp\{i\varphi\} \tag{A.54}$$

where φ is the carrier-phase tracking error. The real-valued correlation function of the internal reference signal used for correlation with the incoming signal is denoted as R and we have a two-dimensional vector of unknowns

$$\mathbf{x} = \begin{pmatrix} a \\ \tau \end{pmatrix} \tag{A.55}$$

As a boundary condition, we require that the imaginary part of τ vanishes. For simplicity we choose as linearization point

$$\mathbf{x}_0 = \begin{pmatrix} a_0 \\ 0 \end{pmatrix} \tag{A.56}$$

We measure two complex-valued correlator values (with a code phase offset of τ_1 and τ_2). The design matrix is given by

$$A = \begin{pmatrix} R(\tau_1) & a_0 R'(\tau_1) \\ R(\tau_2) & a_0 R'(\tau_2) \end{pmatrix} \tag{A.57}$$

Assuming both observations are uncorrelated and each have unit weight, the normal matrix is given by

$$N = A^* A = \begin{pmatrix} R(\tau_1)^2 + R(\tau_2)^2 & a_0(R(\tau_1)R'(\tau_1) + R(\tau_2)R'(\tau_2)) \\ \bar{a}_0(R(\tau_1)R'(\tau_1) + R(\tau_2)R'(\tau_2)) & |a_0|^2 (R'(\tau_1)^2 + R'(\tau_2)^2) \end{pmatrix} \tag{A.58}$$

and its inverse by

$$N^{-1} = \begin{pmatrix} \dfrac{R'(\tau_1)^2 + R'(\tau_2)^2}{(R(\tau_2)R'(\tau_1) - R(\tau_1)R'(\tau_2))^2} & -\dfrac{R(\tau_1)R'(\tau_1) + R(\tau_2)R'(\tau_2)}{\bar{a}_0(R(\tau_2)R'(\tau_1) - R(\tau_1)R'(\tau_2))^2} \\ -\dfrac{R(\tau_1)R'(\tau_1) + R(\tau_2)R'(\tau_2)}{a_0(R(\tau_2)R'(\tau_1) - R(\tau_1)R'(\tau_2))^2} & \dfrac{R(\tau_1)^2 + R(\tau_2)^2}{|a_0|^2 (R(\tau_2)R'(\tau_1) - R(\tau_1)R'(\tau_2))^2} \end{pmatrix} \tag{A.59}$$

In case of a real and symmetric correlation function and if $\tau_1 = -\tau_2$, then

$$N^{-1} = \begin{pmatrix} \dfrac{1}{2R(\tau_1)^2} & 0 \\ 0 & \dfrac{1}{2|a_0|^2 R'(\tau_1)^2} \end{pmatrix} \tag{A.60}$$

In that case, forcing the imaginary part of the code-phase error estimate to vanish will not influence the other parameters. In fact, the complex-valued estimated value for the code phase is given by

$$
\begin{aligned}
\Delta\hat{\tau} = (N^{-1}A^*\mathbf{k}')_2 &= \frac{\bar{a}_0}{2|a_0|^2 R'(\tau_1)^2}(R'(\tau_1)k_1' + R'(-\tau_1)k_2') \\
&= \frac{1}{2a_0R'(\tau_1)}(k_1' - k_2') = \frac{1}{2a_0R'(\tau_1)}(l_1 - l_2)
\end{aligned}
\tag{A.61}
$$

The difference of both observations is a conventional early-minus-late difference and dividing it by the complex-valued signal amplitude brings the early-minus-late difference into the real part of this expression, which is the estimated code phase.

A.1.7 Discussion

Using complex variables in the LSQ adjustment potentially simplifies expressions as the carrier-phase error dependence involving sine/cosine terms can be avoided by using a complex-valued signal amplitude and observations. This reduces the dimension of the involved matrices by a factor of 2. The expressions are especially useful if the estimated parameters are uncorrelated because, in that case, the requirement of real-valued parameters is very easily formulated. Otherwise, those conditions have to be introduced as boundary conditions after forming the normal matrix.

A.2 Representing Digital GNSS Signals

In a navigation receiver, complex- and real-valued signal samples may be used to represent the received navigation signals. Both options are equivalent and will be discussed in this appendix.

The navigation signal is received at RF level as a single (i.e. real-valued) signal. It is down-converted to a lower frequency by one of the following methods:

1. One or more real mixers plus lowpass filters;
2. Filtering and direct RF (bandpass) sampling;
3. One or more complex (I/Q) mixers.

Methods 1 and 2 result in a real-valued signal, whereas Method 3 produces a complex-valued signal. Some variants of the presented methods exist, such as a real mixer plus IF/4 down conversion, which also produces a complex-valued signal. After down conversion, the signal is either centered around an IF or it is located directly at baseband.

A.2.1 Complex-Valued Input Signal

The samples of a complex-valued navigation signal consist of two numbers (the real and the imaginary part). The samples may be generated by two independent ADCs, which are synchronized with each other. They capture the I- and the Q-component of the output from a complex mixer bringing the navigation signal from the RF to the IF or even to baseband.

For complex-sampling, the Nyquist criterion reads that the sampling frequency must be higher than the dual-sided signal bandwidth. The received samples are random variables composed of a deterministic part r_μ and a stochastic-noise part N_μ,

$$S_\mu = r_\mu + N_\mu \tag{A.62}$$

The deterministic part is assumed to be of the form

$$r_\mu = ac(t_\mu - \tau)\exp\{i(\omega t_\mu - \varphi)\} \tag{A.63}$$

This form is assumed to be valid for short time spans and during that time span the four signal parameters a, τ, ω, and ϕ are assumed to be constant. The sampled noise can be assumed to be white and is then modeled as

$$\left\langle N_\mu \bar{N}_\upsilon \right\rangle = 2\delta_{\mu,\upsilon} \tag{A.64}$$

In case the input signal samples are complex-valued, the total signal power P_{sig} is given by

$$P_{sig} = a^2 \tag{A.65}$$

The total noise power P_{noise} for a complex-valued input signal is two, as can be seen from (A.64), and the sample rate equals the (dual-sided) noise bandwidth B. The (dual-sided) noise power spectral density N_0 is thus

$$N_0 = \frac{P_{noise}}{B} = \frac{2}{f_s} \tag{A.66}$$

and the C/N_0 value is

$$C/N_0 = \frac{P_{sig}}{N_0} = \frac{f_s a^2}{2} \tag{A.67}$$

or equivalently the following expressions are helpful

$$a^2 = \frac{2C/N_0}{f_s} = \frac{2C/N_0}{B} \quad \Leftrightarrow \quad af_s = \sqrt{2C/N_0 B} \tag{A.68}$$

A.2.2 Real-Valued Input Signal

To represent a real-valued navigation signal, each signal sample consists of one number generated by one ADC. It captures the I- *or* the Q-component of a complex- or real-mixer output. The ADC may also be connected directly to the amplified and filtered RF signal.

For real sampling, the Nyquist criterion states that the sampling frequency must be higher than two times the dual-sided signal bandwidth. The received samples are random variables composed of a deterministic part r_μ and a stochastic-noise part N_μ,

$$S_\mu = r_\mu + N_\mu \tag{A.69}$$

The real-valued signal samples share the same signal model as the complex samples and their relation is expressed as

$$S_{\mu;real} = \mathrm{Re}\{S_{\mu;complex}\} \tag{A.70}$$

Consequently, the deterministic part is assumed to be of the form

$$r_\mu = ac(t_\mu - \tau)\cos(\omega t_\mu - \varphi) \tag{A.71}$$

and the sampled noise can be assumed to be white; that is,

$$\langle N_\mu N_\upsilon \rangle = \delta_{\mu,\upsilon} \tag{A.72}$$

In case the input signal samples are real-valued, the total signal power P_{sig} is given by

$$P_{sig} = \frac{a^2}{2} \tag{A.73}$$

The (single-sided) noise bandwidth B is given by half of the sample rate f_s (Nyquist criterion). Note, for a real-valued signal, only positive frequency values of its spectrum are considered. The complex-conjugated, but otherwise symmetric, part of the spectrum located at negative frequencies is irrelevant. Because we assume that the real part of the noise samples has unit variance, the total noise power P_{noise} for a real-valued input signal is one, as can be seen from (A.72). The (single-sided) noise power spectral density N_0 is

$$N_0 = \frac{P_{noise}}{B} = \frac{2}{f_s} \tag{A.74}$$

The C/N_0 value is defined as the ratio between received signal power and noise spectral density

$$C/N_0 = \frac{P_{sig}}{N_0} = \frac{f_s a^2}{4} \tag{A.75}$$

or equivalently the following expressions are occasionally helpful

$$a^2 = \frac{4C/N_0}{f_s} \Leftrightarrow \frac{af_s}{2} = \sqrt{2C/N_0 B} \tag{A.76}$$

A.2.3 Comparing Real- and Complex-Valued Signals

Complex- and real-valued signals are two equivalent ways of representing the same navigation signal, which is ultimately defined by its frequency content (or, loosely speaking, by its spectrum). A real-valued signal can be represented by only the positive frequency portion of its signal spectrum, although the spectrum (defined via the Fourier transform) of a real-valued signal also consists of negative frequencies. However, the full spectrum of real-valued signals is symmetric with respect to the zero frequency. On the contrary, the spectrum of a complex-valued signal has positive and negative frequency components and is, in general, not symmetric.

By using the Hilbert transform, a real-valued signal can be transformed into a complex-valued signal of half the sample rate [3]. Converting a complex-valued signal into an equivalent real-valued signal implies doubling the sample rate.

Let's assume a complex-valued signal whose spectral density is confined within the frequency range from 0 to f_s, and vanishes outside. This signal consists of noise samples with a variance of two. Converting it into a real-valued signal consisting of noise samples with a variance of one is a two-step procedure:

- Increase the sample rate by a factor of two and interpolate in time in a way that frequencies from f_s to $2f_s$ vanish (or perform an equivalent procedure in the frequency domain).
- Cut off the imaginary part (in the time domain) and take the real part.

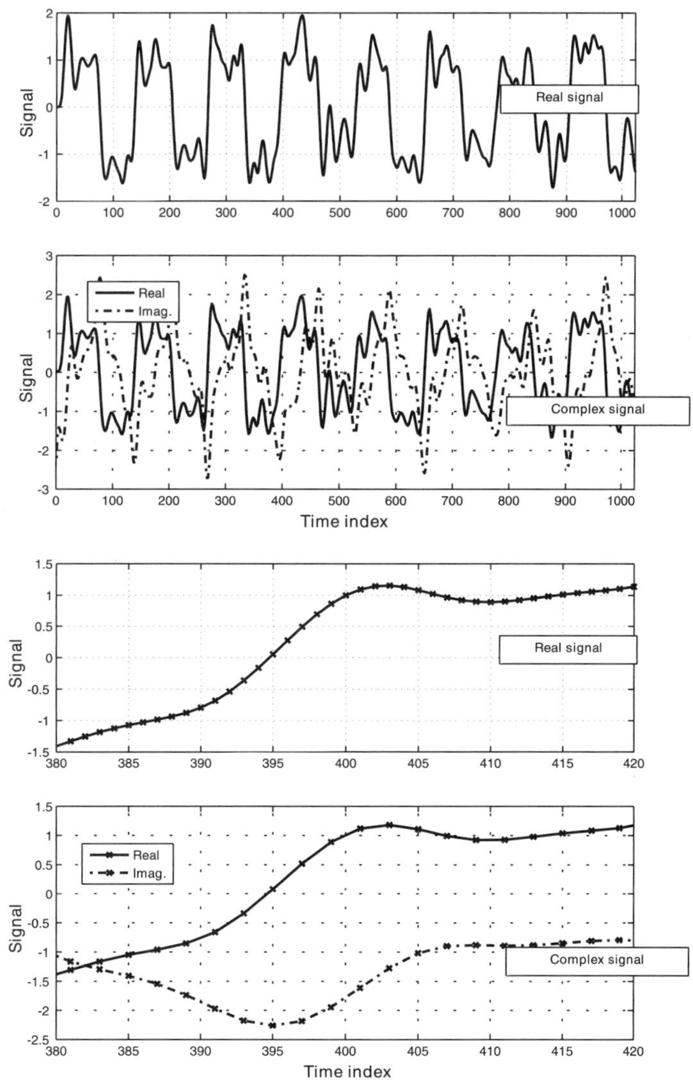

Figure A.1 Example of a complex and real representation of the same signal (lower figure zooms into upper figure).

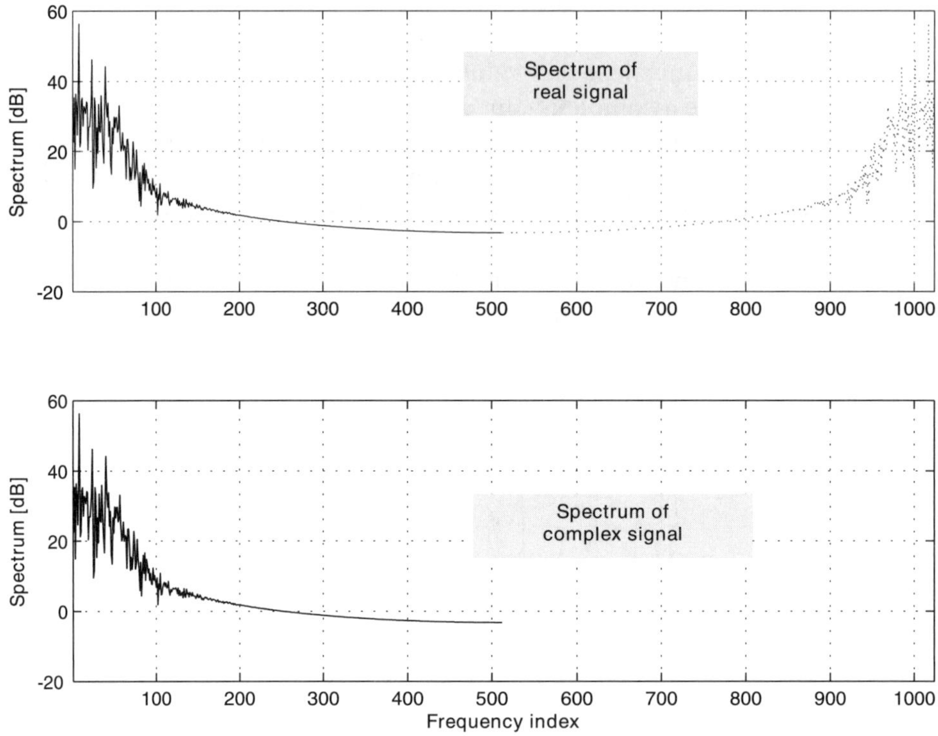

Figure A.2 Spectrum of the complex and real representation of the signal of Figure A.1.

Converting a real-valued signal with a sample rate f_s and with noise samples of unit variance into a complex-valued signal having a noise-sample variance of two is accomplished by a three-step procedure that is related to the Hilbert transform:

- Add a vanishing imaginary signal component (i.e., make the signal complex).
- Remove frequencies from $f_s/2$ to f_s (or equivalently from $-f_s/2$ to 0).
- Reduce sample rate from f_s to $f_s/2$ by decimation.

Figures A.1 through A.3 illustrate the conversion procedure. They are based on a real-valued signal with a length of 1,024 samples. The complex-valued signal has a length of 512 samples and uses half of the sample rate of the real-valued signal. Figures A.1 and A.2 use a signal with low frequency components; Figure A.3 is based on a signal with low and high frequency components.

From Figure A.1, one sees that the real part of the complex-valued signal is virtually identical to the real-valued signal (apart from the different sample rate). This is because the spectrum of the real-valued signal is concentrated at low frequencies. In general, the interpolated real-valued samples depend on the real and imaginary component of the complex-valued signal. For example, in Figure A.3, one sees that a simple linear interpolation of the real-part of the complex-valued signal is not sufficient to get the real-valued signal. Frequency domain interpolation has to be employed.

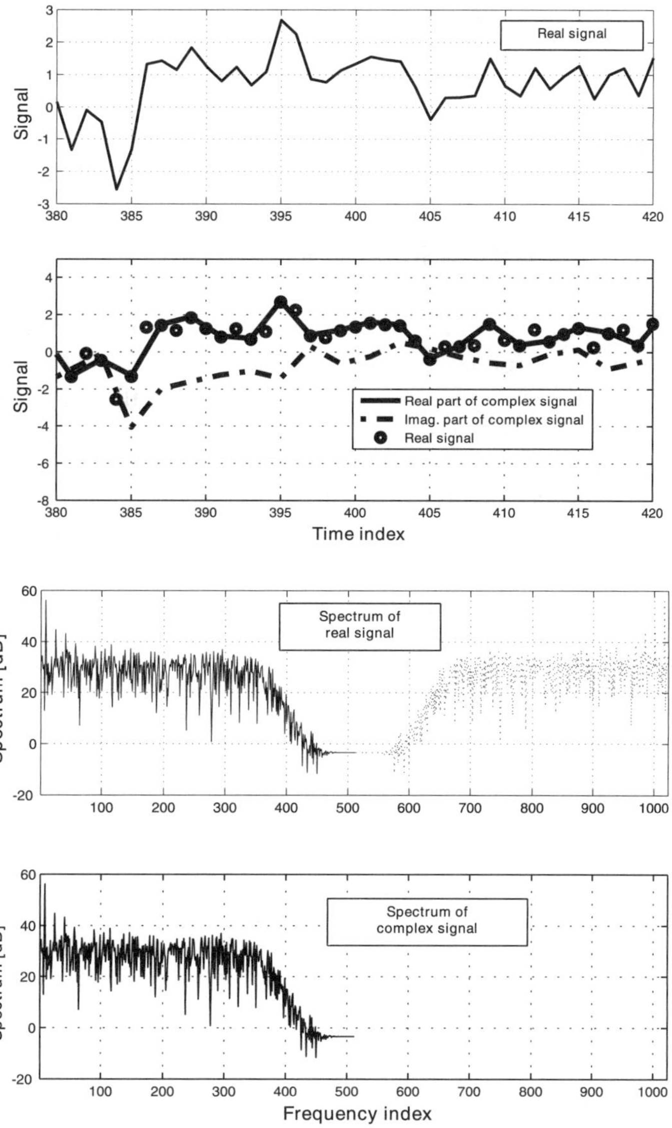

Figure A.3 Example of a complex and real representation of a signal with high frequency contributions.

The imaginary part of the complex-valued signal can also be obtained by calculating the Hilbert transform of the real-valued signal and a subsequent reduction of the sample rate from f_s to $f_s/2$.

The spectra shown in Figures A.2 and A.3 look virtually identical. The negative frequency components of the real-valued signal spectrum are plotted on the right-hand side, above the Nyquist frequency.

It should be noted that the conversion from the real to the complex representation is not completely unique because the frequency component of the real-valued signal corresponding to the negative frequency of $-f_{s,real}/2$ has no counterpart in the complex representation. The spectrum of the complex-valued signal ranges from 0 to $f_{s,real}(L-1)/(2L)$, where L is the number of samples representing the complex-valued signal. For the purpose of navigation signal processing, this can be ignored as long as the main lobes of the navigation signal are not located at that frequency. In that case, a possible conversion would, more or less, only affect the noise part of the signal, which can be ignored.

Neither of the conversions change a or n_0.

A.3 Correlation Function Invariance

This appendix outlines a proof, that the cross-correlation function of two wide-sense stationary stochastic processes is invariant under the sampling process. It implies that the sampling frequency does not influence the shape of the cross-correlation function. Aliasing effects average out because of the wide-sense stationary property.

Let $S(t)$ and $R(t)$ be two complex-valued continuous stochastic processes. Both processes are wide-sense stationary. This implies that first- and second-order moments are constant in time [4].

The continuous-time cross-correlation function is defined as

$$\left\langle S(t-\tau)R(t)\right\rangle_{\mathrm{S,R}} = R_{S,R}(\tau) \tag{A.77}$$

Discrete-time stochastic processes are obtained via sampling the continuous-time processes, like

$$R_\mu = R(t_\mu) \tag{A.78}$$

with the sample index μ

$$t_\mu = \frac{\mu}{f_s} \tag{A.79}$$

and f_s being the sample rate.

Both processes are wide-sense stationary, thus

$$\left\langle \frac{1}{2T}\int_{t=-T}^{T} S(t-\tau)R(t)dt \right\rangle_{\mathrm{S,R}} = \frac{1}{2T}\int_{t=-T}^{T} \left\langle S(t-\tau)R(t)\right\rangle_{\mathrm{S,R}} dt = R_{S,R}(\tau) \tag{A.80}$$

The same relationship also holds for the discrete-time process

$$\left\langle \frac{1}{2L+1}\sum_{\mu=-L}^{L} S(t_\mu-\tau_\mu)R(t_\mu) \right\rangle_{\mathrm{S,R}} = \frac{1}{2L+1}\sum_{\mu=-L}^{L} \left\langle S(t_\mu-\tau_\mu)R(t_\mu)\right\rangle_{\mathrm{S,R}} = R_{S,R}(\tau_\mu) \tag{A.81}$$

for an arbitrary sample rate f_s.

It demonstrates that the cross-correlation function can be derived from discrete samples regardless of the sampling frequency. The sampling frequency, however, determines the timely resolution of the obtained correlation function.

Loosely speaking, sampling needs to be *ergodic* in the sense that the sampling process needs to capture sufficient information on the stochastic properties of both processes. Ergodic sampling does not necessarily imply that the sample rate is larger than the Nyquist rate.

An obvious question arises, why aliasing effects—which occur if sub-Nyquist sampling is used—do not influence this result? In the following example, it will be shown that aliasing effects exist, but they average out when the correlation function is formed.

First, the Fourier transform of a stochastic process is given as

$$\tilde{R}(\omega) = \frac{1}{\sqrt{2\pi}} \int_{t=-\infty}^{\infty} R(t)\exp\{-i\omega t\}dt \tag{A.82}$$

For a specific angular frequency ω, the quantity $\tilde{R}(\omega)$ is a complex-valued random variable.

For two stochastic wide-sense stationary processes, the correlation between two Fourier transforms is

$$\begin{aligned}
\left\langle \tilde{S}(\omega)\tilde{R}(\omega')\right\rangle_{\mathrm{S,R}} &= \frac{1}{2\pi} \int_{t,t'=-\infty}^{\infty} \left\langle S(t)R(t')\right\rangle_{\mathrm{S,R}} \exp\{-i\omega t - i\omega' t'\}dtdt' \\
&= \frac{1}{2\pi} \int_{t,t'=-\infty}^{\infty} \left\langle S(t'-t)R(t')\right\rangle_{\mathrm{S,R}} \exp\{-i\omega(t'-t) - i\omega' t'\}dtdt' \\
&= \frac{1}{2\pi} \int_{t,t'=-\infty}^{\infty} R_{S,R}(t)\exp\{-i\omega(t'-t) - i\omega' t'\}dtdt' \\
&= \frac{1}{2\pi} \int_{t,t'=-\infty}^{\infty} R_{S,R}(t)\exp\{-it'(\omega+\omega') + i\omega t\}dtdt' \\
&= \delta(\omega+\omega') \int_{t=-\infty}^{\infty} R_{S,R}(t)\exp\{i\omega t\}dt
\end{aligned} \tag{A.83}$$

where the relation

$$\int_{t'=-\infty}^{\infty} \exp\{-i\omega t'\}dt' = 2\pi\delta(\omega) \tag{A.84}$$

has been used to represent Dirac's delta function. Overall, the correlation between the two Fourier transforms vanishes if both frequencies do not add up to zero.

If only discrete-time values τ_v of the cross-correlation function are considered

$$\tau_v = \frac{v}{f_s} \tag{A.85}$$

then the expected value of the timely averaged *continuous-time correlation function* for the shift index v is given by

$$\begin{aligned}
&\int\limits_{t=-\infty}^{\infty} \left\langle S(t-\tau_\upsilon)R(t)\right\rangle_{\mathrm{S,R}} dt \\
&= \frac{1}{2\pi} \int\limits_{t=-\infty}^{\infty} \int\limits_{\omega,\omega'=-\infty}^{\infty} \left\langle \tilde{S}(\omega)\tilde{R}(\omega')\right\rangle_{\mathrm{S,R}} \exp\{i\omega(t-\tau_\upsilon)+i\omega' t\} d\omega d\omega' dt \\
&= \int\limits_{\omega,\omega'=-\infty}^{\infty} \left\langle \tilde{S}(\omega)\tilde{R}(\omega')\right\rangle_{\mathrm{S,R}} \exp\{-i\omega\tau_\upsilon\}\delta(\omega+\omega') d\omega d\omega' \qquad \text{(A.86)} \\
&= \int\limits_{\omega=-\infty}^{\infty} \left\langle \tilde{S}(\omega)\tilde{R}(-\omega)\right\rangle_{\mathrm{S,R}} \exp\{-i\omega\tau_\upsilon\} d\omega \\
&= \sum_{m=-\infty}^{\infty} \int\limits_{\hat{\omega}=-\pi f_s}^{\pi f_s} \left\langle \tilde{S}(\hat{\omega}+2\pi m f_s)\tilde{R}(-\hat{\omega}-2\pi m f_s)\right\rangle_{\mathrm{S,R}} \exp\{-i\hat{\omega}\tau_\upsilon\} d\hat{\omega}
\end{aligned}$$

Here the auxiliary angular frequency $\hat{\omega}$ has been introduced,

$$\omega = \hat{\omega} + 2\pi m f_s \qquad \text{(A.87)}$$

The index m is used to count the continuous-time frequencies ω, which are all aliased after sampling onto the same discrete-time frequency $\hat{\omega}$.

On the other hand, the expected value of the timely averaged *discrete-time correlation function* is given by

$$\begin{aligned}
&\sum_{\mu=-\infty}^{\infty} \left\langle S(t_\mu-\tau_\upsilon)R(t_\mu)\right\rangle_{\mathrm{S,R}} \\
&= \frac{1}{2\pi} \sum_{\mu=-\infty}^{\infty} \int\limits_{\omega,\omega'=-\infty}^{\infty} \left\langle \tilde{S}(\omega)\tilde{R}(\omega')\right\rangle_{\mathrm{S,R}} \exp\{i\omega(t_\mu-\tau_\upsilon)+i\omega' t_\mu\} d\omega d\omega' \\
&= \frac{1}{2\pi} \int\limits_{\omega,\omega'=-\infty}^{\infty} \left\langle \tilde{S}(\omega)\tilde{R}(\omega')\right\rangle_{\mathrm{S,R}} \exp\{-i\omega\tau_\upsilon\} \sum_{\mu=-\infty}^{\infty} \exp\{i\omega t_\mu + i\omega' t_\mu\} d\omega d\omega' \qquad \text{(A.88)} \\
&= \int\limits_{\omega,\omega'=-\infty}^{\infty} \left\langle \tilde{S}(\omega)\tilde{R}(\omega')\right\rangle_{\mathrm{S,R}} \exp\{-i\omega\tau_\upsilon\} \sum_{k=-\infty}^{\infty} \delta(\omega'-(2\pi k f_s-\omega)) d\omega d\omega' \\
&= \sum_{k=-\infty}^{\infty} \int\limits_{\omega=-\infty}^{\infty} \left\langle \tilde{S}(\omega)\tilde{R}(2\pi k f_s-\omega)\right\rangle_{\mathrm{S,R}} \exp\{-i\omega\tau_\upsilon\} d\omega
\end{aligned}$$

Here the definition

$$\sum_{\mu=-\infty}^{\infty} \exp\{i\omega t_\mu\} = 2\pi \sum_{k=-\infty}^{\infty} \delta(\omega - 2\pi k f_s) \qquad \text{(A.89)}$$

for the frequency comb function has been used.

The expression (A.88) can be further simplified by introducing the auxiliary frequency $\hat{\omega}$ and we obtain

$$\begin{aligned}&\sum_{\mu=-\infty}^{\infty} \left\langle S(t_\mu - \tau_\upsilon)R(t_\mu)\right\rangle_{\mathbf{S,R}} \\ &= \sum_{m,k'=-\infty}^{\infty} \int_{\hat{\omega}=-\pi f_s}^{\pi f_s} \left\langle \tilde{S}(\hat{\omega}+2\pi m f_s)\tilde{R}(-2\pi k' f_s - \hat{\omega})\right\rangle_{\mathbf{S,R}} \exp\{-i\hat{\omega}\tau_\upsilon\}d\hat{\omega}\end{aligned} \tag{A.90}$$

For two wide-sense stationary processes, (A.86) equals (A.90) because

$$m \neq k' \quad \Rightarrow \quad \left\langle \tilde{S}(\hat{\omega}+2\pi m f_s)\tilde{R}(-2\pi k' f_s - \hat{\omega})\right\rangle_{\mathbf{S,R}} = 0 \tag{A.91}$$

of (A.83).

Overall, the frequency domain picture of aliasing effects on the cross-correlation function can be described as follows: aliasing effects occur during sampling, causing correlation of frequency components, which do not correlate in the continuous-time case. However, the ensemble average of the aliasing effects averages to zero, because wide-sense stationary processes have a diagonal covariance matrix in the frequency domain.

A.4 Useful Formulas

This appendix summarizes some useful formulas for the reader's convenience.

A.4.1 Fourier Transform

Definitions and normalization constants of three different Fourier transforms are summarized here. We consider the continuous Fourier transform and the discrete Fourier transform with finite and infinite length.

A.4.1.1 Continuous-Time

Let $x(t)$ be a complex valued function of time t in seconds, then it is related to its *angular frequency* Fourier transform $\tilde{x}(\omega)$(unit [rad/s]) via

$$x(t) = \frac{1}{\sqrt{2\pi}} \int_{-\infty}^{\infty} \tilde{x}(\omega)\exp\{i\omega t\}d\omega \tag{A.92}$$

and

$$\tilde{x}(\omega) = \frac{1}{\sqrt{2\pi}} \int_{-\infty}^{\infty} x(t)\exp\{-i\omega t\}dt \tag{A.93}$$

Dirac's delta function is expressed as

$$\int_{-\infty}^{\infty} \exp\{-i\omega t\}dt = 2\pi\delta(\omega), \quad \int_{-\infty}^{\infty} \exp\{i\omega t\}d\omega = 2\pi\delta(t) \tag{A.94}$$

The function $x(t)$ is related to its *frequency* Fourier transform $\tilde{x}(f)$ (unit [1/Hz]) via

$$x(t) = \int_{-\infty}^{\infty} \tilde{x}(f)\exp\{2\pi ift\}df \tag{A.95}$$

and

$$\tilde{x}(f) = \int_{-\infty}^{\infty} x(t)\exp\{-2\pi ift\}dt \tag{A.96}$$

Dirac's delta function is expressed as

$$\int_{-\infty}^{\infty} \exp\{-2\pi ift\}dt = \delta(f), \quad \int_{-\infty}^{\infty} \exp\{2\pi ift\}df = \delta(t) \tag{A.97}$$

The Fourier transform $\tilde{x}(f)$ (unit [1/Hz]) is related to the angular frequency Fourier transform $\tilde{x}(\omega)$ (in radians per seconds) via

$$\tilde{x}(\omega) = \sqrt{2\pi}\tilde{x}(f) \tag{A.98}$$

A.4.1.2 Discrete Infinite Length Signal

Let x_μ be an infinite series of complex numbers, obtained by sampling a complex-valued signal with a sample rate f_s, then x_μ is related to its Fourier transform $\tilde{x}(f)$ (unit [1/Hz]) via

$$x_\mu = \frac{1}{f_s} \int_{f=-f_s/2}^{f_s/2} \tilde{x}(f)\exp\left\{\frac{2\pi if}{f_s}\mu\right\}df \tag{A.99}$$

and

$$\tilde{x}(f) = \sum_{\mu=-\infty}^{\infty} x_\mu \exp\left\{-\frac{2\pi if}{f_s}\mu\right\} \qquad {}^{f_s}\!/_2 < f \le {}^{f_s}\!/_2 \tag{A.100}$$

Dirac's delta function equivalent, the Kronecker delta and the Dirac comb, are expressed as

$$\frac{1}{f_s} \int_{f=-f_s/2}^{f_s/2} \exp\left\{\frac{2\pi if}{f_s}\mu\right\}df = \delta_{\mu,0}, \quad \sum_{\mu=-\infty}^{\infty} \exp\left\{-\frac{2\pi if}{f_s}\mu\right\} = f_s \sum_{k=-\infty}^{\infty} \delta(f - kf_s) \tag{A.101}$$

A.4.1.3 Discrete Finite Length Signal

Let x_μ be a finite series of complex numbers ($0 \leq \mu < L$), then x_μ is related to its discrete Fourier transform $\tilde{x}_k$ via

$$x_\mu = \frac{1}{L}\sum_{k=0}^{L-1} \tilde{x}_k \exp\left\{2\pi i \frac{\mu k}{L}\right\} \tag{A.102}$$

and

$$\tilde{x}_k = \sum_{\mu=0}^{L-1} x_\mu \exp\left\{-2\pi i \frac{\mu k}{L}\right\} \tag{A.103}$$

Dirac's delta function equivalent, the Kronecker delta, is expressed as

$$\sum_{\mu=0}^{L-1} \exp\left\{2\pi i \frac{\mu k}{L}\right\} = L\delta_{k,0}, \quad \sum_{k=0}^{L-1} \exp\left\{-2\pi i \frac{\mu k}{L}\right\} = L\delta_{\mu,0} \tag{A.104}$$

A.4.2 Correlation Function

Often, two signals are correlated with each other and the correlation is performed over a limited time span on equidistantly spaced epochs. Although the result cannot be simplified in a strict mathematical sense, the resulting expression is commonly approximated by the cross-correlation function of the two signals. This argument is formalized in the following discussion.

Let $x(t)$ and $y(t)$ be two functions. We define the (approximate) correlation function of the two functions as

$$R_{x,y}(\tau) \approx \frac{1}{L}\sum_{\mu=1}^{L} x(t_\mu - \tau)y(t_\mu) \tag{A.105}$$

where L is the number of samples involved in the summation. The specific values of the epochs t_μ are context-specific but they are usually equally spaced. The range of possible τ values is much smaller than the time span covered by t_μ,

$$|\tau| \ll t_L - t_1 \tag{A.106}$$

The following identity holds approximately

$$R_{y,x}(\tau) \approx \frac{1}{L}\sum_{\mu=1}^{L} y(t_\mu - \tau)x(t_\mu) \approx \frac{1}{L}\sum_{\mu=1}^{L} x(t_\mu + \tau)y(t_\mu) \approx R_{x,y}(-\tau) \tag{A.107}$$

If $x'(t)$ and $y'(t)$ denote the first derivatives of $x(t)$ and $y(t)$, then the correlation functions involving derivatives are approximated as

$$\begin{aligned} R_{x',y}(\tau) &\approx \frac{1}{L}\sum_{\mu=1}^{L} x'(t_\mu - \tau)y(t_\mu) = -\frac{\partial}{\partial\tau}\frac{1}{L}\sum_{\mu=1}^{L} x(t_\mu - \tau)y(t_\mu) \approx -R'_{x,y}(\tau) \\ R_{x,y'}(\tau) &\approx \frac{1}{L}\sum_{\mu=1}^{L} x(t_\mu - \tau)y'(t_\mu) \approx \frac{\partial}{\partial\tau}\frac{1}{L}\sum_{\mu=1}^{L} x(t_\mu)y(t_\mu + \tau) \approx R'_{x,y}(\tau) \end{aligned} \tag{A.108}$$

For the second derivates, the following identities hold approximately

$$R_{x'',y}(\tau) \approx R_{x,y''}(\tau) \approx -R_{x',y'}(\tau) \approx R''_{x,y}\ (\tau) \tag{A.109}$$

For completeness, the following approximate identities are listed here

$$\begin{aligned}
&\frac{1}{L}\sum_{\mu=1}^{L} x(t_\mu - \tau_1)y(t_\mu - \tau_2) \approx R_{x,y}(\tau_1 - \tau_2)\\
&\frac{1}{L}\sum_{\mu=1}^{L} x'(t_\mu - \tau_1)y(t_\mu - \tau_2) \approx R_{x',y}(\tau_1 - \tau_2) \approx -R'_{x,y}\ (\tau_1 - \tau_2)\\
&\frac{1}{L}\sum_{\mu=1}^{L} x''(t_\mu - \tau_1)y(t_\mu - \tau_2) \approx R_{x'',y}(\tau_1 - \tau_2) \approx R''_{x,y}\ (\tau_1 - \tau_2)
\end{aligned} \tag{A.110}$$

For the auto-correlation function of a complex-valued function, the following expressions are approximately valid

$$\begin{aligned}
&R_{\bar{x},x}(\tau) \approx \frac{1}{L}\sum_{\mu=1}^{L} \bar{x}(t_\mu - \tau)x(t_\mu) \approx \overline{\frac{1}{L}\sum_{\mu=1}^{L} x(t_\mu)\bar{x}(t_\mu + \tau)} \approx \overline{R_{\bar{x},x}(-\tau)}\\
&\Rightarrow \operatorname{Im}\{R_{\bar{x},x}(0)\} \approx 0
\end{aligned} \tag{A.111}$$

for the first derivative

$$\begin{aligned}
&R'_{\bar{x},x}\ (\tau) = \frac{\partial}{\partial\tau}R_{\bar{x},x}(\tau) \approx \frac{\partial}{\partial\tau}\overline{R_{\bar{x},x}(-\tau)} = -\overline{R'_{\bar{x},x}\ (-\tau)}\\
&\Rightarrow \operatorname{Re}\{R'_{\bar{x},x}\ (0)\} \approx 0
\end{aligned} \tag{A.112}$$

and for the second derivative

$$\begin{aligned}
&R''_{\bar{x},x}\ (\tau) = \frac{\partial^2}{\partial\tau^2}R_{\bar{x},x}(\tau) \approx \frac{\partial^2}{\partial\tau^2}\overline{R_{\bar{x},x}(-\tau)} = \overline{R''_{\bar{x},x}\ (-\tau)}\\
&\Rightarrow \operatorname{Im}\{R''_{\bar{x},x}\ (0)\} \approx 0
\end{aligned} \tag{A.113}$$

A.4.3 Correlation with an Auxiliary Function

Let $x(t)$ and $y(t)$ be two functions. We define the (approximate) correlation function of the two functions including an auxiliary function $h(t)$ as

$$
\begin{aligned}
R_{x,y}(\tau;h) &\approx \frac{1}{L}\sum_{\mu=1}^{L} x(t_\mu-\tau)y(t_\mu)h(t_\mu) \\
&= \frac{R_{x,y}(\tau)}{L}\sum_{\mu=1}^{L}\frac{x(t_\mu-\tau)y(t_\mu)}{R_{x,y}(\tau)}h(t_\mu) \\
&\approx \frac{R_{x,y}(\tau)}{L}\sum_{\mu=1}^{L}\frac{x(t_\mu)y(t_\mu)}{R_{x,y}(0)}h(t_\mu) \\
&\approx R_{x,y}(\tau)\frac{1}{t_L-t_0}\int_{t=t_0}^{t_L} W(t)h(t)dt
\end{aligned}
\tag{A.114}
$$

with

$$
W(t)=\frac{x(t)y(t)}{R_{x,y}(0)} \tag{A.115}
$$

This formula is based on the assumption that the product of the two signals fulfills

$$
\int_{t=t_0}^{t_L}(x(t-\tau)y(t)R_{x,y}(0)-x(t)y(t)R_{x,y}(\tau))h(t)dt \approx 0 \tag{A.116}
$$

sufficiently well. Because this is a nontrivial assumption for the functions x, y and h, its validity has to be verified explicitly.

The function $W(t)$ identifies the timely occurrence of the cross-correlation energy. If $x(t)$ and $y(t)$ are realizations of two wide-sense stationary processes, then $W(t)$ is well approximated by one.

The formula for correlation with an auxiliary function is summarized as

$$
R_{x,y}(\tau;h)\approx R_{x,y}(\tau)\frac{R_{x,y}(0;h)}{R_{x,y}(0)} \tag{A.117}
$$

and indicates the required *independence* of the auxiliary function from the correlation function.

A.4.4 Correlation with Doppler

The following example is based on Section A.4.3 with h being a complex sinusoidal function. Let $x(t)$ and $y(t)$ be two functions. We define the (approximate) correlation function of the two functions including a Doppler offset ω as

$$
R_{x,y}(\tau,\omega)=R_{x,y}(\tau;\exp\{i\omega t\})\approx R_{x,y}(\tau)\frac{R_{x,y}(0;\exp\{i\omega t\})}{R_{x,y}(0)}=R_{x,y}(\tau)\kappa(\omega) \tag{A.118}
$$

Similar to Section A.4.3, it is assumed that the correlation function is independent from the complex sinusoidal function. The Doppler correlation function $\kappa(\omega)$ evaluates to

$$\begin{aligned}\kappa(\omega) &= \frac{R_{x,y}(0;\exp\{i\omega t\})}{R_{x,y}(0)} \\ &\approx \frac{1}{(t_L - t_0)} \int_{t=t_0}^{t_L} W(t)\exp\{i\omega t\}dt = \frac{1}{R_{x,y}(0)(t_L - t_0)} \int_{t=t_0}^{t_L} x(t)y(t)\exp\{i\omega t\}dt\end{aligned} \tag{A.119}$$

For $x(t)$ and $y(t)$ being realizations of wide-sense stationary processes (i.e., $W(t) \approx 1$) the function $\kappa(\omega)$ is well approximated as

$$\kappa(\omega) \approx \frac{i(e^{i\omega t_0} - e^{i\omega t_L})}{(t_L - t_0)\omega} \tag{A.120}$$

For a symmetrically chosen summation interval of duration T

$$t_L = -t_0 = T/2 \tag{A.121}$$

and $x(t)$ and $y(t)$ being realizations of wide-sense stationary processes, the function $\kappa(\omega)$ is

$$\kappa(\omega) \approx \operatorname{sinc}\left(\frac{\omega T}{2}\right) \tag{A.122}$$

If x and y equal the Gaussian pulse sequence defined in (1.44), then the correlation energy function is

$$\begin{aligned}W(t) &\approx \frac{c_0^2 \exp\left\{-\frac{(t-a-d)^2}{2b^2}\right\}^2 + c_0^2 \exp\left\{-\frac{(t-a+d)^2}{2b^2}\right\}^2}{R_{c,c}(0)} \\ &\approx \frac{\exp\left\{-\frac{(t-a-d)^2}{b^2}\right\} + \exp\left\{-\frac{(t-a+d)^2}{b^2}\right\}}{2b\sqrt{b}}\end{aligned} \tag{A.123}$$

under the parameter assumptions of Section 1.9.4.

The function $\kappa(\omega)$ is for the Gaussian pulse sequence approximated as

$$\kappa(\omega) \approx \exp\{i\omega a\}\exp\left\{-\frac{b^2\omega^2}{4}\right\}\cos(d\omega) \tag{A.124}$$

A.4.5 Correlation in Continuous Time

The correlation function between two sampled signals can be approximated by the correlation function of the underlying continuous-time signals. The Fourier transform of the continuous-time correlation function is the cross-power density spectrum and is the product of the Fourier transform of both signals. Fourier transforms and spectra of many modernized continuous-time GNSS signals can be found in the thesis by Ávila Rodríguez [5].

The discrete sum yielding the correlation function is approximated by an integral as

$$\begin{aligned} R_{x,y}(\tau) &\approx \frac{1}{L}\sum_{\mu=1}^{L} x(t_\mu - \tau)y(t_\mu) = \frac{1}{L\Delta t}\sum_{\mu=1}^{L} x(t_\mu - \tau)y(t_\mu)\Delta t \\ &\approx \frac{1}{t_L - t_1}\int_{t=t_1}^{t_L} x(t-\tau)y(t)dt \end{aligned} \tag{A.125}$$

where $\Delta t = 1/f_s$ is the sampling period.

If both signals are wide-sense stationary, the integration boundaries can be shifted. The integration is extended to infinity and the signals are expressed via the Fourier transforms,

$$\begin{aligned} R_{x,y}(\tau) &\approx \lim_{T\to\infty}\frac{1}{2T}\int_{t=-T}^{T} x(t-\tau)y(t)dt \\ &= \lim_{T\to\infty}\frac{1}{4\pi T}\int_{t=-T}^{T}\int_{\omega',\omega''=-\infty}^{\infty} \tilde{x}(\omega')\tilde{y}(\omega'')\exp\{i\omega'(t-\tau)+i\omega''t\}d\omega' d\omega'' dt \\ &= \lim_{T\to\infty}\frac{1}{4\pi T}\int_{\omega',\omega''=-\infty}^{\infty} \tilde{x}(\omega')\tilde{y}(\omega'')\exp\{-i\omega'\tau\}\int_{t=-T}^{T}\exp\{it(\omega'+\omega'')\}dt d\omega' d\omega'' \end{aligned} \tag{A.126}$$

Exchanging limes and integration yields the following expression

$$\lim_{T\to\infty}\frac{1}{2T}\int_{t=-T}^{T}\exp\{it(\omega'+\omega'')\}dt = \begin{cases} 1 & \omega' = -\omega'' \\ 0 & \omega' \neq -\omega'' \end{cases} \tag{A.127}$$

and the correlation function simplifies to

$$R_{x,y}(\tau) \approx \frac{1}{2\pi}\int_{\omega=-\infty}^{\infty} \tilde{x}(-\omega)\tilde{y}(\omega)\exp\{i\omega\tau\}d\omega \tag{A.128}$$

Therefore the angular frequency domain Fourier transform of the correlation function is

$$\tilde{R}_{x,y}(\omega) \approx \frac{1}{\sqrt{2\pi}}\tilde{x}(-\omega)\tilde{y}(\omega) \tag{A.129}$$

Because of different Fourier transform normalization factors, the normal (i.e., in [1/Hz]) Fourier transform of the correlation function is

$$\tilde{R}_{x,y}(f) \approx \tilde{x}(-f)\tilde{y}(f) \tag{A.130}$$

A.4.6 Probability Density Functions

In the following discussion, a number of probability density functions are listed which are used in Chapter 5 on signal detection. The formulas are based on Kay's book [6]. For different distributions, the probability density function $p(x)$ and the right-tail probability $Q(x)$ (also termed complementary cumulative distribution function) will be listed. Both functions are related via

$$Q(x) = \int_{x}^{\infty} p(t)dt \tag{A.131}$$

A.4.6.1 Normal Distribution

The normal distribution function for a one-dimensional real-valued random variable with mean μ and a variance σ^2 is defined as

$$p_{N;\mu,\sigma^2}(x) = \frac{1}{\sqrt{2\pi\sigma^2}} \exp\left\{-\frac{1}{2\sigma^2}(x-\mu)^2\right\} \tag{A.132}$$

The corresponding right-tail probability function cannot be expressed in closed form, but for $x - \mu >> 4\sigma$ the following approximation is valid

$$Q_{N;\mu,\sigma^2}(x) = \int_{x}^{\infty} p_{N;\mu,\sigma^2}(t)dt \approx \frac{\sigma}{(x-\mu)\sqrt{2\pi}} \exp\left\{-\frac{(x-\mu)^2}{2\sigma^2}\right\} \tag{A.133}$$

For $\mu = 0$ and $\sigma = 1$ we may omit the sub-script

$$Q(x) = Q_{N;0,1}(x) \tag{A.134}$$

A.4.6.2 Central Chi-Squared Distribution

The central chi-squared distribution arises as the probability distribution of the sum of v squared normal distributed independent random variables y_k with zero mean and unit variance, if y_i is $N(0,1)$ then x given by

$$x = \sum_{k=1}^{\upsilon} y_k^2 \sim \chi_{\upsilon}^2 \tag{A.135}$$

is chi-square distributed with v degrees of freedom. The probability density function is defined for positive values of x as

$$p_{\chi^2;\upsilon}(x) = \frac{1}{2^{\frac{\upsilon}{2}}\Gamma\left(\frac{\upsilon}{2}\right)} x^{\frac{\upsilon}{2}-1} \exp\left(-\frac{1}{2}x\right) \tag{A.136}$$

and vanishes for negative values. The mean and variance are

$$\begin{aligned} \langle x \rangle &= \upsilon \\ \operatorname{var}\langle x \rangle &= \left\langle \left(x - \langle x \rangle\right)^2 \right\rangle = 2\upsilon \end{aligned} \tag{A.137}$$

and the right-tail probability is given as

$$Q_{\chi^2;\upsilon}(x) = \begin{cases} 2Q(\sqrt{x}) & \upsilon = 1 \\ \exp(-x/2)\displaystyle\sum_{k=0}^{\upsilon/2-1} \frac{\left(x/2\right)^k}{k!} & \upsilon = 2 \\ 2Q\left(\sqrt{x}\right) + \dfrac{\exp(-x/2)}{\sqrt{\pi}} \displaystyle\sum_{k=1}^{(\upsilon-1)/2} \frac{(k-1)!(2x)^{k-1/2}}{(2k-1)!} & \upsilon \geq 2 \end{cases} \tag{A.138}$$

A.4.6.3 Noncentral Chi-Squared Distribution

The noncentral chi-squared distribution arises as a result of summing the squares of ν-squared independent Gaussian random variables with nonzero means, if y_i is $N(\mu_k,1)$ then x given by

$$x = \sum_{k=1}^{\upsilon} y_k^2 \sim \chi_{\upsilon,\lambda}^2 \tag{A.139}$$

has a noncentral chi-squared distribution with ν degrees of freedom and the noncentrality parameter λ defined as

$$\lambda = \sum_{k=1}^{\upsilon} \mu_k^2 \tag{A.140}$$

The probability density function is defined for positive values of x as

$$p_{\chi^2;\upsilon,\lambda}(x) = \frac{1}{2}\left(\frac{x}{\lambda}\right)^{\frac{\upsilon-2}{4}} \exp\left\{-\frac{1}{2}(x+\lambda)\right\} I_{\frac{\upsilon}{2}-1}(\sqrt{\lambda x}) \tag{A.141}$$

and vanishes for negative values of x. The symbol I_r denotes the modified Bessel function of the first kind and order r. The mean and variance are

$$\begin{aligned} \langle x \rangle &= \upsilon + \lambda \\ \operatorname{var}\langle x \rangle &= \left\langle \left(x - \langle x \rangle\right)^2 \right\rangle = 2\upsilon + 4\lambda \end{aligned} \tag{A.142}$$

The right-tail probability cannot be expressed in a closed form. However, for a large number of degrees of freedom ν, the right-tail probability can be approximated by using the central limit theorem as

$$Q_{\chi^2;\upsilon,\lambda}(x) \approx Q\left(\frac{x-\upsilon-\lambda}{\sqrt{2\upsilon+4\lambda}}\right) \tag{A.143}$$

A.4.6.4 Noncentral Chi-Squared Distribution for Complex-Valued Random Variables

If z_k are complex -independent random variables whose real and imaginary part are independent and if Re{z_k} is N(Re{μ_k}, 1), and Im{z_k} is N(Im{μ_k}, 1), then x given by

$$x = \sum_{k=1}^{v} \left|z_k\right|^2 \sim \chi^2_{2v,\lambda} \tag{A.144}$$

has a noncentral chi-squared distribution with $2v$ degrees of freedom and the noncentrality parameter λ as

$$\lambda = \sum_{k=1}^{v} \left|\mu_k\right|^2 \tag{A.145}$$

The same formulas for the density function and moments as in Section A.3.6.3 apply.

References

[1] Leick, A., *GPS Satellite Surveying*, New York: Wiley, 2004.

[2] Kay, S. M., *Fundamentals of Statistical Signal Processing: Estimation Theory*, Englewood Cliffs: Prentice Hall, 1993.

[3] Diniz, P. S. R., E. A. B. da Silva, and S. L. Netto, *Digital Signal Processing, System Analysis and Design*, Cambridge, U.K.: Cambridge University Press, 2002.

[4] Porat, B., *Digital Processing of Random Signals, Theory & Methods*, Englewood Cliffs, NJ: Prentice-Hall, 1994.

[5] Ávila Rodríguez, J. Á., *On Generalized Signal Waveforms for Satellite Navigation*. University of Federal Armed Forces Munich, Werner-Heisenberg-Weg 39, D-85577 Neubiberg, http://www.unibw.de/unibib/digibib/ediss/bauv, 2008.

[6] Kay, S. M., *Fundamentals of Statistical Signal Processing: Detection Theory*, Englewood Cliffs, NJ: Prentice-Hall, 1998.

Abbreviations

i.i.d.	independent and identically distributed
ipexSR	Institute of Geodesy and Navigation PC-Based Experimental Software Receiver
ACRLB	Asymptotic CRLB
ADC	Analog-to-Digital Conversion
AEP	Application Environment Profile
AGC	Automatic Gain Control
AGNSS	Assisted GNSS
AltBOC	Alternate BOC
API	Application Programming Interface
ASIC	Application-Specific Integrated Circuit
BOC	Binary Offset Carrier
BPSK	Binary Phase Shift Keying
CCRW	Code Continuous Reference Waveform
CDMA	Code Division Multiple Access
CF	Core Framework
CORBA	Common Object Requesting Broker Architecture
COTS	Commercial Off The Shelf
CRLB	Cramér-Rao Lower Bound
CW	Continuous Wave
DAC	Digital-to-Analog Conversion
DC	Direct Current
DLL	Delay Lock Loop
DoD	Department of Defense
DOP	Dilution of Precision
DSP	Digital Signal Processor
EDGE	Enhanced Data Rates for GSM Evolution
EGNOS	European Geostationary Navigation Overlay Service
EIB	Element Interconnect Bus
EPLRS	Enhanced Position Location Reporting System
FDMA	Frequency Division Multiple Access
FFT	Fast Fourier Transform
FIR	Finite Impulse Response
FLL	Frequency Lock Loop
FM	Frequency Modulation
FPGA	Field-Programmable Gate Array
GATE	Galileo Test and Development Environment
GFLOPS	Giga Floating-Point Operations Per Second

GIG	Global Information Grid
GLRT	Generalized Likelihood Ratio Test
GNSS	Global Navigation Satellite System
GNU	GNU's Not Unix (free software packages)
GOPS	Giga Operations Per Second
GPP	General Purpose Processor
GPS	Global Positioning System
GSM	Global System for Mobile Communications
HDTV	High-Definition Television
HF	High Frequency
HPA	High-Power Amplifier
HSDPA	High-Speed Downlink Packet Access
HSUPA	High-Speed Uplink Packet Access
ICD	Interface Control Document
IEEE	Institute of Electrical and Electronics Engineers
IGS	International GNSS Service
IMU	Inertial Measurement Unit
INS	Inertial Navigation System
IP	Internet Protocol
JAN-TE	Joint Airborne Networking–Tactical Edge
JCRLB	Joint CRLB
JTRS	Joint Tactical Radio System
LAN	Local Area Network
LNA	Low-Noise Amplifier
LORAN-C	LOng RAnge Navigation (-C)
LSB	Least Significant Bit
LSQ	Least-squares
MAP	Maximum A Posteriori Probability
MBOC	Multiplexed BOC
MCRBL	Modified CRLB
MEMS	Microelectromechanical system
MFLOPS	Mega Floating-Point Operations Per Second
MIPS	Million Instructions Per Second
ML	Maximum-Likelihood
MLE	Maximum-Likelihood estimator
MMAE	Minimum Mean Absolute Error
MMSE	Minimum Mean Squared Error
Mod-SDR	Modular SDR
MSB	Most Significant Bit
MSE	Mean Squared Error
MUOS	Mobile User Objective System
MVUE	Minimum Variance Unbiased Estimator
NCO	Numerically Controlled Oscillator
NMEA	National Marine Electronics Association
NSGU	Navigation Signal Generation Unit
OCXO	Oven Controlled Crystal Oscillator
ORB	Object Request Broker

OS	Open Service
OS	Operating System
PaC-SDR	Parameter-Controlled SDR
PC	Personal Computer
PCI	Peripheral Component Interconnect
PDA	Personal Digital Assistant
PL	Pseudolite
PLD	Programmable Logic Device
PLL	Phase Lock Loop
PND	Personal Navigation Device
POSIX	Portable Operating System Interface
PPE	Power Processor Element
PRN	Pseudorandom noise
PSD	Power Spectral Density
PVT	Position, Velocity and Time
QPSK	Quadratur Phase Shift Keying
R&D	Research and Development
RF	Radio Frequency
RINEX	Receiver INdependent EXchange Format
RMS	Root Mean Square
RTCM	Radio Technical Commission for Maritime services
RTK	Real-Time Kinematic
SATCOM	SATellite COMmunications
SCA	Software Communications Architecture
SDR	Software-Defined Radio
SI	Système International d'Unités
SINCGARS	SINgle Channel Ground and Airborne Radio System
SIS	Signal In Space
SISNET	Signal-In-Space NETwork
SMA	Sub-Miniature-A
SNR	Signal-to-Noise Ratio
SPE	Synergistic-Processor Element
SPI	Serial Peripheral Interface
SPS	Standard Positioning Service
SRW	Soldier Radio Waveform
SSP	Synchronous Serial Port
TCP	Transmission Control Protocol
TCRBL	True CRLB
TCXO	Temperature Controlled Crystal Oscillator
TDMA	Time Division Multiple Access
TOA	Time of Arrival
TTFF	Time to First Fix
UHF	Ultra-High Frequency
UMP	Uniform Most Powerful
UMPC	Ultra-Mobile PC
UMPUT	UMP Unbiased Test
UMTS	Universal Mobile Telecommunications System

USB	Universal Serial Bus
USRP	Universal Software Radio Peripheral
VDLL	Vector DLL
VLBI	Very Long Baseline Interferometry
WiMAX	Worldwide Interoperability for Microwave Access
WLAN	Wireless LAN
WNW	Wideband Networking Waveform

List of Symbols

The following conventions are generally used for mathematical symbols: *Lower case* variables denote deterministic quantities (e.g. r, $\mathbf{s}$) and *upper case* symbols denote random variables (e.g. R, $\mathbf{S}$). Estimates of deterministic but unknown quantities like $\hat{\tau}$, $\hat{\mathbf{p}}$ are themselves random variables but retain lower case symbols. A bold symbol like $\mathbf{n}$ denotes a vector having the elements n_μ. Matrices use italic upper case letters, like A or I.

Unless otherwise noticed, SI units are used. Code phases, delays or pseudoranges are expressed in [s], distances in [m], angles and carrier phases in [rad], frequencies in [Hz] and angular frequencies in [rad/s].

Commonly Used Symbols

Symbol	*Description*
$\mathbf{n}$	Noise (part of received signal samples)
$\mathbf{p}$	Low rate pseudorange parameters
$\mathbf{q}$	High rate pseudorange parameters
$\mathbf{r}$	Deterministic signal (part of received signal samples)
$\mathbf{s}$	Received signal samples
$\mathbf{x}$	Position parameters
ξ	Nuisance parameters
a	Signal amplitude
c	Speed of light [299,792,458 m/s]
$c(t)$	Baseband representation of the navigation signal
C/N_0	Carrier to noise power density ratio [Hz]
f_s	Sample rate in [Samples/s]
i	Imaginary unit
I	Fisher information matrix
L	Number of samples used for coherent integration
$L(\mathbf{s})$	Likelihood ratio
$N(\mathbf{q}, \mathbf{I})$	Normally distributed random variables, with mean $\mathbf{q}$ and covariance matrix I
$P(X > \gamma)$	Probability that the random variable X is larger than γ
$p(\mathbf{n})$	Probability density function for the random variables $\mathbf{N}$
$R_{x,y}(\tau)$	Cross-correlation function of signal x and y
t_μ	Time of reception for sample s_μ in [s]
T_{coh}	Coherent integration time in [s]
τ	Code phase in [s]

χ_{freq}	Nonuniformity of signal power distribution in time
$\delta(x)$	Dirac's delta function
$\delta_{\mu,\upsilon}$	Kronecker's Delta (1 for $\mu=\upsilon$, 0 otherwise)
$\kappa(\omega)$	Doppler correlation function
σ^2	Variance of the noise samples
φ	Carrier phase in [rad]
ω	Angular Doppler frequency in [rad/s]

Operations

Symbol	*Description*
$\langle f(\mathbf{N})\rangle_{\mathbf{N}} = \int f(\mathbf{n})\, d\mu(\mathbf{n}) = \int f(\mathbf{n}) p(\mathbf{n}) \prod_{\mu} dn_{\mu}$	Expected value of the function f with respect to the random variables **N** with the probability density function $p(\mathbf{n})$
$\mathrm{var}\langle \mathbf{S}\rangle_{\mathbf{N}} = \left\langle (\mathbf{S}-\langle \mathbf{S}\rangle_{\mathbf{N}})\ (\mathbf{S}-\langle \mathbf{S}\rangle_{\mathbf{N}})^{*}\right\rangle_{\mathbf{N}} = \left\langle \mathbf{S}\mathbf{S}^{*}\right\rangle_{\mathbf{N}} - \langle \mathbf{S}\rangle_{\mathbf{N}} \langle \mathbf{S}\rangle_{\mathbf{N}}^{*}$	(Co-)variance of a vector of random variables **S**, **S** depends on **N**
$\mathrm{var}\langle X\rangle_{\mathbf{N}} = \left\langle \lvert X\rvert^{2}\right\rangle_{\mathbf{N}} - \lvert\langle X\rangle_{\mathbf{N}}\rvert^{2}$	Variance of a random variable X being a function of **N**
$\mathrm{cov}\langle \mathbf{S},\mathbf{T}\rangle_{\mathbf{N}} = \left\langle (\mathbf{S}-\langle \mathbf{S}\rangle_{\mathbf{N}})\, (\mathbf{T}-\langle \mathbf{T}\rangle_{\mathbf{N}})^{*}\right\rangle_{\mathbf{N}}$	Covariance of two vectors of random variables **S** and **T**; **S** and **T** depend on **N**
$\overline{x}$	Complex conjugate of x
$\mathbf{x}^{T}$	Transposed vector of **x**
$\mathbf{x}^{*}$	Hermitian conjugate (transposed and complex conjugate) of vector (or matrix) **x**
$N \sim Q(a, b)$	Random variable N is from distribution $Q(a,b)$
$x \approx y$	x approximately equal to y
$x \propto y$	x proportional to y
$x \overset{!}{=} y$	x should be equal to y
$\mathbf{x}^{T} \bullet \mathbf{y} = \sum_{\mu} x_{\mu} y_{\mu}$	Dot product operation
$c'(t) = dc(t)/dt$	First derivative of $c(t)$
<=>	Equivalence

About the Author

Thomas Pany is a senior research engineer at IFEN GmbH in Germany, a company that provides a complete portfolio of leading edge GNSS testing products, including GNSS test receivers. He studied physics and began working with GPS receivers in 1997. He performed RTCM data quality monitoring in a network of permanent GPS reference stations and received a Ph.D. from the Graz University of Technology in Austria for his work on modeling GPS and synthetic aperture radar signal propagation within the troposphere. In 2001, he began working at the University of Federal Armed Forces in Munich, and, over eight years, he has developed a number of different GNSS software receivers. This book derives from his work during that time. He has authored over 100 papers in the field of positioning, GNSS receivers, Galileo signal structure, and GPS science.

Index

The GNSS Technology and Applications Series

Elliott Kaplan and Christopher Hegarty, Series Editors

A-GPS: Assisted GPS, GNSS, and SBAS, Frank van Diggelen

Applied Satellite Navigation Using GPS, GALILEO, and Augmentation Systems, Ramjee Prasad and Marina Ruggieri

Digital Terrain Modeling: Acquisition, Manipulation, and Applications, Naser El-Sheimy, Caterina Valeo, and Ayman Habib

Geographical Information Systems Demystified, Stephen R. Galati

GNSS Applications and Methods, Scott Gleason and Demoz Gebre-Egziabher

GNSS Markets and Applications, Len Jacobson

GNSS Receivers for Weak Signals, Nesreen I. Ziedan

Introduction to GPS: The Global Positioning System, Second Edition, Ahmed El-Rabbany

Navigation Signal Processing for GNSS Software Receivers, Thomas Pany

Principles of GNSS, Inertial, and Multisensor Integrated Navigation Systems, Paul D. Groves

Server-Side GPS and Assisted-GPS in Java™, Neil Harper

Spread Spectrum Systems for GNSS and Wireless Communications, Jack K. Holmes

Understanding GPS: Principles and Applications, Second Edition, Elliott Kaplan and Christopher Hegarty, editors

Ubiquitous Positioning, Robin Mannings

Wireless Positioning Technologies and Applications, Alan Bensky

For further information on these and other Artech House titles, including previously considered out-of-print books now available through our In-Print-Forever® (IPF®) program, contact:

Artech House Publishers
685 Canton Street
Norwood, MA 02062
Phone: 781-769-9750
Fax: 781-769-6334
e-mail: artech@artechhouse.com

Artech House Books
16 Sussex Street
London SW1V HRW UK
Phone: +44 (0)20 7596 8750
Fax: +44 (0)20 7630 0166
e-mail: artech-uk@artechhouse.com

Find us on the World Wide Web at: www.artechhouse.com